Genetics of Complex
Human Diseases

A LABORATORY MANUAL

ALSO FROM COLD SPRING HARBOR LABORATORY PRESS

RELATED LABORATORY MANUALS

Genetic Variation: A Laboratory Manual
Proteomics: A Cold Spring Harbor Laboratory Course Manual

RELATED HANDBOOKS

Lab Math: A Handbook of Measurements, Calculations, and Other Quantitative
Skills for Use at the Bench
Statistics at the Bench: A Step-by-Step Handbook for Biologists

RELATED WEBSITE

 Cold Spring Harbor Protocols
www.cshprotocols.org

Genetics of Complex Human Diseases

A LABORATORY MANUAL

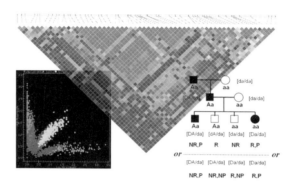

EDITED BY

Ammar Al-Chalabi
MRC Centre for
Neurodegeneration Research
King's College London

Laura Almasy
Southwest Foundation for
Biomedical Research
San Antonio, Texas

www.cshprotocols.org

COLD SPRING HARBOR LABORATORY PRESS
Cold Spring Harbor, New York • www.cshlpress.com

Genetics of Complex Human Diseases
A LABORATORY MANUAL

Publisher	John Inglis
Acquisition Editor	Alexander Gann
Director of Development, Marketing, & Sales	Jan Argentine
Developmental Editor	Judy Cuddihy
Project Coordinator	Mary Cozza
Permissions Coordinator	Carol Brown
Production Editors	Kathleen Bubbeo, Rena Steuer
Desktop Editors	Susan Schaefer, Lauren Heller
Production Manager	Denise Weiss
Book Marketing Manager	Ingrid Benirschke
Cover Designer	Mike Albano

Front cover artwork: (*Top right*) Triangle plot showing the linkage disequilibrium relationships of single-nucleotide polymorphisms (SNPs) as visualized using Haploview software (Chapter 8, Al-Chalabi). (*Lower left*) Quality control of SNP genotyping with data shown from the Sequenom MassArray reaction (Chapter 16, Edenberg and Liu). (*Lower right*) Confounding of recombination and penetrance in a pedigree segregating an autosomal dominant susceptibility locus (Chapter 3, Borecki and Rice).

Library of Congress Cataloging-in-Publication Data

Genetics of complex human diseases / edited by Ammar Al-Chalabi, Laura Almasy.
   p. cm.
 ISBN 978-0-87969-882-9 (hardcover : alk. paper) -- ISBN 978-0-87969-883-6
(pbk. : alk. paper)
 1. Medical genetics. I. Al-Chalabi, Ammar II. Almasy, Laura.
 RB155.G38985 2010
 616'.042--dc22

                2009023930

10 9 8 7 6 5 4 3 2 1

To Cathy, Thomas, and William — A. A.

To Fred — L. A.

Contents

Preface, ix

1   Introduction, 1
Ammar Al-Chalabi and Laura Almasy

2   "Statistics 101"—A Primer for the Genetics of Complex Human Disease, 5
Janet Sinsheimer

3   Linkage Analysis of Discrete Traits, 17
Ingrid B. Borecki and John P. Rice

4   Epidemiologic Considerations in Complex Disease Genetics, 27
John Gallacher

5   Variance Component Methods for Analysis of Complex Phenotypes, 37
Laura Almasy and John Blangero

6   Multiple Testing and Power Calculations in Genetic Association Studies, 49
Hon-Cheong So and Pak C. Sham

7   Introduction to Genetic Association Studies, 61
Cathryn M. Lewis and Jo Knight

8   Genome-Wide Association Studies, 73
Ammar Al-Chalabi

9   Introduction to Linkage Disequilibrium, the HapMap, and Imputation, 89
Benjamin M. Neale

10   Meta-Analysis of Genome-Wide Association Studies, 95
Paul I.W. de Bakker, Benjamin M. Neale, and Mark J. Daly

11   Gene–Environment Interaction and Common Disease, 105
Ruth J.F. Loos and Nicholas J. Wareham

12   Family-Based Genetic Association Tests, 119
Eden R. Martin and Evadnie Rampersaud

13   Copy-Number Variation and Common Human Diseases, 131
Dheeraj Malhotra and Jonathan Sebat

14   Oncogenomics, 143
     Simon J. Furney, Gunes Gundem, and Nuria Lopez-Bigas

15   When the Genetic Code Is Not Enough—How Sequence Variations Can Affect Pre-mRNA Splicing and Cause (Complex) Disease, 165
     Brage Storstein Andresen and Adrian R. Krainer

16   Laboratory Methods for High-Throughput Genotyping, 183
     Howard J. Edenberg and Yunlong Liu

17   Gene Set Analysis and Network Analysis for Genome-Wide Association Studies, 195
     Inti Pedroso and Gerome Breen

Index, 205

Preface

This book grew out of the Genetics of Complex Human Diseases course taught at Cold Spring Harbor Laboratory. The course has been given there every other year since 2004. It has been fascinating to hear speakers at the international cutting edge of statistical and complex disease genetics, and to see the subject matter change as the field has evolved. This book is intended to be a reflection of the course, without being an exact copy of the curriculum. We hope that it conveys the spirit of the course in containing chapters that are written by well-recognized authors with a clear teaching style, who offer a combination of detail for those who want the reasons behind the conclusions and an overview for those new to the subject.

Of course, it is impossible to write such a book without a great deal of help from many people, and thanks are due to the authors who have done a remarkable job of simplifying a complicated subject, and the speakers in the course, who every other year donate their time to pass on their knowledge to others. We also thank all of those at Cold Spring Harbor Laboratory Press who guided us through the publishing process: John Inglis, Publisher, and Alex Gann, Editorial Director; Mary Cozza, Project Coordinator, and Judy Cuddihy, Developmental Editor; and Production staff Kathy Bubbeo, Rena Steuer, Susan Schaefer, and Lauren Heller. Without their efforts, this book would not exist.

AMMAR AL-CHALABI
LAURA ALMASY
May 2009

Genetics of Complex Human Diseases

A LABORATORY MANUAL

1 | Introduction

Ammar Al-Chalabi[1] and Laura Almasy[2]

[1]MRC Centre for Neurodegeneration Research, King's College London, London SE5 8AF, United Kingdom; [2]Southwest Foundation for Biomedical Research, San Antonio, Texas 78227

WHY GENETICS MATTERS

A central principle underlying modern genetics research is that by knowing which genetic variations contribute to disease phenotypes we will understand the mechanism of disease causation and therefore be able to intervene and treat or prevent a disease. We are by no means at the end of our quest to understand the genome and its relationship to disease or other phenotypes, but we have made sufficient progress that some diseases are now yielding their secrets to us.

In putting this book together, we have aimed to bring together the tools that geneticists use to find disease genes with the genetic concepts and statistical theories that underpin them. Anyone with an interest in human genetics or who uses genetic techniques in research will find this book useful, particularly if studying diseases with complex inheritance. Although some statistics and mathematics are used, their use is explained in detail both as an overview in an introductory chapter and specifically within each chapter. Even so, statistical ability is not necessary for an understanding of the chapters. The themes covered are diverse but with an emphasis on association studies, as these are the basis for many current study designs for complex diseases. On the other hand, classical methods such as linkage are included because they are useful for the investigation of many phenotypes and are the basis of subsequent methods, and an understanding of them is required for successful assessment of the existing research. The chapters are a blend of practical guides in undertaking genetic studies and a review of the subject, making this a valuable resource for those who want to know both the how and the why.

A BRIEF HISTORY OF MODERN GENETICS

Keeping Up with Genetic Ideas

Genetics is advancing at an extremely rapid rate, and a good measure of this is the rate at which significant milestones have been passed. Although there have been many important steps in the journey to our present genetic understanding, major leaps of knowledge have occurred either because of new approaches (mathematical, conceptual, or technological) or because existing ideas were overturned or subsumed.

Mendel, Darwin, and the Wave–Particle Debate on Heredity

In 1859, Darwin published his theory of evolution, which relied on the concept of heredity (Darwin 1859). The year 1865 saw the first truly modern scientific analysis of heredity when Gregor Mendel,

1

an Augustinian monk in the Abbey of St. Thomas in Brno, performed his meticulous experiments crossing and counting 29,000 pea plants (Mendel 1866). In 1905, William Bateson coined the term "genetics" and became the first Professor of Genetics, at Cambridge University in England (Bateson 1907). At this time, there was great debate about whether the mechanism of heredity was based on particles or waves, with Bateson proposing that genes were waves or vibrations, and Karl Pearson (who gave us the chi-square test, regression, and correlation) believing they were discrete particles. (There was an even longer-running and similar debate about the nature of light.) Mendel's laws could only be explained with a particle theory of heredity, whereas the theory of evolution seemed to require gradual changes that were difficult to account for with hereditary particles, but would best be explained by vibrations or waves, which could allow blending of parental traits. On the other hand, a wave theory could not account for evolution, as any blending would result in a loss of diversity with time. This was resolved in 1918 by Ronald Fisher (inventor of the concept of statistical variance), who with J.B.S. Haldane and Sewell Wright showed that a particle-based polygenic theory of inheritance would be consistent with Mendel's laws and with evolution (Fisher 1918). Between 1903 and 1910, Walter Sutton and Thomas Hunt Morgan showed that genes residing on chromosomes were the hereditary units. Thus, the late 1800s and early 1900s gave us the foundations for modern genetics and statistics and are the direct ancestors of the concepts in this book.

The Central Dogma of Molecular Biology

The 1930s and 1940s saw the development of the central dogma of genetics (proposed in 1941, and formally in 1958), which is that information flows from DNA (shown in 1933 to be found in chromosomes) to RNA to proteins and does not flow back (Crick 1970). The central dogma was broken in 1964 with the discovery of retroviruses and in 1982, with the discovery of prions.

DNA and the Genetic Code

In 1953, James D. Watson and Francis H.C. Crick, using crystallographic data from Rosalind Franklin in Maurice Wilkins's lab, resolved the structure of DNA to be a double helix with base pairing (Watson and Crick 1953). By 1967, what we now call the genetic code was cracked by a number of investigators. Today with the knowledge that only 2% of the human genome codes for protein, it is apparent that the genetic code is far from cracked—we still do not fully understand how to read the message encoded in the remaining 98%.

Genomics Arrives

The 1970s saw the arrival of the genomics era, with Walter Fiers's group giving us the first gene sequence in 1972 for the bacteriophage MS2 coat protein, and the first complete RNA genome in 1976, again for MS2, closely followed in 1977 with the first sequence of an entire DNA genome (the bacteriophage ΦX174) from Fred Sanger's group (Min Jou et al. 1972; Fiers et al. 1976; Sanger et al. 1977).

The year 1983 saw the invention of the polymerase chain reaction (PCR) by Kary Mullis (Mullis and Faloona 1987), bringing molecular biology truly into the genomics era. In 1990, the Human Genome Project began formally and was announced as completed in 2001, although it was not until April 14, 2003 that it was 99% complete to 99.99% accuracy (Lander et al. 2001; Venter et al. 2001).

Postgenomics

We have now entered the "postgenomics era." This is characterized by huge international collaborations put together to achieve sufficient statistical power to find common genetic variants contributing to common diseases, studies of rare sequence variants by deep resequencing, harmonization of biobanks and epidemiological studies, efforts to understand the structural variation of the genome, a realization of the importance of intronic and intergenic (so-called "junk") DNA, epigenetics, the new roles of RNA, developments in statistical theory and computing power, and the thousand-dollar genome.

HOW THE STUDY OF COMPLEX DISEASE SITS IN THE GENETIC LANDSCAPE

Complex disease genetics has grown from an understanding that the "one gene, one disease" model is too simplistic and cannot explain familial clustering of diseases that do not show Mendelian inheritance or the complexity of phenotypes seen in health and disease. The topics we have selected for inclusion are by no means all that could have been, but they do form a coherent and logical whole. We begin with an introduction to the basic statistics used by geneticists and a review of the importance of epidemiology. We move to variance components and linkage and family-based tests of association before devoting several chapters to various aspects of genome-wide association studies, their problems, and how to successfully overcome them. We have included the latest ideas in meta-analysis and imputation, gene x environment interaction, and copy-number variation, as well as pathways-based analyses, cancer genetics, and how RNA splicing contributes to complex disease. We finish with a roundup of the latest laboratory technologies.

FINAL THOUGHTS

Undoubtedly, our present understanding of the genome will change dramatically during the next few years. As we unravel the meaning of noncoding DNA and understand the complex interactions between microRNAs, epigenetic signals, and genes, and the control of posttranslational protein modifications, our view of the genome will become a true reflection of the complexity that lies beneath the apparent simplicity of the genetic sequence.

REFERENCES

Bateson, W. 1907. *The Progress of Genetic Research. In Report of the Third 1906 International Conference on Genetics: Hybridization (the cross-breeding of genera or species), the cross-breeding of varieties, and general plant breeding* (ed. W. Wilks). Royal Horticultural Society, London.

Crick, F. 1970. Central dogma of molecular biology. *Nature* 227: 561–563.

Darwin, C. 1859. *On the origin of species by means of natural selection, or the preservation of favoured races in the struggle for life.*

Fiers, W., Contreras, R., Duerinck, F., Haegeman, G., Iserentant, D., Merregaert, J., Min Jou, W., Molemans, F., Raeymaekers, A., Van den Berghe, A., et al. 1976. Complete nucleotide-sequence of bacteriophage MS2-RNA—Primary and secondary structure of replicase gene. *Nature* 260: 500–507.

Fisher, R.A. 1918. The correlation between relatives on the supposition of Mendelian inheritance. *Trans. R. Soc. Edinb.* 52: 399–433.

Lander, E.S., Linton, L.M., Birren, B., Nusbaum, C., Zody, M.C., Baldwin, J., Devon, K., Dewar, K., Doyle, M., FitzHugh, W., et al. 2001. Initial sequencing and analysis of the human genome. *Nature* 409: 860–921.

Mendel, G. 1866. Versuche über Pflanzen-Hybriden. *Verh. Naturforsch. Ver. Brünn* 4: 3–47.

Min Jou, W., Haegeman, G., Ysebaert, M., and Fiers, W. 1972. Nucleotide sequence of the gene coding for the bacteriophage MS2 coat protein. *Nature* 237: 82–88.

Mullis, K.B. and Faloona, F.A. 1987. Specific synthesis of DNA in vitro via a polymerase-catalyzed chain reaction. *Methods Enzymol.* 155: 335–350.

Sanger, F., Air, G.M., Barrell, B.G., Brown, N.L., Coulson, A.R., Fiddes, C.A., Hutchison, C.A., Slocombe, P.M., and Smith, M. 1977. Nucleotide sequence of bacteriophage phi X174 DNA. *Nature* 265: 687–695.

Venter, J.C., Adams, M.D., Myers, E.W., Li, P.W., Mural, R.J., Sutton, G.G., Smith, H.O., Yandell, M., Evans, C.A., Holt, R.A., et al. 2001. The sequence of the human genome. *Science* 291: 1304–1351.

Watson, J.D. and Crick, F.H. 1953. Molecular structure of nucleic acids; a structure for deoxyribose nucleic acid. *Nature* 171: 737–738.

2 "Statistics 101"—A Primer for the Genetics of Complex Human Disease

Janet Sinsheimer

Departments of Human Genetics, Biomathematics, David Geffen School of Medicine at UCLA, and Biostatistics, UCLA School of Public Health, Los Angeles, California 90095

INTRODUCTION

This chapter reviews the basis of probability and statistics used in the genetic analysis of complex human diseases and illustrates their use in several simple examples. Much of the material presented here is so fundamental to statistics that it has become common knowledge in the field and the originators are no longer cited (e.g., Gauss). However, there are a number of statistics texts available if readers would like a more thorough treatment: These include Brase and Brase (2007), Gonick and Smith (1993), Lange (2002, 2003), Milton (1999), Mood et al. (1974), Sham (1998), and Young (1995), and the material presented in this chapter draws on these texts.

SET THEORY BASICS

To start, we need to briefly review the set theory basics that apply to genetics analysis. Let the entire event space be denoted by Ω and the null space (or empty set) by $\emptyset$. Let A and B be two events that are both elements of Ω. The symbol $\cap$ denotes "and" or "intersection" (Fig. 2.1) and the symbol $\in$ indicates when an event is a member, or element, of a set. The outcomes A and B are mutually exclusive when $A \cap B = \emptyset$. If $A \in \Omega$, then the complement of A (not A), which is denoted as $\bar{A}$, is also an element of Ω. The symbol $\cup$ denotes "or" or "union." Our final rule here is that if $A \in \Omega$ and $B \in \Omega$ then $A \cup B \in \Omega$.

PROBABILITY THEORY IN GENETIC ANALYSIS

We can now move on to the probability of an event or set of events. Let $P(A)$ denote the probability of outcome A, in which A can be a single event or a set of events. Likewise, let $P(B)$ denote the probability of outcome B. The probability of both outcomes A and B occurring is denoted by $P(A \cap B)$

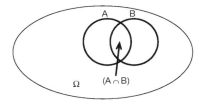

FIGURE 2.1 A Venn diagram illustrating the intersection of two outcomes.

or $P(A, B)$. The probability of outcome A or outcome B occurring is $P(A \cup B)$. The basis of probability comes from the following four rules: (1) $P(\Omega) = 1$; (2) $P(\emptyset) = 0$; (3) $0 \leq P(A) \leq 1$; and (4) if $A \cap B = \emptyset$ then $P(A \cup B) = P(A) + P(B)$. These four rules can be used in combination to provide more probabilistic insights. As an example, the probability of $\bar{B}$ can be derived as follows: $P(\Omega) = 1$, so $P(B \cup \bar{B}) = 1$. These events are mutually exclusive so $P(B) + P(\bar{B}) = 1$ and finally $P(\bar{B}) = 1 - P(B)$.

Conditional Probability

The concept of conditional probability plays an important role in genetic analysis. The conditional probability of event A given event B is

$$P(A|B) = \frac{P(A \cap B)}{P(B)}.$$

Conditional probabilities can be particularly useful when it is difficult to calculate $P(A)$ directly. Using this notation, we can express the probability of A in terms of joint or conditional probabilities,

$$P(A) = P(A \cap B) + P(A \cap \bar{B}) = P(A|B)P(B) + P(A|\bar{B})P(\bar{B}).$$

Here outcome A is partitioned into two parts by the two outcomes B and $\bar{B}$. We can extend this partitioning to any number of outcomes.

Bayes' theorem also relies on conditional probability. Suppose it is difficult to calculate $P(B|A)$ directly; however, $P(B)$, $P(\bar{B})$, $P(A|B)$, and $P(A|\bar{B})$ can be calculated. By Bayes' theorem,

$$P(B|A) = \frac{P(A|B)P(B)}{P(A|B)P(B) + P(A|\bar{B})P(\bar{B})}.$$

Again this theorem can be extended to more than the two outcomes B and $\bar{B}$.

Independence

Another important concept is independence. If A and B are independent, then $P(A \cap B) = P(A)P(B)$. If $P(B) > 0$, then, using the definition of conditional probability, independence can be expressed as $P(A|B) = P(A)$. Intuitively then, independence between events A and B means that knowing that B occurred does not change the probability of A occurring.

Example: Linkage Equilibrium

A haplotype is a set of alleles at genes inherited from the same parent. Two loci C and D, which make up part of a haplotype, are in linkage equilibrium if the haplotype frequency equals the product of the allele frequencies. That is $P(c_i, d_j) = P(c_i)P(d_j)$ for any allele c_i at C and any allele d_j at D. So linkage equilibrium is an example of independence. When a haplotype frequency is not equal to the product of the corresponding allele frequencies, then these loci are in linkage disequilibrium (LD; see Chapter 8 for more information on LD).

Example: Hardy–Weinberg Equilibrium

For a specified locus, the Hardy–Weinberg law ($p^2 + 2pq + q^2 = 1$) allows us to calculate genotype frequencies in terms of the allele frequencies (Hardy 1908; Weinberg 1908). If there is a very large population size, no mutation, no selection, and random mating with respect to the locus, then the Hardy–Weinberg law dictates that the genotype frequencies in a population remain constant from generation to generation and the expected genotype frequencies can be derived from the expected allele frequencies (Hardy–Weinberg equilibrium, HWE). As an example, consider the blood group

MN. If allele M has population frequency p and allele N has frequency q, then under HWE we can calculate the genotype frequencies. The genotype M/N is an unordered genotype that is shorthand for the union of the two ordered M|N and N|M genotypes (the parental origin of the alleles is known). That is, M/N denotes the event that the M is from the mother and the N is from the father or that the N is from the mother and the M is from the father. The two ordered genotypes are mutually exclusive, so using our probability rules we know that $P(\text{M/N}) = P(\text{M from mother, N from father}) + P(\text{N from mother, M from father})$. Random union of gametes implies independence so $P(\text{M/N}) = pq + qp = 2pq$. Similarly, we can show that $P(\text{M/M}) = p^2$, and $P(\text{N/N}) = q^2$ under HWE.

Example: Determining the Frequency of ABO Blood Type by Affection Status

Clarke and coworkers (1959) provide blood types for 521 individuals suffering from duodenal ulcers and 680 unaffected controls. Lange (2002) summarized the data from the ulcer patients, noting that there were 186 with blood type A, 38 with blood type B, 13 with blood type AB, and 284 with blood type O, and for the controls, there were 279 with blood type A, 69 with blood type B, 17 with blood type AB, and 315 with blood type O.

The frequency of blood type O in those individuals with ulcers is the conditional probability

$$P(\text{O} \cap \text{Ulcer}) / P(\text{Ulcer}) = \frac{284/1201}{521/1201} = \frac{284}{521} = P(\text{O} | \text{Ulcer}) = 0.545.$$

Likewise the frequency of blood type O in the unaffected individuals $P(\text{O} | \text{Unaffected}) = 315/680 = 0.465$. Note that these probabilities are not the same as the probability of having an ulcer given blood type O, $P(\text{Ulcer} | \text{O}) = 0.474$, or the probability of not having an ulcer given blood type O, $P(\text{Unaffected} | \text{O}) = 0.526$. Assuming HWE and an iterative gene counting algorithm, which is an example of an expectation–maximization (E-M) algorithm (see Lange 2002), provides the allele frequencies for ulcer patients and the allele frequencies for unaffected controls ($P(\text{O} | \text{Ulcer}) = 0.7363$, $P(\text{A} | \text{Ulcer}) = 0.2136$, $P(\text{B} | \text{Ulcer}) = 0.0501$, $P(\text{O} | \text{Unaffected}) = 0.6853$, $P(\text{A} | \text{Unaffected}) = 0.2492$, $P(\text{B} | \text{Unaffected}) = 0.0655$).

VARIABLES, PARAMETERS, AND DISTRIBUTIONS IN GENETIC ANALYSIS

We start with a few definitions. A random variable is something we can measure, control, or manipulate in research. Examples include continuous values like high-density lipoprotein cholesterol levels or discrete values like disease status. A parameter is any characteristic of a population and a statistic is any characteristic of a sample. Some common parameters for a continuous random variable X are its population mean, μ_x, population variance, σ_x^2, and population standard deviation, σ_x. The relationship between two continuous random variables X and Y can be summarized by their population covariance, σ_{xy}. Some common measures for a dichotomous categorical trait include the proportion and its variance. The data are assumed to be generated from some underlying probabilistic model. This model defines the range of possible values and how likely they are and is called the theoretical distribution. For a random variable X, the probability density is $f(x) = P(X = x)$ for all allowable values x and the cumulative probability distribution is $F(x) = P(X \leq x)$.

Often we cannot directly observe a parameter, and so we use a statistic calculated on a sample to estimate the parameter value and make inferences concerning the population. The observed data from the sample make up the observed distribution. The parameters μ, σ_x^2, σ_x, σ_{xy} are estimated by the sample mean

$$\bar{x} = \frac{\sum_{i=1}^{n} x_i}{n},$$

sample variance

$$v_x = \frac{\sum_{i=1}^{n}\left(x_i - \bar{x}\right)^2}{n-1},$$

sample standard deviation

$$\text{sd}_x = \sqrt{\frac{\sum_{i=1}^{n}\left(x_i - \bar{x}\right)^2}{n-1}},$$

and sample covariance

$$\text{cov}_{xy} = \frac{\sum_{i=1}^{n}\left(x_i - \bar{x}\right)\left(y_i - \bar{y}\right)}{n-1},$$

in which n is the number of individuals in the sample. Another important parameter is the population correlation coefficient, $\rho = \sigma_{xy}/\sigma_x\sigma_y$, which is a standardized covariance between two continuous variables. The population correlation is estimated by the sample correlation coefficient r, which is calculated as $r = \text{cov}_{xy}/\text{sd}_x\text{sd}_y$.

We are often interested in whether the observed distribution could have been drawn from a putative theoretical distribution. We can get a handle on whether this hypothesis is true by determining whether a statistic value for a sample is consistent with a parameter value that is calculated from the putative theoretical distribution.

Theoretical probability distributions depend on whether the variable is discrete or continuous. There are a number of theoretical distributions that pop up frequently in genetic analysis. These distributions include the uniform, the binomial, the normal, and the chi-square. Rather than concentrate too heavily on their formulas, we will examine them intuitively and point out their most important features.

Discrete Uniform Distribution

Let there be a finite number of possible outcomes W; each are equally probable (with the probability of a specific outcome being $1/W$). If there are n total observations from a uniform distribution, we expect n/W of these observations to be in each of the possible categories with variance $(n/W)(1 - 1/W)$. The discrete uniform distribution is found in genetic analysis when quantifying prior beliefs in which there is little or no real evidence to support one hypothesis more than another. As an example, suppose gene mapping has narrowed down the location of a disease conferring mutation to a chromosomal region containing 20 genes. Without additional biological information, it may be a good starting point to suppose that the probability that any one of these genes contains the mutation is 1/20, that is, to assume a discrete uniform distribution.

Normal Distribution

The normal distribution is defined for continuous data (data that can take on any value). The probability density is a bell-shaped curve that is symmetric and defined for the whole real line (Fig. 2.2). Because the density is symmetric and unimodal, the mean, the 50th percentile (median), and the mode (the peak) all have the same value. The bigger σ_x is, the greater the spread of the density. The normal density has inflection points (points in which the second derivative is 0) at $\mu_x - \sigma_x$ and $\mu_x + \sigma_x$. Because the data are continuous, the probability of getting exactly the value x is 0. The probability that the value is less than or equal to x is always positive (the cumulative distribution). Likewise the probability that

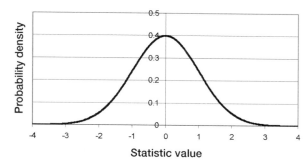

FIGURE 2.2 The standard normal density. Shown is an example of the normal density with μ = 0 and σ = 1 (the standard normal density).

the value is greater than x is always positive. Many of the methods that are used to map quantitative trait loci (e.g., variance component methods for gene mapping [see as examples, Amos 1994; Almasy et al. 1998; Bauman et al. 2005]) use models that implicitly assume that the trait or a transformation of the trait has a distribution that is normal or is a mixture of normal distributions.

The normal distribution is particularly important in statistics because, by the central limit theorem, it is the theoretical distribution of the sampling mean $\bar{x}$. More precisely, as the sample size used to calculate $\bar{x}$ increases, the shape of the sampling distribution of $\bar{x}$ approaches a normal distribution with mean μ_x and standard deviation $\sigma_x / \sqrt{n}$.

Chi-Square Distribution

Another important distribution in genetic analysis is the chi-square distribution. The chi-square distribution is defined for continuous data. If Y is the sum of the squares of N independent variables X_i ($i = 1, \ldots, N$) in which each X_i has a normal density with $\mu = 0$ and $\sigma = 1$ (each has a standard normal density), then Y has a chi-square distribution. The shape of the density depends on N (Fig. 2.3). Because Y is a sum of squared values, Y can never be less than 0. However, Y can be arbitrarily large. Often we assume that a goodness-of-fit comparison of observed and expected results follows a chi-square distribution for categorical data, even though the statistic takes on discrete values. This approximation is reasonable provided we have enough data. Chi-square approximations show up in the contingency table analyses that examine the question of whether genotype frequencies are associated with disease status.

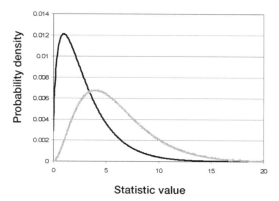

FIGURE 2.3 The chi-square density's shape depends on the degrees of freedom. The black curve is the chi-square density with df = 3. The gray curve is the chi-square density with df = 6.

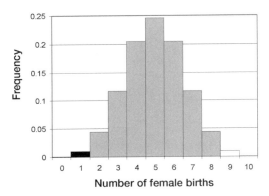

FIGURE 2.4 Example of a binomial density and a graphical depiction of a one-sided p value and a two-sided p value. The density is binomial with $p = 1/2$. The x axis denotes the number of births. There can be 0 to 10 births. Although difficult to see in the graph for the values 0 and 10, each of the 11 possible outcomes has nonzero probability and is represented by a histogram block. The area defined by the black blocks at 0 and 1 represent the p value for the one-sided test. The area represented by the black blocks at 0 and 1 and the white blocks at 9 and 10 represent the p value for the two-sided test.

Binomial Distribution

The binomial distribution is also defined for discrete data. Suppose that there are n independent trials and two possible outcomes, success and failure. Then the probability of exactly s successes (the probability density function) is

$$\frac{n!}{(n-s)!\,s!}\,p^{s}\left(1-p\right)^{n-s},$$

in which p is the probability of a single success. The terms $p^{s}(1-p)^{n-s}$ are present because each of the s successes and $n - s$ failures is independent and so their probabilities multiply. The term $n!/(n-s)!s!$ is the number of ways that these s successes and $n - s$ failures can be arranged. The density is unimodal but not usually symmetric. When $p = 1/2$ the density is symmetric (Fig. 2.4), but when $p < 1/2$ the mode of the density is skewed toward low values of s and when $p > 1/2$ the mode of the density is skewed toward high values of s. The probability of at least s successes (the cumulative distribution) sums the probability densities for 0 to s,

$$\sum_{i=0}^{s}\frac{n!}{(n-i)!\,i!}\,p^{i}\left(1-p\right)^{n-i}.$$

The binomial distribution and its generalization for multiple possible outcomes, the multinomial distribution, underlie a number of statistical genetic tests. The probability of s affected offspring from a random sample of n independent affected–unaffected matings provides a classic example from segregation analysis (Sham 1998).

MAXIMUM LIKELIHOOD ESTIMATION

A likelihood L_{H} is proportional to the probability of the data given the hypothesis H, in which the constant of proportionality is arbitrary. We can think of a likelihood as a surface (a function) whose height varies with the parameter values but the data stay fixed. Therefore, maximum likelihood estimation finds the parameter values that best fit the observed data. These values are called the maximum likelihood estimates (MLEs) of the parameters. Because the logarithm is a monotonic transformation (the logarithm preserves inequalities among the y values, i.e., if $y_{1} < y_{2}$, then $\ln(y_{1}) < \ln(y_{2})$), the MLEs for the likelihood are also the MLEs for natural logarithm of the likelihood (log-likelihood). MLE has several desirable properties. MLEs are often not too difficult to calculate (at least numerically), they are consistent estimates (as the sample size increases the estimate gets better), and they are asymptotically unbiased (expected value of the MLEs equals the true parameter values, at least when the sample size approaches infinity).

The log-likelihood tends to be easier and more stable to maximize. When the number of parameters is small and the log-likelihood amenable, we can maximize the log-likelihood analytically. However, most of the time we maximize the log-likelihood numerically. When the log-likelihood has multiple maxima (i.e., a surface with multiple hills and valleys) and there are a number of parameters, it can be tricky to find the global maximum. To help find the global maximum, it is useful to try several initial values for the estimates when implementing the maximization algorithm.

Example: Maximum Likelihood Estimation for Female Births

Suppose we want to estimate the number of female births in n independent births. We assume that the distribution is binomial with s successes in n trials and the probability of a female at any given birth is p. The likelihood is then $L(p \mid s, n) = kp^s(1-p)^{n-s}$ in which k is an arbitrary positive constant that does not depend on p. Here the maximum of the likelihood represents the point in the likelihood surface that has a slope of 0 and has negative curvature. In this case, the maximum can be found by setting the first derivative of the log-likelihood equal to 0 and solving for p. The MLE is $\hat{p} = s/n$, which is the observed proportion of female births. Any other value $\tilde{p}$ will lead to a lower value for the likelihood. For example, let $s = 3$ and $n = 10$, then $\hat{p} = 0.3$ and $L(\hat{p} = 0.3 \mid 3, 10) = 0.002224k$. When $\tilde{p} = 0.2$, $L(\tilde{p} = 0.2 \mid 3, 10) = 0.001678k$ and when $\tilde{p} = 0.4$, $L(\tilde{p} = 0.4 \mid 3, 10) = 0.001792k$.

HYPOTHESIS TESTING PROVIDES CONFIDENCE IN A RESULT

Along with estimation, we want some measure of confidence in our result. Hypothesis testing is a very common way to provide this confidence. Hypothesis testing starts with a research hypothesis and a contradictory explanation. The contradictory explanation is the null hypothesis H_0, and the research hypothesis (the hypothesis we often believe is the truth) is the alternative hypothesis H_a. The purpose of the hypothesis test is to support or refute H_0. H_0 will be rejected or will fail to be rejected depending on the value of a test statistic. If the test statistic is very unlikely given H_0 is true, then H_0 is rejected. The decision made to reject or not to reject H_0 falls into one of four possible cases. There are two possible correct decisions: (1) H_0 was not rejected and H_0 is actually true or (2) H_0 was rejected and H_a is actually true. There are two possible incorrect decisions: (1) type I errors (false positives) in which H_0 was rejected but H_0 is actually true or (2) type II errors (false negatives) in which H_0 was accepted but H_a is actually true. Although the stochastic nature of data dictates that a correct decision cannot be made all the time, a desirable hypothesis testing procedure has both low type I and type II error rates.

p Values

As part of a hypothesis test, we determine how unusual a test statistic is if H_0 is true. One such measure is the p value, which is the probability of a test statistic that is as extreme or more extreme than the observed test statistic if H_0 is true. Before applying a test, a significance level (α) should be designated. The significance level is a cutoff, and H_0 is rejected if the p value is less than α. In other words, H_0 is rejected when the p value is considered too small to have occurred by chance under H_0, in which too small is defined by being less than a prechosen α. The significance level depends on how willing we are to make a type I error, and, although there are some popular values, the value of α is an arbitrary matter of choice. See Chapter 6 for more discussion of type I and type II error rates and significance thresholds in the context of power and multiple testing in association studies.

Before moving on to an example, there are a few caveats regarding p values that deserve mention. First, the p value is not the same as the posterior probability that the null hypothesis is true, $P(H_0 \mid \text{data})$. Second, just because a p value is small it does not mean that the null hypothesis is impossible. As an example, a p value of 0.05 means that if we repeated the experiment 100 times and the null hypothesis is true, then we would expect to see data as extreme or more extreme than observed 5 of the 100 times. This relates to our final caveat, if multiple experiments are performed, then the rejection of any individual test must be made less frequent if an overall experimental-wide α is desired. That is, we need to take into account multiple testing in our decision-making process. Some researchers advocate monitoring the expected proportion of false rejections of H_0 among all rejections, the false discovery rate (Benjamini and Hochberg 1995).

Example: Female Births

Suppose we want to determine if females are less likely to be born than males in a population. Our null hypothesis (H_0) is $p = 1/2$, and in this case our alternative hypothesis, H_a, is $p < 1/2$ (a one-sided test). We collect data on 10 births (too small a sample size in practice but useful for this illustration). The null distribution is binomial, with 10 trials and $p = 1/2$. Let $\alpha = 0.05$. The test statistic is the number of female births (s). We find that only 1 of these 10 children are female ($s = 1$). Test statistics as extreme or more extreme for this one-sided test are $s = 1$ and $s = 0$. The p value is 0.01074 (Fig. 2.4).

Alternatively we could have conducted a two-sided test. The research question is whether the probability of a female birth differs from the probability of a male birth. H_0 remains $p = 1/2$, but H_a is now $p \neq 1/2$, so that too many or too few female births are cause for rejection of the null hypothesis. The observed data remain 1 female birth out of 10 births. The null distribution is still binomial with 10 trials and $p = 1/2$, but now test statistics that are as extreme or more extreme than the observed test statistic are 0, 1, 9, and 10 and the p value is 0.02148 (Fig. 2.4).

Example: Association between ABO Blood Type and Ulcer Susceptibility

We return to the Clarke et al. (1959) data. The null hypothesis here is that there is no association between ABO blood type and duodenal ulcers. This is equivalent to a hypothesis of independence—that the joint probability of ABO blood type and ulcer status can be calculated as the product of the marginal probability of ABO blood type and the marginal probability of ulcer status. The alternative hypothesis is that the ABO blood type and duodenal ulcers are associated. Let $\alpha = 0.05$. We now need to specify the test statistic and its distribution under the null hypothesis. Let N_{ij} be the number of individuals with blood type i and affection status j, let N_j be the number of individuals with affection status j, let $N_{i.}$ be the number of individuals with blood type i, and let $N_{..}$ be the total number of individuals in the study. Then the expected number of individuals with blood type i and affection status j is

$$E\left(N_{ij}\right) = \frac{N_{\cdot j}N_{i\cdot}}{N_{..}}$$

under the null hypothesis of no association. As an example, the expected number of individuals with type A blood and ulcers is

$$E\left(N_{11}\right) = \frac{N_{\cdot1}N_{1\cdot}}{N_{..}} = \frac{521{*}465}{1201} = 201.72 \,,$$

which is different from the observed number (186). But could this difference be caused by chance alone or is there evidence of an association? A statistic that measures the relative difference between the observed number of individuals with each blood type and affection status from their expected values is

$$t = \sum_{j=1}^{2} \sum_{i=1}^{4} \frac{\left(N_{ij} - N_{.j}N_{i.}/N_{..}\right)^2}{N_{.j}N_{i.}/N_{..}} \ .$$

When there is no association (under the null hypothesis), t has an asymptotically chi-square distribution with three degrees of freedom (df). Using the Clarke et al. (1959) data, $t = 8.824$. The probability of a t that is greater than or equal to 8.824 under the null hypothesis of no association (the p value) is 0.0317. Thus, we reject the null hypothesis and conclude that there is evidence for an association between ABO blood type and ulcer susceptibility.

Likelihood Ratios, Likelihood Ratio Test Statistics, and Logarithm of the Odds Scores

Often hypothesis testing uses a likelihood ratio (LR). We can compare the likelihood for different models or hypotheses by calculating the likelihood ratio, LR $= L_{H1}/L_{H2}$, in which L_{H1} is the likelihood for hypothesis 1 and L_{H2} is the likelihood for hypothesis 2. When LR > 1, the data are more likely under H_1 than under H_2, and, when LR < 1, the data are more likely under H_2 than under H_1. Often, as in our female birth example, we do not have a simple alternative hypothesis. We can also form a LR for a null hypothesis whose state space is nested within a larger hypothesis space. Let θ denote the parameters, Ω_0 denote the null hypothesis space for these parameters, and Ω_A denote the alternative space. Let these two spaces be mutually exclusive ($\Omega_0 \cap \Omega_A = \emptyset$). Then the likelihood ratio compares the maximum of the likelihood for the union of the spaces to the maximum of the likelihood for just the null space,

$$LR = \frac{\max_{\theta \in \Omega_0 \cup \Omega_A} L(\theta)}{\max_{\theta \in \Omega_0} L(\theta)} \ .$$

Notice that now LR ≥ 1.

Twice the natural logarithm of the likelihood ratio is called the likelihood ratio test statistic (LRT). The dfs depend on the number of additional parameters maximized in H_a versus H_0. Under the null hypothesis, the distribution of the LRT is often (but not always) asymptotically chi-square with these dfs (Self and Liang 1987). A logarithm of the odds (LOD) score is similar to the LRT, except that a LOD score is the base 10 logarithm of the LR.

Example: Hypothesis Testing Using an LRT for Female Births

Rather than using the binomial distribution, we can conduct a hypothesis test using the LRT assuming that, under H_0, the LRT has an asymptotic chi-square distribution with df $= 1$. As before, let $\alpha = 0.05$ and let the test be two-sided. H_0 is $p = 1/2$ and H_a is $p \neq 1/2$. The likelihood under H_0 is

$$\max_{p \in \Omega_0} L(p) = K \left(\frac{1}{2}\right)^{10} = K * 0.000977.$$

The maximum likelihood under H_a is

$$\max_{p \in \Omega_0 \cup \Omega_A} L(p) = K \left(\frac{1}{10}\right) \left(\frac{9}{10}\right)^{9} = K * 0.03874.$$

LR $= 39.682$, LRT $= 7.3613$, and the p value $= 0.00666$. Note that this p value from the large sample approximation to the distribution differs from the p value we calculated assuming that the data came from a binomial distribution (0.02148). When the sample size is small, we are better off using the exact p value derived from using the binomial distribution rather than relying on an asymptotic approximation.

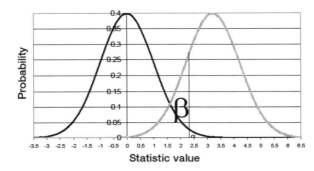

FIGURE 2.5 Graphical depiction of the null and alternative densities. The black curve represents density under the null hypothesis and the gray curve represents the density under the alternative hypothesis. The vertical line at 2.32 marks the boundary of the one-sided significance level ($\alpha = 0.01$, area to the right of the line under the null density curve) and the type II error ($\beta = 0.19$, area to the left of the line under the alternative density curve). The power is 0.81.

Power

The power of a test is the probability of rejecting the null hypothesis given an alternative hypothesis is true. Power and the probability of a type II error (β) sum to 1. Power depends on the sample size and the specifications of the alternative hypothesis. Power also depends on the maximum type I error that can be tolerated, the significance level. Holding everything else constant, the smaller α is, the lower the power and correspondingly the greater the chance of a type II error (Fig. 2.5).

Whenever possible, power estimations should be performed before embarking on a genetic analysis because they will produce information relevant to study design. As an example, the relative power of various genome-wide association study (GWAS) designs can be compared and balanced against the type I error rates. Reducing the significance level will reduce the number of false inferences of association, but it may produce unacceptable decreases in power in a GWAS. In such circumstances, one possible approach is to modify the overall nature of the testing, accepting that false positives will be generated if the "true" results are also to be found, and then seeking replication to distinguish between the true and false results. How to optimize GWAS designs is an active area of research (e.g., see Dudbridge et al. 2006; Elston et al. 2007; Kraft and Cox 2008; Moskvina and Schmidt 2008).

SUMMARY

Statistical genetics is a vast discipline. We have covered only the bare bones of probability and statistics related to the discipline of this book. Besides the references given here, each chapter in this volume cites background references that provide additional mathematical and statistical underpinnings.

REFERENCES

Almasy, L. and Blangero, J. 1998. Multipoint quantitative-trait linkage analysis in general pedigrees *Am. J. Hum. Genet.* **62**: 1198–1211.

Amos, C.I. 1994. Robust variance-components approach for assessing genetic linkage in pedigrees. *Am. J. Hum. Genet.* **54**: 535–543.

Bauman, L.E., Almasy, L., Blangero, J., Duggirala, R., Sinsheimer, J.S., and Lange, K. 2005. Fishing for pleiotropic QTLs in a polygenic sea. *Ann. Hum. Genet.* **69**: 590–611.

Benjamini, Y. and Hochberg, Y. 1995. Controlling the false discovery rate—A practical and powerful approach to multiple testing. *J. Roy. Stat. Soc. B* **57**: 289–300.

Brase, C.H. and Brase, C.P. 2007. *Understanding basic statistics,* 4th ed. Houghton Mifflin, New York.

Clarke, C.A., Price-Evans, D.A., McConnell, R.B., and Sheppard, P.M. 1959. Secretion of blood group antigens and peptic ulcers. *Brit. Med. J.* **1**: 603–607.

Dudbridge, F., Gusnanto, A., and Koeleman, B.P. 2006. Detecting multiple associations in genome-wide studies. *Hum. Genomics* **2:** 310–317.

Elston, R.C., Lin, D., and Zheng, G. 2007. Multistage sampling for genetic studies. *Annu. Rev. Genomics Hum. Genet.* **8:** 327–342.

Gonick, L. and Smith, W. 1993. *The cartoon guide to statistics*. Harper Perennial, New York.

Hardy, G.H. 1908. Mendelian proportions in a mixed population. *Science* **28:** 49–50.

Kraft, P. and Cox, D.G. 2008. Study designs for genome-wide association studies. *Adv. Genet.* **60:** 465–504.

Lange, K. 2002. *Mathematical and statistical methods for genetic analysis*, 2nd ed. Springer, New York.

Lange, K. 2003. *Applied probability*. Springer, New York.

Milton, J.S. 1999. *Statistical methods in the biological and health sciences*. McGraw-Hill, New York.

Mood, A.M., Graybill, F.A., and Boes, D.C. 1974. *Introduction to the theory of statistics*, 3rd ed. McGraw-Hill, New York.

Moskvina, V. and Schmidt, K.M. 2008. On multiple testing correction in genome wide association studies. *Genet. Epidemiol.* **32:** 567–573.

Self, S.G. and Liang, K.-Y. 1987. Asymptotic properties of maximum likelihood estimators and likelihood ratio tests under nonstandard conditions. *J. Am. Stat. Assoc.* **82:** 605–610.

Sham, P. 1998. *Statistics in human genetics*. Arnold Publishing, London.

Weinberg, W. 1908. Uber den Nachweis der Verrebung bein Menschem. *Jahreshefte Verein* **64:** 368–382.

Young, A. 1995. *Course manual for the Wellcome Trust Summer School: Genetic analysis of multifactorial disease*. Wellcome Trust, London.

3 | Linkage Analysis of Discrete Traits

Ingrid B. Borecki[1] and John P. Rice[2]

[1]Division of Statistical Genomics, Department of Genetics, Washington University School of Medicine, St. Louis, Missouri 63110; [2]Department of Psychiatry, Washington University School of Medicine, St. Louis, Missouri 63110

INTRODUCTION

Linkage analysis is an important tool for the investigation of the genetic basis of disease traits and has led to the discovery of the single genes underlying such diseases as cystic fibrosis, Duchenne muscular dystrophy, Huntington's disease, and a wide variety of metabolic disorders. For certain kinds of traits, linkage analysis allows the localization of trait loci on a known genetic map, which can be followed up with fine mapping based on linkage disequilibrium (LD) and ultimate identification of the gene. Successes using this approach have been largely confined to Mendelian monogenic disorders or oligogenic disorders (e.g., breast cancer, Alzheimer's disease). Complex diseases (e.g., asthma, schizophrenia) involving several to many loci, as well as environmental factors, may require additional strategies to locate risk loci.

UNDERSTANDING LINKAGE

Linkage describes a physical relationship among loci that are in close proximity to each other along a DNA strand. Mendel's second law of independent assortment predicts that two loci will segregate independently in forming the gametes, and this is true for loci that are on different chromosomes or are quite far apart on the same chromosome. However, chiasmata (non-sister chromatid crossover during meiosis) can form and recombination occurs during the process of gametogenesis. The closer two loci are, the more likely the alleles on each chromatid will segregate together. When two loci show this departure from independent assortment, they are said to be linked.

To illustrate, let us consider a doubly heterozygous parent at two loci, A and B (AB/ab), as depicted in Figure 3.1A. Provided there is no recombination, the resulting gametes are of the parental type. With one recombination between the two loci (Fig. 3.1B), recombinant gametes are formed (Ab/aB). With two recombinations (Fig. 3.1C), the result is apparently the parental types; without intervening markers, it is not possible to appreciate that a double recombination has occurred. With three recombinations (Fig. 3.1D), again a recombinant type is formed. In general, recombinant types will be observed when there are an odd number of recombinations between genotyped loci.

Let us define θ as the frequency of an odd number of recombinations. When two loci are on different chromosomes or far apart on the same chromosome, $\theta = 1/2$, which is equivalent to independent assortment. In linkage analysis, we attempt to find loci for which $\theta < 1/2$. The recombination fraction

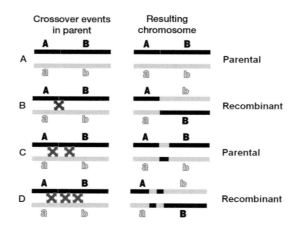

FIGURE 3.1 Gametic products of recombination from a doubly heterozygous parent AB/ab. *Left* column illustrates recombinations in the parent during gametogenesis and *right* column shows the resulting chromosomes transmitted to offspring.

θ is roughly equivalent to the distance between two loci measured in centiMorgans, for small distances. The correspondence between genetic distance (in cM) and physical distance (in base pairs) varies considerably across the genome and by sex, although a rough equivalence is 1 cM $\sim$ 1 Mb (Kong et al. 2002).

The goal of linkage studies is to estimate the recombination fraction between two loci to locate them relative to one another. The simplest method for accomplishing this is simply to count the number of recombinant gametes out of the total sampled for known mating types. However, fundamental to recognizing these types is knowledge of the *phase of linkage* in the parents—that is, the alleles carried together on the same parental chromosome. In the example described above, we assumed the given *phase* to be AB/ab. However, reported genotypes will simply appear as AaBb, and the alternative phase Ab/aB is equally frequent under linkage equilibrium. The designation of recombinant and parental gametes is conditional on both the mating type and the phase of linkage.

To estimate the recombination fraction θ and to combine information over families, maximum likelihood theory is used to calculate the probability of observing the data at hand while estimating the recombination fraction. The likelihood can be written as a function of the probabilities of recombinant (R)/nonrecombinant (NR) gametes, conditional on mating type. In the example shown in Figure 3.2, we assume that the phase of linkage in the parents can be ascertained from grandparental information. Given the phase, it is possible to designate each of the six offspring, with only the last one being an R type. The NR types occur with probability $(1 - \theta)$, and the R types occur with probability θ. The joint probability of the sibship under a linkage hypothesis is $(1 - \theta)^5 * \theta$, because there are five NRs and one R, whereas the same probability under independent assortment is simply $(1/2)^6$. Taking the logarithm (to the base 10) of the probability ratio provides a measure of the weight of support for the linkage hypothesis versus the null alternative of no linkage and is known as the LOD (logarithm of the odds) score (Morton 1955). The LOD score for this family is $\log_{10}[(1 - \theta)^5 * \theta/(1/2)^6]$. In general, the LOD is

$$Z(\theta) = \log_{10}[(L(\text{pedigree} \mid \theta \text{ estimated})/(L(\text{pedigree} \mid \theta = 1/2))] \,. \tag{1}$$

If the phase is not known, an average over the two possible phases is taken:

$$Z(\theta) = \log_{10}[(1/2)(\text{LR for phase A}) + (1/2)(\text{LR for phase B})] \,, \tag{2}$$

where LR is the likelihood ratio, as described above. Because of the log transformation, the joint probability of a collection of families is simply the sum of their specific LOD scores. Note that to

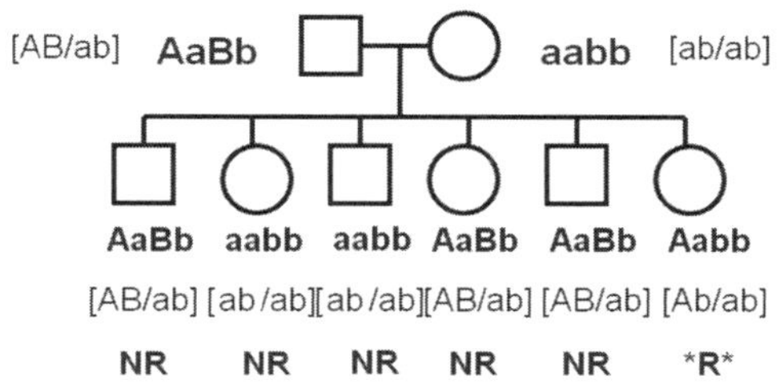

- **If A and B are linked:**
     $(1-\theta)$ $(1-\theta)$ $(1-\theta)$ $(1-\theta)$ $(1-\theta)$ θ $= (1-\theta)^5\,\theta$
- **If A and B are NOT linked:**
     ½   ½   ½   ½   ½   ½ $= (½)^6$

FIGURE 3.2 Sample probability calculation for a phase-known pedigree under a linkage hypothesis and under the null. The linkage phase for each individual is shown in brackets.

obtain a numerical value for the score, one must specify a particular value for the recombination fraction. Typically, an investigator may do a grid search hypothesizing a variety of values (e.g., 0, 0.05, 0.1, 0.2, 0.3, 0.4) for the recombination fraction; the value with the highest LOD score is the most likely one. Alternatively, maximum likelihood estimation methods can be used to precisely estimate the value of θ with the highest likelihood, transcending the discrete grid search. See Chapter 2 for an introduction to probability and maximum likelihood.

LINKAGE ANALYSIS OF A DISEASE PHENOTYPE

Thus far, we have been discussing linkage analysis between markers in which there is a perfect correspondence between genotype and phenotype. However, when the investigator's interest is in a particular disease, an additional step is necessary. With a disease trait, only the phenotype is observed, and the objective is to locate and identify the underlying trait locus. To carry out the linkage analysis, it is necessary essentially to impute the underlying trait locus genotype. This is done by relying on assumptions regarding the mode of inheritance of the locus underlying the trait.

First, we assume the disease or trait of interest is caused by a single locus described by the single major locus (SML) model. That is, for a two-allele locus influencing a dichotomous phenotype, we denote the two alleles by D and d, with p the frequency of D and $q = 1 - p$ the frequency of d. Under the assumption of Hardy–Weinberg equilibrium (HWE), the probabilities for the three genotypes DD, Dd, and dd are p^2, $2pq$, and q^2, respectively. The penetrances f_{DD}, f_{Dd}, and f_{dd} are defined as the probability that individuals of genotype DD, Dd, and dd, respectively, are affected. Accordingly, the SML model can be described in terms of the four parameters p, f_{DD}, f_{Dd}, and f_{dd}. We further assume that all familial resemblance is due to that single locus. Thus, the phenotype of an individual depends only on his/her genotype. That is, if $(X_1, \ldots, X_m)$ denotes the phenotypes in a family of size m, and $(g_1, \ldots, g_m)$ denotes their genotypes, then $P(X_i \mid g_1, \ldots, g_m) = P(X_i \mid g_i)$ and $P(X_i, X_j \mid g_i, g_j) = P(X_i \mid g_i)P(X_j \mid g_j)$, for individuals i and j, where $P(\ )$ denotes the probability of the event in parentheses. Let θ denote the recombination fraction between the trait locus (D) and a marker M. Then the likelihood of a family depends on the five parameters $p, f_{DD}, f_{Dd}, f_{dd}$, and θ. Under the assumptions of the SML model, the likelihood of large pedigrees can be calculated efficiently using the Elston and Stewart algorithm (1971). This method was implemented in the original LINKAGE package (Lathrop et al. 1984); Cottington and colleagues (1993) then developed FASTLINK with more efficient calculations. Other packages, some with computational enhancements, include MENDEL (Lange et al. 1988) and LODLINK and VITESSE (O'Connell and Weeks 1995).

The LOD curve $Z(\theta)$ is defined as

$$Z(0) = \log_{10} \frac{L(\text{data} \mid \theta, f_{DD}, f_{Dd}, f_{dd})}{L(\text{data} \mid \theta, 1/2, f_{DD}, f_{Dd}, f_{dd})}, \quad (3)$$

where the LOD score $Z(\theta)$ is summed over all families in the study for a particular value of θ. Note that in a nuclear family, one parent must be a double heterozygote to provide evidence for linkage. In the example in Figure 3.2 with two genotyped markers, it is the transmissions from the doubly heterozygous father that are being scored; the doubly homozygous mother is uninformative for linkage. For a disease locus with alleles D and d, and a marker locus with alleles A and a, the parent whose genotype is Ddaa always transmits a to his offspring and it is impossible to deduce whether a recombination has occurred. Moreover, if the father's and mother's genotypes are DdAa and ddaa, respectively, and the offspring is DdAa, there is no information for linkage because the phase of the doubly heterozygous parent is not known. If we know the father's haplotype is DA/da, then we can directly count the number of R and NR offspring. If we observe k recombinants in n matings, the LOD score of the test as the (base 10) log of the likelihood ratio is

$$\text{LR} = \frac{L(k/n \mid \theta = \theta_0)}{L(k/n \mid \theta = 1/2)} = \frac{\theta_0^k (1-\theta_0)^{(n-k)}}{(1/2)^n}. \quad (4)$$

We then see that in the particular case that $k = 0$, the maximum likelihood estimate for θ_0 is $\theta_0 = 0$, and hence $\text{LR} = 2^n$ so that the LOD score is $\log_{10}(\text{LR}) = n \log_{10}(2)$. Each observation ($n$) adds about 0.3 (i.e., $\log_{10}$ of 2) to the observed LOD score.

Dominant, recessive, and additive models (or anything in between) can be posited, along with assumed penetrances. A dominant model can be captured by the values $f_i = (0, 1, 1)$ corresponding to the genotypes i = DD, Dd, and dd, whereas a recessive model is $(0, 0, 1)$. An additive model places the value of a heterozygote exactly in the middle between the two homozygotes $f_i = (0, 0.5, 1)$. Incomplete penetrance is specified by a penetrance < 1 for a risk genotype, whereas sporadic cases or phenocopies can be accommodated by a penetrance > 0 for a normal genotype. Incomplete penetrance allows for individuals who carry one or more risk alleles but never become affected, whereas phenocopies allow for individuals who are affected because of environmental factors or loci other than the one currently under consideration.

These exact assumptions play a critical role in the calculation of the probability of family data and LOD scores. In the example shown in Figure 3.3, a three-generation pedigree is depicted with affected individuals shaded and genotypes at a hypothetical marker locus A. A dominant mode of inheritance may be a reasonable initial assumption because there are affected individuals in each generation. Under this mode of inheritance, each affected individual must have at least one copy of the underlying disease susceptibility allele (D). If we further assume that the disease allele is relatively rare in the population, then those affected individuals are more likely to be heterozygotes than homozygotes (if q is the frequency of the risk allele and small, and the frequency of the normal allele is $p = 1 - q$, then $2pq \gg q^2$). Assuming full penetrance and no phenocopies, the genotypes for the disease locus shown under each individual can be imputed. Using the grandparents to establish phase, we know that the disease allele is traveling with the A allele in the affected father. The offspring can be designated as R or NR as before, and all subjects with the risk allele are penetrant (P). Simply relaxing one of these assumptions—allowing for incomplete penetrance—already changes the picture (as shown below the dotted line in Fig. 3.3). In that case, it is possible that the middle two offspring actually carry the risk allele, but are nonpenetrant (NP) and thus not affected. That changes the recombinant status of those offspring. These possibilities are weighted by the precise value of the penetrance between 0 and 1. This example serves to show the reliance on assumptions regarding the genetic model of this so-called "parametric" linkage analysis because of the need to specify the trait genetic model.

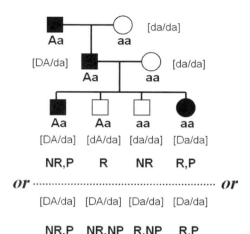

FIGURE 3.3 Confounding of recombination and penetrance in a pedigree segregating an autosomal dominant susceptibility locus.

There are several variations on the basic genetic model that may need to be accommodated in linkage analysis. Not all diseases of interest are present at birth; therefore it may be necessary to take age at onset into account. That is, there could be individuals carrying the risk genotype, but who are not sufficiently through the risk period in order to be affected (e.g., Huntington's chorea with an average onset in middle age). Likewise, the older an individual is and remains not affected, the greater the chances that their genotype does not contain risk allele(s). This can be modeled by defining different penetrances for each genotype by age strata (user defined). Similarly, variations in disease prevalence by sex may necessitate sex-specific penetrances. For rare diseases, it may also be important to model a spontaneous mutation rate. However, the challenge is that these values must be provided by the investigator, usually in the absence of actual data to inform their choices. To devise feasible sets of parameter estimates (including disease allele frequency and appropriate penetrance functions), it is possible to constrain the analysis to those sets that lead to a population prevalence consistent with that observed in the reference population.

GENETIC MAPS AND MULTIPOINT ANALYSIS

Different types of genetic markers have been used to create a linkage map of the human genome. Most linkage studies have been performed using ~300–400 microsatellites, or polymorphisms with variable-number tandem repeats (VNTRs), uniformly distributed across the genome at a density of ~1 in every 5–10 cM. These markers were favored because of their high polymorphic information content; a large number of alleles ensures a high frequency of heterozygosity, which is necessary for informative analysis as described above. However, microsatellites are technically somewhat difficult to type reliably (often distinguishing genotypes varying in the number of copies of a dinucleotide repeat by a factor of 2), and, with the discovery and characterization of an abundant class of diallelic single-nucleotide polymorphisms (SNPs), linkage maps are now constructed using these polymorphisms. SNPs are relatively easy and inexpensive to type on dense arrays, and what SNPs lack in polymorphism, they make up for in density. Usually ~3000 SNPs will provide a map of equal information content to traditional microsatellite maps. However, care must be taken to select a set of markers not in LD; if the parents are not genotyped, the LOD score is inflated and the type I error rate (i.e., false positives) is increased for affected pair-based analyses of the type described below.

The human linkage map was created using LOD score analysis, ascertaining order and recombination fraction (θ) among the markers. Note that θ is a probability, and it is necessary to relate θ to genetic distance. For three markers, the θ between the flanking markers is less then the sum of the θs between the contiguous markers. We say that two markers are 1 cM apart if the probability of recombination is 0.01. The human genome is approximately 3330 cM in length. Under the assumption that crossovers in different intervals are independent (no interference), then Haldane (1919) noted the map distance in cM, x, is given by $x = -(1/2)\ln(1 - 2\theta)$ for $0 \leq \theta < 1/2$. The inverse is given by $\theta = (1/2)[1 - \exp(-2|x|)]$. When performing multipoint analyses, the θ values are transformed to distances, so that only $N - 1$ are needed when analyzing N markers and others are computed using distances and then applying the inverse of the Haldane function. There are several other such map functions (cf. Ott 1999) that allow for interference. They all have the property that $x \approx \theta$ when θ is small.

Multipoint analysis can be performed by moving the location of the trait locus across the linear genetic map, in the same way that we gridded across values of θ above, and calculating LOD scores based on all the marker/haplotype information in the family on the chromosome. Again, the location with the highest LOD score is the most likely location of the trait locus. It should be noted that the extended haplotype information afforded by the multipoint approach allows for the calculation of more accurate odds, and thus provides more information to exclude or confirm linkage. There is a computational challenge associated with multipoint analysis in that although the computer programs can analyze pedigrees with hundreds of individuals, it is not practical to do so using hundreds of markers at once. An alternative algorithm, as implemented in GENEHUNTER (Kruglyak and Lander 1995) or MERLIN (Abecasis et al. 2002) can handle hundreds of markers, but only in families of size less than 20 individuals. A typical approach for large pedigrees is to use a sliding window of markers. Alternatively, Markov chain Monte Carlo–based estimation approaches, implemented in programs such as SimWalk and LOKI, can accommodate both large pedigrees and a large number of markers.

WHAT ARE THE EFFECTS OF MODEL MISSPECIFICATION?

The sensitivity of parametric LOD score linkage analysis to misspecification of the genetic model has been investigated, and the results are nicely summarized in Xu et al. (1998). By far, the most influential parameter is the dominance (Clerget-Darpoux et al. 1986), which does not influence the type I error rate (false positives), but can severely diminish evidence for linkage. In contrast, incorrect specification of the penetrances or the disease allele frequency can produce variations in the expected LOD score, with more severe drops associated with values further from the truth, but still retaining the ability to detect trait loci, at least under some simple monogenic scenarios.

Another noteworthy feature of linkage analysis, depicted in Figure 3.4, is that the disease state (D) and the marker (M) are the two measured entities (shown in squares), leaving the underlying trait locus (G) as a latent, unmeasured factor (shown in a circle) that we wish to locate and identify. This leads to a situation in which recombination fraction (θ) or location is confounded with penetrance. Location can be estimated fixing the penetrance or penetrance can be estimated fixing the

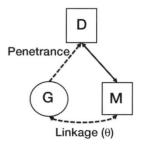

FIGURE 3.4 The linkage paradigm, where D is disease state, M is marker, G is underlying trait locus, and θ is recombination fraction.

location, but both cannot be estimated simultaneously. Thus, incorrectly specified penetrances also can affect the estimate of location, which could adversely impact follow-up fine mapping or positional cloning efforts.

NONPARAMETRIC LINKAGE ANALYSIS USES IDENTITY BY DESCENT

In the parametric approach described above, investigators would first perform segregation analysis to estimate allele frequency and penetrances before linkage analysis. This led to the mapping of many Mendelian diseases. Although the use of parametric methods for complex diseases has been advocated (Greenberg 1989), the lack of success using the SML model for more complex traits moved the field to the use of sib-pair methods (cf. Kruglyak and Lander 1995).

These affected sib-pair (ASP) methods are often described as a "nonparametric" or "model-free" because they do not rely on assumptions or estimates of the parameters describing the mode of inheritance. However, the equivalence between ASP tests and parametric tests has been shown in some cases (Knapp et al. 1994). Furthermore, implicit model assumptions exist, and the power of such tests is thus influenced by the appropriateness of these assumptions (Whittemore 1996).

ASP analyses test for excess sharing of marker alleles identical by descent (IBD) in affected–affected sib-pairs. As noted earlier, the SML model may be parameterized in terms of five parameters:

$$\{f_{DD}, f_{Dd}, f_{dd}, q, \text{and } \theta\} . \tag{5}$$

Suarez et al. (1978) have noted that the IBD distribution of sib-pairs, conditioned on their phenotype, does not depend on q. Risch (1990) used an alternative parameterization $\{K, \lambda_s, \lambda_o, \theta\}$, where $\lambda_R = K_R/K$, with K the population prevalence and K_R the risk to a type of relative R of an affected individual. Here λ_s and λ_o are risk ratios to siblings and offspring, respectively. Moreover, Risch noted the IBD distribution in affected sib-pairs depends only on $\{\lambda_s, \lambda_o, \theta\}$. With multipoint marker data, all three parameters $\{\lambda_s, \lambda_o, x\}$, where x is chromosome location, can be estimated when a trait locus is present. Under the null hypothesis, the parameter space is degenerate, so that likelihood theory does not apply. However, Hauser et al. (1996) provide simulation results in terms of $\lambda = \lambda_o = \lambda_s$, where λ is the locus-specific recurrence risk ratio.

The important point here is that these statistics, as implemented in programs such MERLIN, are in fact the LOD score maximized over these parameters (the MOD score). The likelihood is P(marker data | both sibs affected and parents unknown phenotype), and maximizing the likelihood is equivalent to maximizing the LOD score (Clerget-Darpoux 1986).

Risch (1990) used the MOD score for affected sib-pairs and their parents. He parameterized the model in terms of IBD at the marker locus and notes that if IBD can be determined, then the MOD score does not depend on the marker allele frequencies. Pairs where parents are missing or the IBD cannot be deduced can also be used with this method, although the MOD score is a function of the marker allele frequencies. Holmans (1993) restricted the range of the admissible parameter space in Risch's approach to yield a more powerful test. He further notes that even if the marker frequencies are known, there can be more power to use $2N$ pairs with untyped parents than to use N pairs with typed parents (with a total of $4N$ typings in each case). Judicious use of these guidelines within the context of a particular study can lead to powerful tests of linkage and localization of trait loci.

HOW TO INTERPRET LOD SCORES

There are historical roots for the interpretation of LOD scores (Morton 1955). A value of 3 is taken as evidence of linkage (essentially the odds favoring a specific linkage hypothesis are 1000:1),

TABLE 3.1 Correspondence between landmark LOD scores and asymptotic p values

LOD	p value
0.834	0.05
2.0	0.001
2.5	0.0007
3.0	0.0001
3.6	0.00006

whereas at the opposite end of the spectrum, a value of –2 is taken as evidence against linkage (with odds 1:100 that the loci are linked). It is easy enough to convert a LOD score to a chi-square with 1 degree of freedom (representing the hypothesis test for the estimated value of θ vs. its restricted value of 1/2 indicating no linkage) by changing from the $\log_{10}$ to the natural log:

$$2(\log_e 10) * Z = 4.6 * Z \approx \chi_1^2, \tag{6}$$

where Z is the LOD score at the maximum of $Z(\theta)$. The correspondence between common landmarks on both scales—odds ratios and significance values—is shown in Table 3.1. Although a LOD score of 3 represents odds 1000:1, it is actually associated with a p value of 1:10,000. In practice, investigators do not simply test one linkage relationship, but rather conduct a genome-wide scan to locate trait loci, which naturally leads to a multiple testing issue. Lander and Kruglyak (1995) calculated that to achieve genome-wide significance for linkage, assuming an infinitely dense genetic map, a critical value of LOD ~ 3.6 is necessary. That is, 1 in 20 genome screens would provide a LOD score of 3.6 that is false. LOD scores ≥ 2 can be viewed as suggestive, requiring additional supporting data or replication in an independent study.

GENETIC HETEROGENEITY IN MAPPING COMPLEX DISEASES

Recent efforts in gene mapping have increasingly focused on more complex phenotypes and diseases, in that simple Mendelian genetic models may not be sufficient to describe their genetic inheritance. One issue of great importance is genetic heterogeneity; that is, independent variants give rise to the same clinical phenotype. For example, this includes situations in which variants or mutations influencing any of the factors involved in a metabolic pathway could lead to failure of that pathway. In fact, this has been one of the strongest justifications for studying single extended pedigrees, as it is likely that genetic heterogeneity will be limited in a particular lineage and only one or a few mutations can be responsible for the phenotype of interest. Unrecognized genetic heterogeneity can lead to substantial losses of power to detect any underlying trait loci. Genetic heterogeneity presenting in the form of different mutations within the same gene influencing disease risk (also known as allelic heterogeneity) is not nearly so problematic in the linkage context; cosegregation of markers with disease alleles in families (even different ones across families) would still be expected to produce a linkage signal that could be traced to the sought-after disease locus. However, it is possible that a linkage signal could be degraded in rare situations with unusual combinations of disease alleles' different dominance characteristics.

There are several strategies that can be used to confront genetic heterogeneity (here, meaning different disease loci). Perhaps the most important requires attention to phenotypic assessment and definition of "affected" status. Furthermore, age at onset also can be an important factor. Generally, it is assumed that cases with an earlier onset are more likely to be the result of penetrant susceptibility alleles and less likely to be sporadic cases. Careful clinical observations can help partition a sample into more homogeneous subgroups. For example, only ~20% of all breast cancers are familial, and the disease is relatively common such that not all families with an affected woman would be segregating a susceptibility locus. However, focusing on families with multiple affected members with early premenopausal onset, as well as those in which ovarian cancer also occurred, led to the identification of *BRCA1*, a major susceptibility locus accounting for approximately one-half of all familial cases (Hall et al. 1990). Therefore, careful choice of subsets of families may greatly increase the power to detect disease susceptibility loci, even if there is a reduction in sample size.

Examination of the family-specific LOD scores can provide a clue suggestive of heterogeneity if there are families strongly supporting linkage while others do not. Examination of the differences between these groups of families potentially can provide insight as to external stratifying factors that may result in more homogeneous subsets (Faraway 1993). This partitioning can be achieved statistically using an admixture test as suggested by Ott (1991). In this approach, a proportion α of the families are linked while $1 - \alpha$ are not ($\theta = 0.5$). This additional parameter, α, is estimated along with the recombination fraction.

WHAT IS THE UTILITY OF LINKAGE ANALYSIS IN AN ERA OF GENOME-WIDE ASSOCIATION STUDIES?

Linkage analysis requires family data to trace the cosegregation of loci across generations, and it can provide localization of trait loci. How well these loci are located depends on the locus-specific effect size, the number of informative families, the size of the families, and the extent of heterogeneity. With the ability to locate trait loci via linkage disequilibrium mapping using genome-wide association studies (GWAS), the question is: How useful are linkage studies? One of the most prominent and persistent challenges of the GWAS approach lies in the statistical interpretation of the results accounting for a vast multiple comparisons problem. Many types of filters with which to reduce the number of statistical tests are used (e.g., restricting attention to candidate genes, coding sequences, and rapidly evolving sequences). Linkage analysis can provide valuable priors of the likely locations of trait loci that can be formally used in a Bayesian framework to improve interpretability (see, e.g., Lewinger et al. 2007). Combined linkage-association analyses also hold promise in that they bring complementary information to the task. Moreover, the current prevailing approach is focused on identifying common variants that can be found on common haplotypes. Investigators are slowly turning their attention to the possibility that collections of rare, or at least less common (<1%), variants may account for complex diseases, with different combinations leading to clinically similar phenotypes. This type of heterogeneity could be effectively addressed with linkage-association studies conducted in large pedigrees segregating a small set of variants.

In conclusion, many disease loci have been successfully identified using linkage analysis, and careful use of a variety of methods, appropriate to the problem and data at hand, will be necessary to understand the genetic underpinning of conditions of biomedical importance.

REFERENCES

Abecasis, G.R., Cherny, S.S., Cookson, W.O., and Cardon, L.R. 2002. Merlin—Rapid analysis of dense genetic maps using sparse gene flow trees. *Nat. Genet.* **30:** 97–101.

Clerget-Darpoux, F., Bonaiti-Pellie, C., and Hochez, J. 1986. Effects of misspecifying genetic parameters in lod score analysis. *Biometrics* **42:** 393–399.

Cottingham, Jr., R.W., Idury, R.M., and Schäffer, A.A. 1993. Faster sequential genetic linkage computations. *Am. J. Hum. Genet.* **53:** 252–263.

Elston, R. and Stewart, J. 1971. A general model for the genetic analysis of pedigree data. *Hum. Hered.* **21:** 523–542.

Faraway, J.J. 1993. Distribution of the admixture test for the detection of linkage under heterogeneity. *Genet. Epidemiol.* **10:** 75–83.

Greenberg, D.A. 1989. Inferring mode of inheritance by comparison of lod scores. *Am. J. Med. Genet.* **34:** 480–486.

Hall, J.M., Lee, M.K., Newman, B., Morrow, J.E., Anderson, L.A., Huey, B., and King, M.C. 1990. Linkage of early-onset familial breast cancer to chromosome 17q21. *Science* **250:** 1684–1689.

Hauser, L., Boehnke, M., Guo, S., and Risch, N. 1996. Affected-sib-pair interval mapping and exclusion for complex genetic traits. *Genet. Epidemiol.* **13:** 117–137.

Heath, S.C. 1977. Markov chain segregation and linkage analysis for oligogenic models. *Am. J. Hum. Genet.* **6**: 748–760.

Holmans, P. 1993. Asymptotic properties of affected-sib-pair linkage analysis. *Am. J. Hum. Genet.* **52**: 362–374.

Knapp, M., Seuchter, S.A., and Baur, M.P. 1994. Linkage analysis in nuclear families. 2: Relationship between affected sib-pair tests and lod score analysis. *Hum. Hered.* **44**:.44–51.

Kong, A., Gudbjartsson, D.F., Sainz, J., Jonsdottir, G.M., Gudjonsson, S.A., Richardsson, B., Sigurdardottir, S., Barnard, J., Hallbeck, B., Masson, G., et al. 2002. A high-resolution recombination map of the human genome. *Nat. Genet.* **31**: 241–247.

Kruglyak, L. and Lander, E.S. 1995. Complete multipoint sib-pair analysis of qualitative and quantitative traits. *Am. J. Hum. Genet.* **57**: 439–454.

Lander, E. and Kruglyak, L. 1995. Genetic dissection of complex traits: Guidelines for interpreting and reporting linkage results. *Nat. Genet.* **11**: 241–247.

Lange, K., Weeks, D., and Boehnke, M. 1988. Programs for pedigree analysis: MENDEL, RISHER, and dGENE. *Genet. Epidemiol.* **51**: 235–249.

Lathrop, M., Lalouel, J.M., Julier, C., and Ott, J. 1984. Strategies for multi-locus linkage analysis in humans. *Proc. Natl. Acad. Sci.* **81**: 3443–3446.

Lewinger, J.P., Conti, D.V., Baurley, J.W., Triche, T.J., and Thomas, D.C. 2007. Hierarchical Bayes prioritization of marker associations from a genome-wide association scan for further investigation. *Genet. Epidemiol.* **31**: 871–882.

Morton, N.E. 1955. Sequential tests for the detection of linkage. *Am.*

J. Hum. Genet. **7**: 277–318.

O'Connell, J.R. and Weeks, D.E. 1995. The VITESSE algorithm for rapid exact multilocus linkage analysis via genotype set-recoding and fuzzy inheritance. *Nat. Genet.* **11**: 402–408.

Ott, J. 1991. *Analysis of human genetic linkage*, rev. ed. Johns Hopkins University Press, Baltimore.

Risch, N. 1990. Linkage strategies for genetically complex traits. III. The effect of marker polymorphism on analysis of affected relative pairs. *Am. J. Hum. Genet.* **46**: 242–253.

Sobel, E. and Lange, K. 1996. Descent graphs in pedigree analysis: Applications to haplotyping, location scores, and marker sharing statistics. *Am. J. Hum. Genet.* **58**: 1323–1337.

Sobel, E., Sengul, H., and Weeks, D.E. 2001. Multipoint estimation of identity-by-descent probabilities at arbitrary positions among marker loci on general pedigrees. *Hum. Hered.* **52**: 121–131.

Sobel, E., Papp, J.C., and Lange, K. 20002. Detection and integration of genotyping errors in statistical genetics. *Am. J. Hum. Genet.* **70**: 496–508.

Suarez, B.K., Rice, J.P., and Reich, T. 1978. The generalized sib-pair IBD distribution: Its use in the detection of linkage. *Ann. Hum Genet.* **42**: 87–94.

Whittemore, A.S. 1996. Genome scanning for linkage: An overview. *Am. J. Hum. Genet.* **59**: 704–716.

Xu, J., Meyers, D.A., and Pericak-Vance, M.A. 1998. Lod score analysis. In *Approaches to gene mapping in complex human diseases* (ed. J.L. Haines and M.A. Pericak-Vance), pp. 253–272. John Wiley, New York.

WWW RESOURCES

http://www.broad.mit.edu/ftp/distribution/software/genehunter/ Kruglyak and Lander 1995. GENEHUNTER.

http://www.broad.mit.edu/ftp/distribution/software/genehunter/ Abecasis et al. 2002. MERLIN.

http://www.genetics.ucla.edu/software/mendel Lange et al. 1998. MENDEL.

http://www.genetics.ucla.edu/software/simwalk Sobel and Lange 1996; Sobel et al. 2001, 2002. SimWalk.

http://linkage.rockefeller.edu/soft/linkage/ Cottington et al. 1993. FASTLINK.

http://linkage.rockefeller.edu/soft/linkage/ Lathrop et al. 1984. LINK-AGE.

http://www.stat.washington.edu/thompson/Genepi/Loki.shtml, http://loki.homeunix.net Heath 1997. LOKI.

http://watson.hgen.pitt.edu/register/soft_doc.html, http://watson.hgen.pitt.edu/register O'Connell and Weeks 1995. VITESSE.

4 | Epidemiologic Considerations in Complex Disease Genetics

John Gallacher

Department of Primary Care and Public Health, Cardiff University,
Cardiff CF14 4YS, United Kingdom

INTRODUCTION

Association studies are becoming an increasingly popular study design not only for testing of candidate genes, but also for genome-wide localization of loci influencing complex disease. The rationale behind these types of studies as well as the statistical methods and study design for various kinds of association analyses are discussed in detail in Chapters 7, 8, and 12. This chapter will cover some of the practical concerns in conducting large-scale association studies using epidemiologic samples.

THE CONDUCT OF ASSOCIATION STUDIES

Governance

Although there are generic governance issues covering population-based research, specific issues must be addressed for genetic association studies, including informed consent, feedback of results, and data security (Lowrence and Collins 2007).

In an era of genome-wide association studies and pooled analyses combining data from multiple studies, it is difficult to be specific regarding the hypotheses that will be tested using any specific biosample or specimen. Unless the samples will be destroyed once a specified range of hypotheses has been tested, it is necessary to obtain an explicit broad-based consent from study participants that will allow the testing of unspecified emerging hypotheses. Ethics committees are now less likely to consider retrospective implicit consent as adequate.

Population-based association studies focused on complex diseases rarely provide clinically useful feedback for individual participants. This is caused by the large number of alleles, each conferring an extremely low degree of risk, that is likely to be identified. Generally it is not helpful to the participant and the participant's physician to provide feedback of unknown clinical significance, and such feedback is an unnecessary expense to the study. As a result, feedback of individual genetic results is not normal practice in these studies.

Data security is an increasingly important public issue, with particular concerns surrounding the identifiability of genetic information. Safeguards such as separating administrative from research databases and encryption are essential to maintain public confidence in the research process.

Choice of Design

Association studies are essentially observational epidemiologic studies in which the independent variable (exposure) of interest is genetic. Although association studies are very flexible in their application, they require careful design if they are to provide interpretable findings. The selection of design depends on the research question of interest. Nevertheless, because of the intricate interplay between the logistics of conducting an association study and the complexity of most research questions, the design of specific association studies invariably involves some level of compromise.

The epidemiologic designs of choice are the case–control study or the cohort study (Szklo and Nieto 2007). In a case–control study, individuals with the outcome of interest, such as a complex disease (cases), are compared with disease-free individuals (controls). Cases and controls may be compared with respect to their past exposure to etiological agents. In other words, participating individuals are initially classified by disease status rather than exposure. In a cohort study, a defined group (the cohort) is assessed for exposure to etiological agents at recruitment and then followed over time to record the incidence and progress of disease. In other words, participating individuals are initially classified by exposure rather than disease status.

Case–Control Studies. For "gene hunting studies" and other research questions in which the mechanism of interest is entirely genetic, the case–control design has much to commend it. Because haplotype is determined at meiosis, it is extremely unlikely to be affected by the environment or the outcome of interest and so the possibility of spurious association or reverse causality is remote. These are strong reasons for considering associations of haplotype with complex disease to be of etiological significance. Case–control studies are also operationally efficient. Cases and controls can be selected to represent extreme tails of the risk distribution and so facilitate identifying associations. As haplotype is extremely robust, exposure data (genotype or haplotype) and outcome data (complex disease) may be collected simultaneously, thus minimizing the duration of the study and, therefore, its infrastructure requirements.

Cohort Studies. The cohort study is the design of choice for investigating research questions that are not entirely genetic, such as joint effects between genetic and environmental exposures. These are widely known as gene–environment interaction studies (discussed in detail in Chapter 11), although in this context the term "interaction" does not necessarily describe the nature of the joint effect (i.e., the extent to which the combination of genetic and environmental exposure amplifies or masks their individual effects). In a gene–environment interaction study, although haplotype may be assessed at any point in the life course, this is generally not adequate for environmental exposures. Although requiring far more infrastructure and being more resource- and time-intensive, cohort studies provide a much richer potential source of information because they can report on many outcomes simultaneously.

Nesting. In practice the flexibility of the cohort design and the efficiency of the case–control design are combined by nesting case–control analyses within cohorts. In this design, as cases accrue within the cohort they can be used for case–control analyses. The remainder of the cohort (non-cases) are used as a pool from which to select controls. Nesting within a cohort increases the likelihood that nongenetic exposure data will be comparable between case and control groups.

Marker Selection

The advent of high-throughput genome-wide scanning and the prospect of cohort sequencing mean that there will invariably be more markers than subjects available for analyses in association studies. This results in multiple statistical comparisons, increasing the likelihood of generating false-positive findings. Strategies for reducing the likelihood of accepting false-positive findings include prioritizing the markers in the analysis, increasing the level of formal statistical significance in hypothesis testing, replicating findings in other studies, and considering prior evidence of linkage that may support association findings (Ioannidis et al 2001; Tabor et al. 2002; Burton et al. 2008). See Chapter 6 for further discussion of multiple testing.

Prioritizing markers may involve assessing the available evidence on the genetic or biological plausibility of a proposed association. However, frequently the evidence for both is scarce. Nevertheless, knowledge of haplotype substructure and linkage disequilibrium as well as relevant biomolecular mechanisms is informative in the efficient selection of markers for an analysis. In practice, though, these decisions in terms of genotyping have already been made by the manufacturers of high-throughput chips, but selection can be made in interpreting the output. See Chapter 16 for more on high-throughput genotyping technologies.

Sample Selection

For case–control studies, the eligible case population is usually clinically determined by the definition of the phenotype or outcome of interest within age and sex parameters. A critical issue, therefore, is the selection of controls. In principle, controls should be genetically comparable to the cases except in terms of the haplotype of interest. There are many strategies for selecting controls, ranging from using unaffected family members to controls who are unrelated but matched with cases to varying degrees on key factors. Each strategy has its own strengths and limitations. For studies of complex disease, using family members increases the level of background comparability but reduces the power of the study to discriminate between cases and controls (i.e., increases the chances of a false-negative finding due to the possibility that relatives carry some of the risk alleles and are unaffected either because they do not carry all of the necessary alleles or because they lack some environmental exposure). The use of unrelated controls increases the power of the study to identify disease-related associations, but it also increases the likelihood that the association will be spurious (i.e., increases the chance of a false-positive result). To minimize these concerns, controls are often selected to match cases as closely as possible in age, sex, relevant environmental exposures (e.g., cigarette smoking), and general ethnic background. When controls are being chosen from a sample with existing genome-wide genotypes, it may be possible to improve on matching by general ethnic background by using the marker data to select controls matched for genetic background, minimizing the possibility of population stratification. (See below and Chapters 7 and 8 for more on population stratification.)

For cohort studies, the priority is to establish a range of exposure. Although most communities of any size will meet this criterion for most exposures, because exposure structures will vary between communities, sampling from a range of communities is preferable where possible. A genetically homogeneous population is also preferred, but this is difficult to achieve except in isolated communities.

Cohort studies have classically striven for the recruitment of population samples that are representative of the population at risk. This ideal enables data of public health interest (e.g., disease incidence) as well as data of etiological interest (e.g., relative risks) to be collected. However, representative population sampling is rarely successful because it requires extremely high response rates of ~90%. Furthermore, populations vary in their risk profiles, and a sample that is representative of one population may not be representative of another. More recently, because the logistical requirements for public health and etiological interest are distinct, these goals have been separated and studies are designed to address one or the other. Etiologic studies typically require larger sample sizes to detect weak associations, whereas public health studies require representative samples to profile the population accurately.

Genetic association studies are primarily etiological studies of biological mechanisms. Representative recruitment is rarely necessary because the mechanisms of interest will be operating in participants and nonparticipants alike (i.e., the findings will be generalizable regardless of representativeness). Complete follow-up, however, is essential. Incomplete ascertainment of outcomes is likely to lead to biased estimates of relative risk (odds ratio or risk ratio). For example, if participants with the outcome are disproportionately represented in the follow-up, the association with disease may be inflated relative to the actual risk, whereas if healthy participants are disproportionately followed, associations may be deflated. This may occur if there is an unknown confounder that influences follow-up rates, such as a subpopulation with different health behaviors in which there is also a different allele frequency. For cohort studies, therefore, participants should be selected to ensure a range of exposure and successful follow-up.

Sample Size

Association studies are well known for their inconsistent findings. Historically, this has been due largely to lack of power due in part to small sample sizes (Burton et al. 2008). Underpowered studies are more likely to report false-negative (an association is present but not detected) findings. They may also generate spurious false-positive associations, which have an increased likelihood of being reported compared with false negatives, so-called publication bias. The false-positive findings are then, unsurprisingly, not replicated. Therefore, sufficient power is an important issue in study design.

Although association studies may be more efficient than linkage studies for complex disease, the numbers of cases required for definitive results remain in the thousands rather than the hundreds (Risch 2000; Burton et al. 2008). Sample size calculations at this level of complexity are specialized and will take into account worst-case scenarios, which, depending on the hypothesis in question, are likely to involve the detection of small associations from multiple statistical tests involving gene–environment interactions.

For genome-wide association studies (GWAS), in which chips testing for 500,000 variants or more are used increasingly (Benjamin et al. 2007; Samani et al. 2007), formal levels of statistical significance vary according to the number of comparisons being made but are typically $p < 10^{-7}$ for main effects and $p < 10^{-10}$ for interactions, rather than the conventional $p < 0.05$ used for single comparisons (Burton et al. 2008b). See Chapter 6 for more on power in association studies.

Random Error

Random error occurs where individual measurements vary unsystematically around the true value (Szklo and Nieto 2007). Under these circumstances, although the point estimate (the summary statistic) is considered to be an unbiased estimate of the true value, random error around that point decreases the statistical power available to the study. Outside of the laboratory, various strategies have been used for the reduction of random error. It is rarely possible to compensate for random error in case–control studies. However, in cohort studies, measurements of exposure may be repeated. If a measurement is repeated within a short period, the differences are likely to reflect measurement error rather than actual changes in exposure. Under these circumstances, both measurements can be used to provide a more accurate measurement of current exposure. The process may be made more efficient by obtaining repeat measurements from a random sample of participants and calculating a general correction factor that may be applied to the sample as a whole.

Bias

Bias is systematic error as opposed to random error (i.e., the point estimate is systematically different from the true value [Szklo and Nieto 2007]). For example, alcohol consumption is generally underreported. Where bias is consistent between cases and controls, however, the effect is cancelled out with respect to the analysis, and the analysis can be considered to show an unbiased estimate of the difference in exposure between groups. Differential bias between cases and controls for non-genetic data is likely when data are collected retrospectively. For example, estimates of stress are likely to be different if you have been diagnosed with heart disease (Gallacher et al. 1984). Other exposures such as exercise or diet are also difficult to assess retrospectively. The most secure solution is to use a cohort design in which concurrent measurement of exposure is possible before the occurrence of disease.

Changes in Exposure

There may be changes in exposure over time for nongenetic factors. For example, smokers may smoke less or quit. This is difficult to measure in case–control studies because recall of past exposure is likely to involve considerable random error and bias. In a cohort study, however, measurement of exposure may be repeated at intervals to obtain a more accurate assessment of exposure during the period of the study.

Quality Control

For biosamples, the challenge for any study of the magnitude required for association studies is one of quality control. Issues to consider when designing sample handling protocols and reviewing sample quality include central versus local processing of samples, needle to processing and storage time, transport, and secure storage (including a secondary archive for samples of high value) (Manolio 2008). For cohort studies, retrieval is an important downstream consideration as most samples will be used for nested case–control analyses and selected for genotyping from within a much larger pool of samples. Extensive piloting of sample collection, processing, storage, and retrieval procedures is a prudent and cost-effective strategy for achieving high-quality biosamples (Elliott and Peakman 2008; Peakman and Elliott 2008).

Paper is unnecessary for self-report data. It is possible for all self-report data, including a record of consent, to be collected, processed, and stored electronically. This reduces the opportunity for human error and enables the data to be available for analysis in virtual real time. Where interview-generated data are desirable, data can still be recorded electronically. Additionally, computerized questionnaires can be designed with internal consistency checks that require participants to verify or reenter information that appears to conflict (e.g., an age of onset or diagnosis that is older than a participant's current age).

THE INTERPRETATION OF ASSOCIATION STUDIES

Understanding Causality

Establishing strong evidence of genetic causality is difficult. Whether through linkage or association designs, statistical association is not proof of causation. Altshuler et al. (2000) provide a thoughtful commentary on this issue, pointing out that there is no generic solution to providing sufficient evidence to establish genetic causality. Consistent statistical evidence, whether by linkage or association (preferably both), is encouraging, but biological functionality is the final criterion. For many conditions in human genetics, the closest simulacrum of functionality will most likely be animal-based model systems.

Causality, Effect Modification, and Confounding

In the absence of experimental evidence of biologic causality, associations from observational studies may indicate relationships that are directly causal (or mediate along a causal pathway) or modify an effect along a causal pathway. Each of these is etiologically informative. However, observational associations may also be spurious and caused by confounding. Confounding of an association between two variables occurs when their association is due to each being causally associated independently with a third, usually unmeasured, variable (Greenland et al. 1999).

Figure 4.1 shows how two genes may operate to affect disease. In Figure 4.1A, gene 1 (G1) and gene 2 (G2) independently influence disease outcome (O). According to the model, G1 and G2 may be considered to be directly causal, although in reality they are part of complex causal pathways; the point here is that the pathways are independent. Figure 4.1B shows how two genes may interact to

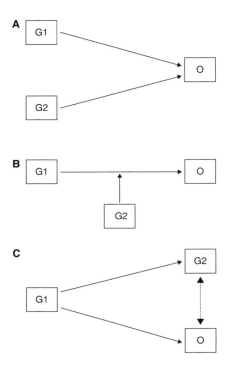

FIGURE 4.1 Possible interpretations of associations between genetic factors (G1 and G2) and complex disease outcomes (O). (*A*) Causal pathway from G1 via G2 to O. (*B*) Effect modification of G1 to O pathway by G2. (*C*) Confounding of association between G2 and O by G1.

affect causality. The association of G1 with O is affected by the presence or absence of G2. In this model, G2 is considered to be an effect modifier of the causal association of G1 with O. However, not all associations reflect causal pathways. In Figure 4.1C, an apparent association between G2 and O is driven by both G2 and O being independently associated with G1, a situation that would occur when G1 and G2 are in linkage disequilibrium. Under these circumstances the association of G2 with O is confounded and may be considered spurious. Complex disease is prone to confounded associations. For example, the inverse association of height with heart disease is not due to either an effect of heart disease on height or height on heart disease but to an effect of early life nutrition on both variables.

In genetic studies, confounding may be caused by a wide variety of factors, including the population stratification of allele frequency, in which cases and controls are differentially populated by genetically distinct groups. Mendelian randomization apart, distinct populations (mating pools), such as those defined by region or ethnicity, are likely to have distinctive allele frequencies irrespective of disease risk, as well as distinctive risks for specific diseases. Allele frequencies (risk-related or otherwise) may be considered to be stratified by population. Although this will result in differential allele frequencies in which the association is causally informative, it will also result in differential allele frequencies for haplotypes that are unrelated to disease. In this latter scenario, a detected association is spurious as a result of confounding by the stratification of allele frequencies between populations.

Population stratification may be addressed in several ways. The most rigorous approach is to design out the problem by using a genetically homogeneous population from which to select cases and controls. However, the more restrictive the definition of eligible controls, such as selecting controls from within the same family as cases, the smaller the pool of eligible controls. Furthermore, the closer the genetic definition of controls to cases, the smaller their genetic distinctiveness and the greater the difficulty of discriminating between disease- and non-disease-related haplotypes. One solution is to adjust statistically for population of origin. This approach is widely used as it requires minimal additional data collection. A more sophisticated solution is to estimate differences in allele frequency between unlinked (non-disease-related) alleles between cases and controls. This method may be used to identify the presence and degree of population stratification and to adjust for it. Population stratification and methods of correcting for it in association studies are discussed in more detail in Chapters 7 and 8.

Other sources of confounding include gene–gene interactions (epistasis), in which the expression of one genotype is affected by another genotype, and gene–environment interactions, in which phenotype is dependent on genotype and exposure to an environmental factor. The distinctions made in Figure 4.1 show that care is required to identify the analytic model being used in the analysis before confounding or causality is inferred.

LARGE-SCALE DATA RESOURCES

A major advantage of association studies is their ability to exploit large sample sizes to detect weak associations. A major limitation in our scientific infrastructure is the ability to generate sufficiently large sample sizes cost-effectively. It is increasingly recognized that stand-alone research projects and studies are no longer cost-effective or efficient for investigating complex disease. In response, large-scale data resources (LSDRs) are being constructed that, because of their size, will allow a wide variety of hypotheses to be tested definitively.

A short-term but effective strategy is the pooling of data from completed studies for secondary analysis, also known as meta-analysis (Ioannidis et al. 2002; Attia et al. 2003; Salanti et al. 2005). This procedure makes efficient use of the available data but is limited to reducing the variability in reported data. This approach is valuable for comparing studies with different confounding structures to arrive at an overall estimate of the likely effect of the exposure.

A further strategy for LSDR development that focuses on efficiency is the pooling of existing data and biosamples from individual studies for further genotyping. The Wellcome Trust Case Control Consortium (WTCCC) is the most widely known example of this strategy (http://www.wtccc.org.uk). The collaboration comprises an analysis of 2000 samples from each of seven diseases (type 1 diabetes, type 2 diabetes, coronary heart disease, hypertension, bipolar disorder, rheumatoid arthritis, and Crohn's disease). For each disease, the case samples have been ascertained from studies widely distributed across the United Kingdom and were compared with a common set of 3000 nationally ascertained controls. A total of 17,000 samples were typed for 500,000 single-nucleotide polymorphisms (SNPs).

A longer-term strategy is the development of biorepositories, widely known as biobanks, that store specimens that can be linked to individual health-related information. Biorepositories are not studies in which data and samples are collected to test specific a priori hypotheses. They are hypothesis-testing resources that are not targeted to specific questions and are established to test emerging hypotheses.

Biobanks are fast becoming national status symbols, but as a result of their potentially high cost, their content is highly variable. The most highly developed biorepository is UK Biobank (http://www.ukbiobank.ac.uk). UK Biobank provides an idea of the resource required to establish a "bespoke" LSDR. It has recruited more than 250,000 of a target of 500,000 UK residents aged 40–69 yr who have consented to be followed indefinitely for genetics-related research. A wide range of baseline data, including exercise, diet, cognitive function, medical history, and environmental exposures, have been collected under standard conditions at recruitment clinics. Blood and urine specimens were also collected at the recruitment clinic, transported to a central processing facility, and stored at –80°C within 24 h of sampling. The central repository has been designed to handle 19,000 sample aliquots per day (Elliott and Peakman 2008). To ensure virtually complete follow-up, participants are followed electronically via the U.K. National Health Service for disease outcomes. To facilitate recruitment and informed consent, the study also runs its own participant support center, which handles up to 2000 calls from the general public per day. UK Biobank has its own Ethics and Governance Framework, which is overseen by an independent Ethics and Governance Council. Other large-scale studies include the Kadoorie Study of Chronic Disease in China (KSCDC) (Chen et al. 2005) and the European Prospective Investigation into Cancer and Nutrition (EPIC) (http://www. srl.cam.ac.uk/epic/international), which have each recruited more than 500,000 participants; the American Cancer Prevention Study-3, which has a target recruitment of 500,000 participants (http://www.cancer.org/docroot/RES/RES_6_6.asp); and the Million Women Study of 1.3 Million U.K. women (http://www.millionwomenstudy.org/introduction).

A more recent conceptual development not yet widely practiced is hospital biobanking, in which, once consent has been obtained, specimens are collected for research as a routine component of clinical care. These "cases" may then be compared with "controls" selected from existing LSDRs. In the United Kingdom, for example, the 1958 birth cohort provides an increasingly used source of universal controls (http://www.b58cgene.sgul.ac.uk/). Primitive versions of this idea are already practiced in the form of tissue banks. Hospital biobanking is essentially an extension of tissue banking from acute and frequently rare conditions to the chronic complex diseases that have greater public health impact. Although hospital biobanking has much potential in terms of the efficient collection of samples, in practice it is difficult to achieve without high levels of public and political support due to issues of trust and the competing pressures on health-care systems and professionals. The major technical challenge associated with hospital biobanking is ensuring adequate quality control of biosample handling, particularly between centers with varying clinical procedures. In practice, the solution to this will likely be the use of a few high-throughput centers focusing on specific research questions, in which standard procedures can be established and monitored, rather than widespread generalized hospital biobanking.

A major resource for coordinating and accessing biobanks worldwide is the Public Population Project in Genomics (P3G) (http://www.p3gconsortium.org). At present the P3G Observatory Study Catalogue contains descriptions, in a standardized form, of 118 population-based biobanks from across the world that have recruited at least 10,000 participants. In total, the catalogued biobanks encompass a total of more than 11 million preexisting or planned recruits. Working in close collaboration with other projects such as the European Biobank and Biomolecular Resources Research Infrastructure (http://bbmri.eu/bbmri), the standardized descriptors have now been extended to catalogue disease-based biobanks (Burton et al. 2009), opening the way forward for development of standardized minimal descriptors. In addition to providing catalogs and tools enabling researchers to identify studies that may provide them with a source of potential participants for pooled analysis, P3G also operates as a means of harmonizing and integrating biobank content enabling the pooling of data (http://www.p3gconsortium.org). Perhaps most usefully, the DataSHaPER suite of core questionnaire variables and data items provides a tool that will help to enhance the commonality of exposure and outcome data between biobanks (http://www.p3gobservatory.org/datashaper/presentation.htm).

INTEGRATED "FUNGIBLE" STUDIES

The cost effectiveness of large and expensive LSDRs, although potentially greater than commissioning a series of stand-alone studies, each with their own infrastructure costs, remains an important obstacle to their development. However, further development of niched LSDRs is essential if the full range of genetic etiologic questions is to be addressed.

A new generation of LSDR may be envisaged that has two distinctive roles. The first is the use of remote study technology. Modern information technology and biotechnology enable recruitment, consent, assessment, and biosampling to be conducted virtually entirely remotely. Provided consent issues are straightforward, participant contact is required only for specialized measurement, such as imaging. Many genetic hypotheses are particularly suitable for testing remotely. Remote study methodology radically reduces costs as well as increases accessibility to studies by prospective participants.

The second potential role for LSDRs is the integration of etiological, public health, and health service delivery questions into a single project. Although specific emphases and resource allocation may vary over time, sufficiently large projects will produce multiple benefits and so represent more attractive investment options for research funders. These projects may be termed "fungible," reflecting the exchange of benefit that will accrue as the emphasis in the study varies over time and according to scientific and political priorities (Gallacher 2007).

With these ideas in place, it is a small step to realize that their fullest expression would be in the recruitment of entire communities to the research enterprise. For genetic association studies, this would provide an extremely rich source of variation sufficient to allow definitive studies on gene–environment relationships.

CONCLUSION

The potential of genetic association studies to identify relationships between genetic factors and complex disease is substantial. The challenge of interpreting association data, however, is also great and it is likely that the design of association studies will become more complex as research questions become more detailed. The infrastructure needed to conduct these more complicated designs, particularly the requirement of sufficient statistical power, will require the development of more cost-effective methodology if the opportunity provided by the design is to be exploited. Association studies yet have much to contribute.

REFERENCES

Altshuler, D., Daly, M., and Kruglyak, L. 2000. Guilt by association. *Nat. Genet.* **26:** 135–137.

Attia, J., Thakkinstian, A., and D'Este, C. 2003. Meta-analyses of molecular association studies: Methodologic lessons for genetic epidemiology. *J. Clin. Epidemiol.* **56:** 297–303.

Benjamin, E.J., Dupuis, J., Larson, M.G., Lunetta, K.L., Booth, S.L., Govindaraju, D.R., Kathiresan, S., Keaney, Jr., J.F., Keyes, M.J., Lin, J.P., et al. 2007. Genome-wide association with select biomarker traits in the Framingham Heart Study. *BMC Med. Genet.* (suppl. 1) **8:** S11.

Burton, P.R., Hansell, A.L., Fortier, I., Manolio, T.A., Khoury, M.J., Little, J., and Elliott, P. 2008. Size matters: Just how big is BIG?: Quantifying realistic sample size requirements for human genome epidemiology. *Int. J. Epidemiol.* **38:** 263–273.

Burton, P.R., Fortier, I., Deschenes, M., Hansell, A., and Palmer, L. 2009. Biobanks and biobank harmonisation. In *An introduction to genetic epidemiology* (ed. G. Davey-Smith et al.), Chapter 6. Policy Press, Bristol.

Chen, Z., Lee, L., Chen, J., Collins, R., Wu, F., Guo, Y., Linksted, P., and Peto, R. 2005. Cohort profile: The Kadoorie Study of Chronic Disease in China (KSCDC). *Int. J. Epidemiol.* **34:** 1243–1249.

Elliott, P. and Peakman, T.C. 2008. The UK Biobank sample handling and storage protocol for the collection, processing and archiving of human blood and urine. *Int. J. Epidemiol.* **37:** 234–244.

Gallacher, J.E. 2007. The case for large scale fungible cohorts. *Eur. J. Public Health* **17:** 548–549.

Gallacher, J.E., Yarnell, J.W., Elwood, P.C., and Phillips, K.M. 1984. Type A behaviour and heart disease prevalent in men in the Caerphilly study. *Br. Med. J.* **289:** 732–733.

Greenland, S., Pearl, J., and Robins, J.M. 1999. Causal diagrams for epidemiologic research. *Epidemiology* **10:** 37–48.

Ioannidis, J.P., Ntzani, E.E., Trikalinos, T.A., and Contopoulos-Ioannidis, D.G. 2001. Replication validity of genetic association studies. *Nat. Genet.* **29:** 306–309.

Ioannidis, J.P., Rosenberg, P.S., Goedert, J.J., and O'Brien, T.R. 2002. Commentary: Meta-analysis of individual participants' data in genetic epidemiology. *Am. J. Epidemiol.* **156:** 204–210.

Lowrance, W.W. and Collins, F.S. 2007. Ethics. Identifiability in genomic research. *Science* **317:** 600–602.

Manolio, T.A. 2008. Biorepositories—At the bleeding edge. *Int. J. Epidemiol.* **37:** 231–233.

Peakman, T.C. and Elliott, P. 2008. The UK Biobank sample handling and storage validation studies. *Int. J. Epidemiol.* (suppl. 1) **37:** i2–i6.

Risch, N.J. 2000. Searching for genetic determinants in the new millennium. *Nature* **405:** 847–856.

Salanti, G., Sanderson, S., and Higgins, J.P. 2005. Obstacles and opportunities in meta-analysis of genetic association studies. *Genet. Med.* **7:** 13–20.

Samani, N.J., Erdmann, J., Hall, A.S., Hengstenberg, C., Mangino, M., Mayer, B., Dixon, R.J., Meiinger, T., Braund, P., Wichmann, H.E., et al. 2007. Genomewide association analysis of coronary artery disease. *N. Engl. J. Med.* **357:** 443–453.

Szklo, M. and Nieto, J. 2007. *Epidemiology: Beyond the basics*, 2nd ed. Jones and Bartlett, Sudbury, Massachusetts.

Tabor, H.K., Risch, N.J., and Myers, R.M. 2002. Candidate-gene approaches for studying complex genetic traits: Practical considerations. *Nat. Rev. Genet.* **3:** 391–397.

WWW RESOURCES

http://www.b58cgene.sgul.ac.uk/ 1958 Birth Cohort.

http://www.cancer.org/docroot/RES/RES_6_6.asp? Cancer Prevention Study03 (CPS-3).

http://www.millionwomenstudy.org/introduction/ Million Women Study.

http://www.p3gconsortium.org/ Public Population Programme in Genomics (P3G).

http://www.srl.cam.ac.uk/epic/international/ European Prospective Investigation of Cancer (EPIC).

http://www.ukbiobank.ac.uk/ UK Biobank.

http://www.wtccc.org.uk/ Wellcome Trust Case Control Consortium.

Variance Component Methods for Analysis of Complex Phenotypes

Laura Almasy and John Blangero

Department of Genetics, Southwest Foundation for Biomedical Research, San Antonio, Texas 78245

INTRODUCTION

Variance component methods have a long history in human quantitative genetics as well as in agricultural genetics and animal breeding. They are designed for genetic analysis of continuously varying quantitative traits like body mass index (BMI), cholesterol levels, or intelligence quotient (IQ). They can be used to assess the strength of genetic effects on a trait, to localize genes influencing a trait through either linkage or association methods, to assess whether associated variants are likely to be the functional variants behind a given localization signal, to explore whether related traits have shared genetic influences in multivariate analyses, and to characterize the genetic effects on a trait through analyses of gene–gene and gene–environment interaction. (An excellent reference for a thorough explanation of classical variance component methods in genetics is Falconer and Mackay 1996.)

WHAT ARE VARIANCE COMPONENT METHODS?

Conceptually, the idea behind variance component methods is very simple—to decompose the overall variance in a phenotype into particular sources. Assuming that the trait of interest is normally distributed, which is a common assumption in variance component analyses, the distribution of a trait or phenotype can be described in terms of the mean and variance of the trait. Figure 5.1 shows the distribution of height in the 1411 participants of the San Antonio Family Heart Study (SAFHS) (Mitchell et al. 1996). The height of study participants ranges from 132.4 cm to 190.5 cm and the mean is 161.64 cm. Most people are about average and a few people are very short or very tall. The variance describes the spread of the trait values around the mean. The variance in height in the SAFHS is 85.65. (See Chapter 2 for calculating variance.) Asking what the sources of variance in a trait are is essentially asking what makes people different from each other.

The most basic way to group these sources of variance is to divide the overall phenotypic variance (σ_p^2) into genetic (σ_g^2) and environmental (σ_e^2) components:

$$\sigma_p^2 = \sigma_g^2 + \sigma_e^2. \tag{1}$$

Each of these can be further subdivided. Genetic variance is often subdivided into additive and dominance variance and sometimes epistatic variance, which arises from interactions among genes. Environmental variance is typically divided into shared and unshared or unique. Shared environmental variance may reflect influences that are common to members of a nuclear family, to spouses, to sibships, or to larger community units that extend beyond the nuclear family. Unshared or unique environmental variance is specific to each individual and may include factors such as measurement error.

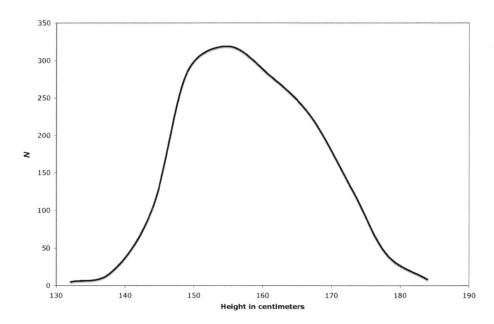

FIGURE 5.1 Distribution of height in the San Antonio Family Heart Study.

ESTIMATING HERITABILITY

Heritability is a measure of the strength of genetic effects on a trait. In its most general sense, heritability (h^2) is defined as the proportion of the phenotypic variance in a trait that is attributable to genetic effects:

$$h^2 = \sigma_g^2/\sigma_p^2 . \qquad (2)$$

This is broad-sense heritability and includes dominance and epistatic interaction effects. However, most human family studies deal with additive genetic or narrow-sense heritability, which is the proportion of the phenotypic variance attributable to additive genetic effects or

$$h^2 = \sigma_a^2/\sigma_p^2 . \qquad (3)$$

The overall phenotypic variance is estimated from the observed distribution of trait values in a sample and is decomposed into genetic and environmental components using the observed covariance in the trait among family members (Ω) and structuring matrices that predict the covariances among family members if they are due to additive genetic effects or to environmental effects:

$$\Omega = 2\Phi\sigma_a^2 + I\sigma_e^2 . \qquad (4)$$

Here Ω is an N-by-N matrix, in which N is the number of individuals in the data set, whose elements are the observed covariances in phenotype for each pair of individuals in the data set (see Chapter 2 for a definition of covariance). The right side of the equation consists of possible sources of covariance among individuals and structuring matrices describing what the covariances among individuals should be if they are caused by that component. In this case, Equation 4 describes a very simple model that includes only aggregate additive genetic effects of an unspecified number of loci at unknown locations in the genome (σ_a^2) and unique, unshared environmental effects (σ_e^2). Each variance component is accompanied by a structuring matrix that predicts the covariance among in-

TABLE 5.1 Family relationships and coefficient of relationship

Degree of relationship	Types of relative pairs	Coefficient of relationship (2Φ)
1	Parent–child, sibling	$1/2 = 0.5$
2	Grandparent–grandchild, half-sibling, avuncular (aunt or uncle with niece or nephew)	$1/4 = 0.25$
3	Great grandparent–grandchild, half-avuncular, grand avuncular, first cousins	$1/8 = 0.125$
4	Great-great grandparent–grandchild, half-grand avuncular, great-grand avuncular, first cousins once removed	$1/16 = 0.0625$

dividuals attributable to that component. In the case of the additive genetic component, the structuring matrix is the coefficient of relationship, 2Φ, which is also twice the kinship coefficient.

The coefficient of relationship can be specified for any two individuals on the basis of their family relationship and requires only knowledge of the pedigree connections between individuals, not their genotypes. It is one-half for first-degree relatives and goes down by a factor of one-half with each degree of relationship, being one-quarter for second-degree relative pairs, one-eighth for third-degree pairs, and so on (Table 5.1). For pairs with more complex types of relationships who are related through multiple lines of descent, as may occur with inbreeding or with marriage loops in a pedigree, this coefficient can also be calculated by tracing the paths between them through all common ancestors, multiplying by one-half for each step along the path and summing across the multiple paths. The coefficient of relationship is also the expected proportion of DNA shared on average by a given relative pair across the whole genome. The basic idea behind this model is intuitively obvious; that is, to the extent that additive genetic effects influence a trait of interest, regardless of how many genes influence the trait, close relatives should be more correlated in their phenotype than are more distant relatives, who should be more correlated than are unrelated individuals.

The structuring matrix for the environmental component is an identity matrix, *I*, which is 1s down the diagonal (i.e., for the individual with themselves) and 0s everywhere else. This implies that the environmental component is unique to each individual and unshared or uncorrelated between individuals. On the basis of the observed covariances in phenotype among individuals in the data set and on these structuring matrices, maximum likelihood techniques are used to estimate the additive genetic and environmental variance components (see Chapter 2 for more on maximum likelihood).

Returning to our example of height in the SAFHS, the maximum likelihood estimate of the additive genetic variance, given the observed covariances among individuals and the kinship coefficients among family members, is 45.39, providing an additive genetic heritability of 45.39/85.65 = 0.53. As a proportion, heritability varies between 0 and 1, with higher values indicating stronger genetic effects.

DEALING WITH SHARED ENVIRONMENTAL EFFECTS

It is important to note that genes are not the only factor shared by family members and that some study designs are susceptible to confounding familial effects with genetic ones, inflating estimates of the additive genetic variance and therefore heritability through unaccounted-for effects of shared environment. In twin studies, a common assumption is that environmental sharing is the same for monozygotic and dizygotic pairs. If this is true, estimating the heritability of a trait by taking the difference in the covariances of the two types of twin pairs results in the environmental variance canceling out. In studies of extended pedigrees, the comparable assumption is that shared environment is unlikely to mimic genetic sharing, which decreases by a factor of one-half with each degree of relationship as shown in Table 5.1. Studies of nuclear families that do not include twins do not have either of these protections and are somewhat more vulnerable to the problem of overestimation of heritability because of the effects of correlations among family members that are due to shared environment rather than shared genes.

One approach to the problem of shared environmental effects is to incorporate them directly into the variance component model. This is easily performed, provided one can specify a structuring matrix that indicates which individuals in the study share the relevant environment. In its simplest form, this could be a matrix of 0s and 1s that specify for each pair of individuals in the study whether they do or do not share the environmental factor. Household is often used in this way, with a matrix indicating which individuals lived in the same household at the time of study, thus serving as a proxy for many difficult-to-measure factors such as diet. One might also use this kind of matrix to model a childhood rearing environment (i.e., which individuals lived together as children) or to allow for correlations between spouses.

Although a simple share/do not share matrix of 0s and 1s is the most common type of environmental sharing incorporated into human variance component studies, there is no reason such a matrix cannot contain continuously varying measures of sharing. One example of this would be a distance matrix in which individuals in the same household have complete sharing (1s in our 0/1 matrix) and individuals in a different household have values decaying toward 0 depending on how far apart the households are. This type of household matrix might be useful as a proxy for environmental exposures such as pollutants.

USING MEASURED ENVIRONMENTAL FACTORS AS COVARIATES

Another important source of trait variation to consider is known environmental factors that can be measured in each study participant. It is more powerful to incorporate a direct measure of an environmental factor than to use indirect measures of whether individuals share or do not share this environmental factor. Using the examples above, if we could measure diet or pollution exposure for each individual, then that would be preferable to using household membership as a proxy for sharing of these factors among family members.

Accounting for the effects of measured environmental factors reduces the unexplained trait variance and effectively magnifies a genetic signal. Covariates are dealt with in variance component analyses as a modification to the trait mean, rather than a component of the variance. Essentially, covariate-specific trait means are used in the calculation of covariances among relatives. In the case of height, it is well known that men are, on average, taller than women. Including sex as a covariate in our example analysis from the SAFHS, we learn that females in this study are, on average, 13.4 cm shorter than males, and, after taking into account mean differences in height between males and females, we reduce the residual trait variance to 39.68, of which 69% can be attributed to additive genetic effects. If we consider the heritability to be a sort of signal-to-noise ratio for genetic effects, including this one covariate increased our ratio by 0.16, from 0.53 to 0.69. Other covariates we might choose to include for height are age, as people do lose some height as they get older, or birth cohort, as there are known secular trends in height.

EFFECTS OF COVARIATE SELECTION ON VARIANCE COMPONENT ANALYSIS

The selection of covariates can have a large effect on the outcome of variance component analyses. Accounting for nongenetic sources of variance can magnify the genetic signal, as shown previously. However, one must consider that it is also possible to choose as covariates traits that absorb genetic variance as well as environmental variance. For example, many individuals with type 2 diabetes also have hypertension, abdominal obesity, high triglyceride levels, and low high-density lipoprotein cholesterol (HDL-C) levels, a clustering of phenotypes described as metabolic syndrome. Because of this, one might choose to include the known correlates blood pressure, triglyceride levels, and HDL-C levels as covariates in genetic analyses of type 2 diabetes. However, because these traits are themselves genetically influenced, including them as covariates raises the possibility that one is correcting out not only environmental factors but also genetic ones, potentially *decreasing* the magnitude of the genetic signals for type 2 diabetes. If hypertension, abdominal obesity, high triglyceride levels, low HDL-C levels, and diabetes commonly occur together because they are influenced by the same genes, then including blood pressure and lipid measurements as covariates in an analysis of type 2 diabetes will likely reduce the power to find genes that influence both these phenotypes and diabetes. There are instances when one may decide to take this route as a deliberate choice—for example, if one is interested in genetic effects on type 2 diabetes that are independent of obesity. However, in general, one should be very cautious about including as a covariate anything that might share overlapping genetic influences with the trait of interest. One way to assess this is to examine the genetic correlations among traits (discussed in the section on Multivariate Analysis).

USING THE LIABILITY THRESHOLD MODEL

Although variance component methods were designed for continuously varying quantitative traits, an extension of this basic model can be used to analyze discrete or categorical traits by assuming that there is an unobserved, continuous, quantitative trait underlying the observed categorical one. This imagined underlying quantitative trait is referred to as the liability and is assumed to be normally distributed. A threshold is placed on this imaginary distribution so that a portion of the distribution equal to the trait prevalence is above the threshold. So if 12% of the population is affected, then the threshold is placed such that 12% of the liability distribution is above the threshold. Covariates, such as age and sex, are modeled as effects on this threshold and allow for different prevalences of the trait in males and females or by age or with smoking or medication use. One conceptual advantage of this model is that it acknowledges differences within affected and unaffected individuals. Some affected individuals are mildly affected and can be thought of as having a liability that is just over the threshold, whereas others are severely affected with a very high liability. Similarly, as the threshold moves with age, some young individuals with higher liabilities who are now unaffected may become affected as they get older.

Of course, it is impossible to measure an individual's liability directly because liability is an unobserved and imaginary trait. We only know which side of the threshold an individual is on given their affection status and where the threshold is for someone of their age, sex, and covariate status. Thus, the analysis is performed by integrating over the possible liability values each individual could have given their observed dichotomous trait status and age, sex, and other covariates. The success of this analysis once again depends on contrasting relatives who are more and less alike in their phenotypes, so it requires the presence of individuals on both sides of the threshold. Such an analysis could not be performed with a sample that contains, for example, only affected individuals. The power of the approach depends in part on the prevalence of the trait. Imagine a relatively rare disease, such as schizophrenia, which has a prevalence of roughly 1%. Knowing someone is affected localizes their liability to a relatively small portion of the curve, the top 1%. But knowing that someone is unaffected tells you almost nothing about their liability as they are somewhere in the bottom 99%. The power of the liability threshold is greatest when the prevalence approaches 50% and affected and unaffected individuals are equally informative (Williams and Blangero 2004).

USING LINKAGE ANALYSIS AND VARIANCE COMPONENTS

The basic model for linkage analysis within a variance components framework is a simple extension of Equation 4, adding in a new locus-specific variance component (σ_{qtl}^2) and a structuring matrix for it (Π) that is a function of observed allele sharing among family members at genotyped markers in a region of interest (Goldgar 1990; Amos 1994; Almasy and Blangero 1998):

$$\Omega = \Pi\sigma_{qtl}^2 + 2\Phi\sigma_a^2 + I\sigma_e^2 . \tag{5}$$

The elements of the Π matrix are the proportion of alleles shared identical by descent (IBD) by each relative pair at a particular location in the genome, which is estimated on the basis of the genotypes at surrounding markers. To be IBD, not only must the two alleles be the same (e.g., both 116 bp for a microsatellite or both G alleles for a single-nucleotide polymorphism [SNP]), but they also must be copies of the same ancestral chromosome. This is the heart of how linkage differs from association. Linkage analysis is not based on which allele any given person or pair of relatives has at a given marker; the genotypes are merely used to mark the flow of chromosomes through pedigrees and to determine how correlated a relative pair is for their alleles on that segment of chromosome. In the region of a gene influencing a trait of interest, relatives who are more correlated in their trait values should have higher IBD allele sharing and relatives who are less correlated phenotypically should have lower IBD sharing. This is true regardless of the type and complexity of the underlying genetic model.

Imagine a gene with extensive allelic heterogeneity. This same quantitative trait locus (QTL) influences the trait of interest in many families, but there are many functional variants. This QTL can still be detected by linkage because, although there may be a different allele in each family, within a family relatives who share the same allele will be more phenotypically alike than relatives who are discordant at the QTL, regardless of which functional allele they carry and whether that particular allele increases or decreases trait values.

IBD sharing is usually represented as a proportion: 0 for pairs that share no alleles, ½ for pairs that share exactly one allele, and 1 for pairs that share both alleles. The Π matrix contains the estimated IBD sharing at a given location for each pair of individuals in the sample. In practice, when parents are ungenotyped or homozygous, we may not be able to determine whether a pair shares 0, 1, or 2 alleles. In this case, the estimated IBD sharing is a weighted average of the probability (P) of sharing one allele and the probability of sharing two alleles:

$$(1/2)P(\text{share 1 allele}) + P(\text{share 2 alleles}) . \tag{6}$$

The power of variance component linkage analysis is a function of the proportion of variance attributable to the QTL (σ_{qtl}^2), the sample size, and the family configuration. For a fixed sample size, linkage power is maximized when the individuals are concentrated into as few pedigrees as possible; larger pedigrees provide more power per person sampled (Blangero et al. 2003). Analytical power formulas can be written down for fixed pedigree configurations (Williams and Blangero 1999), but in practice most studies contain a mixture of different types of pedigrees and power is estimated by simulation. It can be shown that power is greater for quantitative traits than for discrete traits derived from a quantitative measure (e.g., as obesity is from BMI), when QTL effect and sample size and configuration are held constant, unless the quantitative trait is very poorly measured with a high degree of error.

THE IMPORTANCE OF THE ASCERTAINMENT SCHEME FOR STUDY SELECTION

A common rule of thumb often taught is that the ascertainment scheme used to select families for study must be taken into account in segregation analyses but not in linkage analyses. However, not taking into account the way in which families were ascertained can hurt power in variance component analyses. As described previously, the analyses depend on the trait mean and variance, which is being estimated from the sample. If the trait of interest is genetically influenced, family members are correlated with each other for their trait values. So selecting families through an individual with an extreme phenotype (e.g., a BMI > 35) affects the distribution of trait values not only in those probands, but also in their family members. In such a sample, the estimated mean will be higher than the population mean and the variance will be lower than the population variance, because individuals from the lower end of the trait distribution are likely to be underrepresented. Consequently, an individual with a BMI of 40, who would be very extreme compared with the population distribution, is less extreme relative to the sample trait distribution, effectively undervaluing this individual in the analysis. Additionally, correlations among relatives will be underestimated.

The most straightforward ascertainment correction involves conditioning the likelihood of each pedigree on the proband's phenotype (Boehnke and Lange 1984). However, this is only an exact correction when each family was ascertained based on the phenotype of a single individual. When families were ascertained through multiple individuals (e.g., affected sibling pairs), it is possible to condition on both individuals' phenotypes, but this correction is no longer guaranteed to recover the correct population mean and variance and may, in some cases, further reduce analytical power. Another approach that can be used is to fix the trait mean and variance based on measures from epidemiological studies in the same population rather than estimating them. However, if this is performed, one must also fix any covariate effects rather than estimating them. When faced with a complicated ascertainment scheme where it is difficult to identify probands and no appropriate epidemiological data are available for fixing the mean and variance, the good news is that failing to use an ascertainment correction should be conservative. It reduces power but should not increase the false-positive rate.

DEALING WITH NONNORMALITY

Likelihood-based variance component methods typically assume that a trait is normally distributed, like a bell curve. Skewness and kurtosis describe two ways that a distribution may be nonnormal. It may not be symmetrical around the mean with more of the trait values on one side than on the other, in which case it is skewed. Or it may have tails that have too many or too few individuals, which is kurtosis. Examining Figure 5.1, we can see that the distribution of height is slightly skewed, but there is no significant kurtosis. The specific type of nonnormality that can be problematic for variance component analyses is leptokurtosis, when the tails of a distribution are too full and there are more trait values far from the mean than would be expected in a normal distribution. It has been shown that the evidence for linkage can be inflated if such data are analyzed assuming a normal distribution. The increase in the false-positive rate depends on the degree of kurtosis and on the heritability of the trait, but could be two or even three times what is expected.

Fortunately, this situation is easily corrected, and one only need be aware of the issue and take appropriate steps when analyzing leptokurtic traits. Two commonly used corrections are using the t-distribution instead of the normal or calculating a correction constant. The correction constant can be calculated directly for pedigrees of fixed structure, but more commonly it is derived from comparing logarithm of the odds (LOD) scores obtained in simulations under the null of no linkage to the observed LOD score distribution for the trait at hand. These corrections are discussed more fully in Blangero et al. (2000, 2001). Some investigators also choose to use transformations to normalize their data. These transformations could range from taking the natural log of the trait values to rank ordering the trait values and replacing them with a corresponding value from a normal distribution. The use of such transformations is somewhat controversial, with some arguing that changing the distribution of the trait may change the properties and detectability of the underlying genetic signal. One potential safeguard against this is to choose transformations that maintain or enhance the trait heritability. If the goal of the study is gene localization, choosing a transformation that maximizes heritability should not bias any eventual linkage or association results.

MULTIVARIATE ANALYSIS AND PLEIOTROPY

Joint analysis of multiple related phenotypes can be used to answer questions about the nature of the relationship between the traits and to increase power to localize genes influencing the traits (Lange and Boehnke 1983; Almasy et al. 1997). For example, when two traits are known to be correlated, we often would like to know whether this is because they are influenced by the same genes. Identifying networks of related risk factors that share overlapping genetic effects may provide insight into the biology of a disease phenotype. Similarly, showing that two heritable risk factors for the same disease have no overlapping genetic effects suggests that there are at least two independent pathways contributing to disease risk.

As with the variance for a single trait, the overall phenotypic correlation between two traits (ρ_p) can be broken down into a genetic (ρ_g) and an environmental component (ρ_e):

$$\rho_p = \sqrt{h_1^2}\sqrt{h_2^2}\rho_g + \sqrt{1-h_1^2}\sqrt{1-h_2^2}\rho_e \ , \tag{7}$$

in which h_1^2 and h_2^2 are the heritabilities of trait 1 and trait 2. In practice, the genetic and environmental correlations are obtained by estimating the genetic and environmental variance components for each trait (σ_a^2 and σ_e^2) and the covariance between them, using the observed covariances among family members for the two traits and the same structuring matrices as before, 2Φ and I.

The additive genetic correlation ρ_g varies between -1 and 1 and is a measure of pleiotropy, the extent of common genetic effects on the two traits. If $\rho_g = 0$, the two traits are influenced by independent genetic factors. If $\rho_g = -1$ or 1, the genetic influences on the two traits are identical, with the sign indicating whether variants that increase levels of one trait also increase levels of the other ($+1$) or whether factors that increase levels of one trait decrease levels of the other (-1). Likelihood ratio tests can be used to obtain a p value testing the hypothesis of pleiotropy (i.e., whether ρ_g is different from 0). This test of pleiotropy is one way to assess whether a measured cofactor may have overlapping genetic influences with the focal trait in an analysis before deciding whether to use it as a covariate. As with the additive genetic heritability, the additive genetic correlation reflects common effects on the two traits of an unspecified number of as yet unidentified genes.

The linkage models discussed above are also easily expanded to bivariate or multivariate analyses via a QTL variance for each trait and the locus-specific correlations between them (Almasy et al. 1997). For a specific test of pleiotropy, one may choose to fix the locus-specific correlation to 1 or -1, implying that the same functional variant (or variants) in the region affects both traits. In the case in which there are multiple functional variants that comprise a QTL, one may observe a genetic correlation $<|1|$ if some variants influence both traits and some influence only one or if there are gene–environment interactions influencing one trait but not the other.

USING ASSOCIATION ANALYSIS FOR QUANTITATIVE TRAITS

The simplest association analysis for quantitative traits is to test whether the mean trait values differ by genotype, sometimes called a measured genotype test (Boerwinkle et al. 1986). This test is implemented in the same way as are covariate effects, such as age and sex. The genotype of each individual is scored, a regression coefficient is estimated, and a likelihood ratio test is used to assess whether the regression coefficient is different from 0.

Often an additive model of gene action is assumed. For a marker with only two alleles, such as a SNP, an additive model requires a single genotype score with genotypes AA, Aa, and aa being scored as 0, 1, and 2, respectively. This model effectively constrains the trait mean for heterozygotes to be at the midpoint of the mean for the two homozygotes and provides a one degree of freedom association test. Recessive and dominant models, in which the heterozygote mean is constrained to be equal to that of one of the homozygotes and the genotypes are scored 0 or 1, also provide one degree of freedom tests, but are used less commonly. Means may also be estimated separately for each genotype using two 0/1 genotype scores to differentiate the three genotype classes. This does not require any assumptions about the underlying model of gene action. However, it results in a two degree of freedom test and it may lead to parameter estimates that are biologically implausible for many phenotypes (e.g., a situation of overdominance in which the mean for the heterozygote is outside the range of the homozygote means).

These fixed-effects, regression-based association tests for differences in trait mean by genotype are identical to those that might be performed with any statistical analysis software. The advantage to implementing them within a variance component framework is that the nonindependence among family members is accounted for through the additive genetic component in the random effects model of the variance. Ignoring this nonindependence among family members could bias the *p* values of the association tests.

Although the measured genotype test implemented in a variance component framework takes into account the nonindependence among family members, it is still susceptible to the effects of population stratification. A variety of transmission disequilibrium tests for quantitative traits has also been implemented in a variance component framework. These tests protect against association caused by population stratification by separating the genotype score for association with a marker into between- and within-family components and using only the within-family component for the test of association (Fulker et al. 1999; Abecasis et al. 2000a,b; Siegmund et al. 2001).

GENE–GENE AND GENE–ENVIRONMENT INTERACTIONS IN THE VARIANCE COMPONENT ANALYSIS FRAMEWORK

The above models are easily expanded to incorporate or test gene–gene and gene–environment interactions. On the level of aggregate genetic effects at unspecified points in the genome, gene–environment interactions (described in Chapter 11) can be thought of as caused by either differences in magnitudes of genetic effects between environments or differences in which genes influence a trait in different environments. Differences in magnitudes of genetic effects are modeled for dichotomous environments (e.g., smokers and nonsmokers) by specifying environment-specific variance components (σ_a^2, σ_e^2, and σ_q^2 if it is a linkage model). Differences in which genes influence a trait in different environments are modeled with correlations between the genetic and environmental variance components in the two environments (ρ_g and ρ_e).

A simple test of overall, nonlocus-specific, gene–environment interaction is to compare the likelihood of a model in which separate additive genetic components are allowed to differ between environments to the likelihood of a model in which the additive genetic variances are constrained to be equal for a simple one degree of freedom test. The same type of test can also be used in linkage by testing equality of the QTL-specific variances in the two environments. On the level of association, the analogous test would be to model the difference in mean trait values by genotype separately for smokers and nonsmokers and perform a likelihood ratio test using models in which regression coefficients for the SNP effect are estimated separately versus constrained to be equal.

In the simple nonlocus-specific, additive genetic model, gene–environment interaction is also present when the genetic correlation between environments is different from 1, implying different genes influencing the trait in the two environments. Variance component models for gene–environment interaction are described more fully in Blangero (1993). In the case of an environmental factor that varies continuously, the genetic and environmental variance components can be modeled as a function of the environmental measure, with genetic or environmental variances increasing or decreasing per unit of change in the environmental measure, as described in Diego et al. (2003).

Gene–gene interactions, or epistasis, can be modeled on the level of linkage by adding a variance component for epistatic interaction between two loci with an appropriate structuring matrix to a two-QTL linkage model that also contains QTL-specific variance components for the independent effects of each of the loci (Mitchell et al. 1997):

$$\Omega = \Pi_1 \sigma_{q1}^2 + \Pi_2 \sigma_{q2}^2 + \Pi_1 \odot \Pi_2 \, \sigma_{epi}^2 + 2\Phi\sigma_a^2 + I\sigma_e^2 . \tag{8}$$

For additive–additive interaction, the structuring matrix for the epistatic component would be the Hadamard product of the IBD matrices for each of the individual QTLs, $\Pi_1 \odot \Pi_2$. A focal test of gene–gene interaction is then provided by testing whether the new epistatic component of variance, σ_{epi}^2, is greater than 0. Additive–dominance, dominance–additive, and dominance–dominance epistasis types are not often used in human genetic studies, but they can be modeled with the appropriate structuring matrices. For additive components, this is the IBD sharing matrix, Π, whereas for dominance components, it is a locus-specific version of Jacquard's Δ_7, the probability that each pair of individuals shares both alleles IBD. On the level of association, gene–gene interaction can be modeled similar to gene–environment interaction by estimating multiple regression coefficients (as above in gene–environment interaction) and evaluating whether the displacement among trait means by genotype at one locus differs by genotype at a second locus.

IDENTIFYING POTENTIALLY FUNCTIONAL VARIANTS

Ideally, localization of a QTL to a region or gene through linkage or association will be followed by identification of the specific DNA variants that influence a phenotype. Confirmation of functional variants will of course involve laboratory studies of function, such as expression constructs, and potentially animal models. However, statistical genetic techniques may aid in prioritizing variants for these studies. Suppose that there are two functional variants within a particular QTL, a promoter variant that has a relatively small effect on the mean trait values in the population (QTN1 in Fig. 5.2) and a coding change that has a large population-level effect (QTN2). (Remember that the effect size on the population level is a function of both the frequency of a variant and the shift in phenotype values it causes in each individual who carries it. A variant may have a larger population level effect either by being common or by causing a large displacement in the mean phenotype value.) Suppose also that there are many other SNPs in and around this gene that do not affect our phenotype (SNP1–SNP3 in Fig. 5.2). Some of these nonfunctional polymorphisms are in greater or lesser degrees of linkage disequilibrium (LD) with one or the other of the two functional variants. If we rank SNPs for functional studies in order of their p values for association with the phenotype, nonfunctional SNPs that are in strong LD with the coding variant of large effect (SNP2 and SNP3 in Fig. 5.2) will be higher on our list of candidates than QTN1, the promoter variant that is truly functional but has a smaller effect size. This is because the effective effect size for association studies for a given genotyped marker (σ_{mark}^2) is a function of the proportion of variance attributable to a functional variant (σ_{qtn}^2), which we call a quantitative trait nucleotide (QTN), and the correlation between that QTN and the genotyped marker:

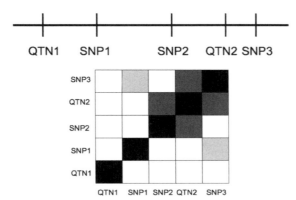

FIGURE 5.2 A chromosomal region with two functional variants (QTN1 and QTN2) and three SNPs and the linkage disequilibrium between these pairs of markers. Darker boxes indicate stronger disequilibrium.

$$\sigma^2_{mark} = \rho\sigma^2_{qtn} , \qquad (9)$$

in which ρ is the correlation between genotypes at the marker and at the QTN, which is also the square root of the common measure of linkage disequilibrium r^2. If our functional variants account for 1% of the trait variance in the case of the weaker QTN1 or 2% in the case of the stronger QTN2, nonfunctional SNPs with r^2 of greater than 0.25 with QTN2 will have a σ^2_{mark} that is greater than the σ^2_{qtn} of QTN1 and will produce stronger p values in an association analysis.

Bayesian methods of multimarker association analysis have been proposed to screen out SNPs whose strong p value in an association analysis is due to LD with another genotyped variant (Blangero et al. 2005). This approach involves obtaining a goodness-of-fit statistic for models including each variant individually and then pairs of variants and then three variants at a time and so on, with the addition of variants continuing until no $n + 1$ variant model fits better than the best n variant model. Then the Bayesian information criterion (BIC) is used to compare these nonnested models. Rather than selecting one model as the "best," all of the models within a certain window of BIC values are retained and Bayesian model averaging is used to obtain a posterior probability for each SNP.

Returning to the example above, with a coding variant of large effect and a promoter variant of smaller effect and other SNPs in LD with them, both of the functional variants and all of the markers in LD with them will do well in the models with individual markers. However, in the two-locus models, once one of the functional variants is in the model, the SNPs in LD with that variant will provide no additional information. Unless another SNP is in complete LD with one of the functional variants ($r^2 = 1$), the models with the true functional variant will provide a better fit than the models with variants that are only in LD with the functional variant. This approach depends on having assayed all of the variants in a region that are present in the samples being analyzed, such that the functional variants are among the genotyped markers, as we will soon have given the growing access to high-throughput sequencing as a routine part of studies. It also depends on the LD between the functional variants and surrounding markers. The degree of LD that can be distinguished depends on the sample size and configuration, but it is not out of the question to be able to pick out a functional variant from markers with an r^2 of 0.95 with that variant.

SUMMARY AND CONCLUSIONS

Variance component methods have a long history in quantitative human, animal, and plant genetics. They can be used to assess the strength of genetic effects on a phenotype of interest, to explore which phenotypes are influenced by the same genes, and to localize, identify, and characterize the genetic variants influencing a trait. They have been used in many successful studies of quantitative risk factors related to human disease (e.g., Mitchell et al. 1996; Comuzzie et al. 1997; Curran et al. 2005; Soria et al. 2005; Goring et al. 2007).

ACKNOWLEDGMENTS

This work was supported in part by National Institutes of Health grants MH59490, MH61622, AA08403, GM31575, HL45522, and HL70751.

REFERENCES

Abecasis, G.R., Cardon, L.R., and Cookson, W.O. 2000a. A general test of association for quantitative traits in nuclear families. *Am. J. Hum. Genet.* **66:** 279–292.

Abecasis, G.R., Cookson, W.O., and Cardon, L.R. 2000b. Pedigree tests of transmission disequilibrium. *Eur. J. Hum. Genet.* **8:** 545–551.

Almasy, L. and Blangero, J. 1998. Multipoint quantitative-trait linkage analysis in general pedigrees. *Am. J. Hum. Genet.* **62:** 1198–1211.

Almasy, L., Dyer, T.D., and Blangero, J. 1997. Bivariate quantitative trait linkage analysis: Pleiotropy versus co-incident linkages. *Genet. Epidemiol.* **14:** 953–958.

Amos, C.I. 1994. Robust variance-components approach for assessing genetic linkage in pedigrees. *Am. J. Hum. Genet.* **54:** 535–543.

Blangero, J. 1993. Statistical genetic approaches to human adaptability. *Hum. Biol.* **65:** 941–966.

Blangero, J., Williams, J.T., and Almasy, L. 2000. Robust LOD scores for variance component-based linkage analysis. *Genet. Epidemiol.* (suppl. 1) **19:** S8–S14.

Blangero, J., Williams, J.T., and Almasy, L. 2001. Variance component methods for detecting complex trait loci. *Adv. Genet.* **42:** 151–181.

Blangero, J., Williams, J.T., and Almasy, L. 2003. Novel family-based approaches to genetic risk in thrombosis. *J. Thromb. Haemost.* **1:** 1391–1397.

Blangero, J., Goring, H.H., Kent, J.W., Jr., Williams, J.T., Peterson, C.P., Almasy, L., and Dyer, T.D. 2005. Quantitative trait nucleotide analysis using Bayesian model selection. *Hum. Biol.* **77:** 541–559.

Boehnke, M. and Lange, K. 1984. Ascertainment and goodness of fit of variance component models for pedigree data. *Prog. Clin. Biol. Res.* **147:** 173–192.

Boerwinkle, E., Chakraborty, R., and Sing, C.F. 1986. The use of measured genotype information in the analysis of quantitative phenotypes in man. I. Models and analytical methods. *Ann. Hum. Genet.* **50:** 181–194.

Comuzzie, A.G., Hixson, J.E., Almasy, L., Mitchell, B.D., Mahaney, M.C., Dyer, T.D., Stern, M.P., MacCluer, J.W., and Blangero, J. 1997. A major quantitative trait locus determining serum leptin levels and fat mass is located on human chromosome 2. *Nat. Genet.* **15:** 273–276.

Curran, J.E., Jowett, J.B., Elliott, K.S., Gao, Y., Gluschenko, K., Wang, J., Abel Azim, D.M., Cai, G., Mahaney, M.C., Comuzzie, A.G., et al. 2005. Genetic variation in selenoprotein S influences inflammatory response. *Nat. Genet.* **37:** 1234–1241.

Diego, V.P., Almasy, L., Dyer, T.D., Soler, J.M., and Blangero, J. 2003. Strategy and model building in the fourth dimension: A null model for genotype x age interaction as a Gaussian stationary stochastic process. *BMC Genet.* (suppl. 1) **4:** S34.

Falconer, D.S. and Mackay, T.F.C. 1996. *Introduction to quantitative genetics.* Longman, Essex, England.

Fulker, D.W., Cherny, S.S., Sham, P.C., and Hewitt, J.K. 1999. Combined linkage and association sib-pair analysis for quantitative traits. *Am. J. Hum. Genet.* **64:** 259–267.

Goldgar, D.E. 1990. Multipoint analysis of human quantitative genetic variation. *Am. J. Hum. Genet.* **47:** 957–967.

Goring, H.H., Curran, J.E., Johnson, M.P., Dyer, T.D., Charlesworth, J., Cole, S.A., Jowett, J.B., Abraham, L.J., Rainwater, D.L., Comuzzie, A.G., et al. 2007. Discovery of expression QTLs using large-scale transcriptional profiling in human lymphocytes. *Nat. Genet.* **39:** 1208–1216.

Lange, K. and Boehnke, M. 1983. Extensions to pedigree analysis. IV. Covariance component models for multivariate traits. *Am. J. Med. Genet.* **14:** 513–524.

Mitchell, B.D., Kammerer, C.M., Blangero, J., Mahaney, M.C., Rainwater, D.L., Dyke, B., Hixson, J.E., Henkel, R.D., Sharp, R.M., Comuzzie, A.G., et al. 1996. Genetic and environmental contributions to cardiovascular risk factors in Mexican Americans. The San Antonio Family Heart Study. *Circulation* **94:** 2159–2170.

Mitchell, B.D., Ghosh, S., Schneider, J.L., Birznieks, G., and Blangero, J. 1997. Power of variance component linkage analysis to detect epistasis. *Genet. Epidemiol.* **14:** 1017–1022.

Siegmund, K.D., Vora, H., and Gauderman, W.J. 2001. Combined linkage and association analysis in pedigrees. *Genet. Epidemiol.* (suppl. 1) **21:** S358–S363.

Soria, J.M., Almasy, L., Souto, J.C., Sabater-Lleal, M., Fontcuberta, J., and Blangero, J. 2005. The F7 gene and clotting factor VII levels: Dissection of a human quantitative trait locus. *Hum. Biol.* **77:** 561–575.

Williams, J.T. and Blangero, J. 1999. Power of variance component linkage analysis to detect quantitative trait loci. *Ann. Hum. Genet.* **63:** 545–563.

Williams, J.T. and Blangero, J. 2004. Power of variance component linkage analysis-II. Discrete traits. *Ann. Hum. Genet.* **68:** 620–632.

6 | Multiple Testing and Power Calculations in Genetic Association Studies

Hon-Cheong So and Pak C. Sham

Department of Psychiatry, LKS Faculty of Medicine, University of Hong Kong, Hong Kong, China

INTRODUCTION

Modern genetic association studies typically involve multiple single-nucleotide polymorphisms (SNPs) and/or multiple genes. With the development of high-throughput genotyping technologies and the reduction in genotyping cost, investigators can now assay up to a million SNPs for direct or indirect association with disease phenotypes. In addition, some studies involve multiple disease or related phenotypes and use multiple methods of statistical analysis. The combination of multiple genetic loci, multiple phenotypes, and multiple methods of evaluating associations between genotype and phenotype means that modern genetic studies often involve the testing of an enormous number of hypotheses. When multiple hypothesis tests are performed in a study, there is a risk of inflation of the type I error rate (i.e., the chance of falsely claiming an association when there is none).

Several methods for multiple-testing correction are in popular use, and they all have strengths and weaknesses. Because no single method is universally adopted or always appropriate, it is important to understand the principles, strengths, and weaknesses of the methods so that they can be applied appropriately in practice.

Statistical power is the probability of rejecting the null hypothesis when it is false (i.e., when a genuine association exists between the tested genetic locus and the disease). Power calculation is crucial in study design both to show the adequacy of the sample size and to improve the efficiency of the study design. It is an essential component in the design and planning of a genetic study, both to ensure that the study has a realistic chance of discovering associations, and to arrive at an efficient design that achieves the desired power at low cost.

MULTIPLE TESTING CAN LEAD TO TYPE I ERRORS

The classical approach to hypothesis testing is to set up a null hypothesis (H_0) and an alternative hypothesis (H_1) and then calculate a test statistic T from the observed data that is used to decide whether the H_0 should be rejected. In association studies, H_0 usually refers to an effect size of 0, whereas H_1 usually refers to an effect size greater than 0. For case–control association studies, a convenient measure of effect size is the logarithm of the odds ratio, log(OR). The p value is defined as the probability of obtaining a test statistic at least as extreme as the value of T calculated from the observed data, given that H_0 is true:

$$p = P(T \geq t \mid H_0). \qquad (1)$$

If the p value is smaller than a preset threshold α (traditionally set at 0.05), then H_0 is rejected and the result is considered to be statistically significant.

The problem of multiple testing is that, when multiple hypotheses are tested in a study, keeping the significance threshold at the traditional value of 0.05 may lead to many false-positive significant results. For example, if 100,000 tests are performed, it is expected that 5% of them or 5000 tests will have $p < 0.05$ by chance, when H_0 is in fact true for all the tests. Most genetic association studies involve multiple comparisons owing to the testing of multiple SNPs, multiple genes, and multiple phenotypes.

THERE ARE THREE PRINCIPAL METHODS FOR MULTIPLE-TESTING CORRECTION

The principal methods for multiple-testing correction are (1) control of the family-wise error rate (Bonferroni or permutation); (2) Bayesian analysis, for predetermined significance threshold; and (3) the false discovery rate methodology. The Bonferroni correction is most commonly used when the magnitude of multiple testing is modest, tests are only weakly dependent (e.g., involving tens or hundreds of distinct genetic variants), and the criteria for the inclusion of variants are not entirely clear. Permutation-based adjusted significance levels are particularly useful when there are strong dependencies among the tests, such as when different test statistics are used for the same data (e.g., tests designed for a different assumed mode of inheritance). Predetermined significance thresholds and false discovery rate methodology are both suitable for very large-scale multiple testing. Predetermined significance thresholds are particularly appealing for scenarios in which the "hypothesis space" is well-defined (e.g., in whole-genome linkage scans and genome-wide association studies using fixed marker panels). However, the fixed thresholds may not be appropriate when the assumptions do not accurately represent reality, and in these scenarios the false discovery methodology is a data-driven alternative that allows the significance thresholds to be set according to the overall pattern of results.

Control of Family-wise Error Rate

Bonferroni Correction

The Bonferroni method simply raises the critical significance level according to the number of independent tests performed in the study. For m independent tests, the significance level is set at $\alpha = 0.05/m$. The justification for this method is straightforward. When H_0 is true for m independent tests, the p values are each distributed as a uniform $(0, 1)$ random variable. Hence the probability of having *at least one* false positive result under H_0, when adopting a critical p value α, is given by

$$P(\text{smallest } p \leq \alpha \mid H_0) = 1 - P[(p_1 > \alpha) \cap (p_2 > \alpha) \cap \cdots \cap (p_n > \alpha)]$$
$$= 1 - (1 - \alpha)^m. \tag{2}$$

This probability is also known as the family-wise error rate (FWER). If it is set to be 0.05, then solving $1 - (1 - \alpha)^m = 0.05$ gives

$$\alpha = 1 - (1 - 0.05)^{1/m} \quad \text{(Sidak correction)}$$
$$\approx 0.05/m \quad \text{(Bonferroni correction)}. \tag{3}$$

However, Bonferroni correction may be conservative when the tests are not independent, which is often the case in association studies. Nyholt (2004) has proposed a correction method based on spectral decomposition of the matrix of pairwise linkage disequilibrium values between SNPs. The eigenvalues are used to deduce the effective number of independent tests performed. A Sidak correction is then performed according to the effective number of independent tests. Although Nyholt's method is computationally fast, it can lead to overcorrection and therefore conservative p values in the presence of a strong haplotype block structure (Salyakina et al. 2005).

Permutation Test

The permutation procedure is a robust but computationally intensive alternative to the Bonferroni correction for multiple-testing correction in the face of dependent tests. To calculate permutation-based p values, the case–control (or phenotype) labels are randomly shuffled and all m tests are re-calculated on the reshuffled data set, with the smallest p value of these m tests being recorded. The procedure is then repeated for a large number of times (typically 1000, or until a certain number [e.g., 5] of the permuted replicates have a p value less than the p value of the actual data) to construct an empirical null frequency distribution of the smallest p values. The p value calculated from the real data can be compared with this null distribution to determine an empirical adjusted p value. If n permutations were performed and the p value from the real data set is smaller than exactly r of the n smallest p values from the permuted data sets, then the empirical adjusted p value is given by

$$\text{Adjusted } p = (r + 1)/(n + 1) . \tag{4}$$

The main idea behind permutation testing is that by random reassignment of the case–control status, we can model what happens under the null hypothesis and hence calculate an empirical adjusted p value. However, the permutation procedure can be computationally intensive. To achieve a very small p value, it is necessary to perform a large number of permutations. Some procedures have been proposed to reduce the heavy computational load, for instance by simulation or fitting analytic distributions to empirical distributions (Dudbridge and Koeleman 2004; Seaman and Muller-Myhsok 2005).

Potential Problems

The Bonferroni method, permutation test (as described above to obtain the empirical significance of the smallest p value from a number of tests), and other methods that control the FWER have been criticized for philosophical reasons. It may seem unreasonable and counterintuitive that the interpretation of a finding depends on the number of other tests also being performed. Also, the general H_0 (that all H_0 are true) is rarely of interest. Methods that control the FWER are prone to making type II errors (i.e., not rejecting the general H_0 when important effects exist).

A Bayesian Perspective

The p value itself does not directly tell us how likely it is that the null hypothesis is true given the observed data. Instead, it is necessary to invoke Bayes' rule to evaluate the probability that H_0 is true when a test is declared significant:

$$
\begin{aligned}
P(H_0 \mid p \le \alpha) &= \frac{P(p \le \alpha \mid H_0)P(H_0)}{P(p \le \alpha \mid H_0)P(H_0) + P(p \le \alpha \mid H_1)P(H_1)} \\[2mm]
&= \frac{\alpha \pi_0}{\alpha \pi_0 + (1-\beta)(1-\pi_0)} .
\end{aligned}
\tag{5}
$$

In this formulation, $1 - \beta$ represents the power of the study, $P(H_0 \mid p \le \alpha)$ can be regarded as the false-positive rate given that a test is declared significant, and π_0 is the prior probability that H_0 is true. Re-expressing the equation in terms of α, we obtain

$$
\alpha = \frac{P(H_0 \mid p \le \alpha)}{1 - P(H_0 \mid p \le \alpha)} \frac{1 - \pi_0}{\pi_0} (1 - \beta) .
\tag{6}
$$

Interestingly, this formula provides a rough justification of the traditional practice of setting α at 0.05. When $\pi_0 \approx 1/2$ (i.e., H_0 and H_1 are considered equally likely before the study) and $1 - \beta \approx 1$ (i.e., the study has high power), the false-positive rate $\approx \alpha$. Setting α at 0.05 implies that we consider a false-positive rate of 5% to be acceptable, which may be quite reasonable for most scientific studies.

Equation 6 also illustrates how α is related to the study power and the prior probability of H_0. When the power $(1 - \beta)$ is low, α has to be set proportionately lower to maintain a fixed false-positive rate. In other words, the p value has to be smaller to produce the same false-positive rate for a study with weak power than one with greater power. Similarly, when the prior probability of H_0 is high, $(1 - \pi_0)/\pi_0$ is low, then again α has to be set proportionately lower to keep the false-positive rate fixed at the desired level.

Performing multiple tests usually reflects a lack of strong prior hypotheses and is therefore associated with a high π_0. The Bonferroni adjustment effectively sets α to be directly proportional to $1/m$, which is equivalent to assuming (a fixed) $\pi_0 = m/(m + 1)$. In other words, we expect only one truly significant result among all tests in all situations. This is clearly not a very reasonable assumption, because the prior hypothesis π_0 should reflect the fact that the marker loci are chosen at random rather than the number of marker loci in a study.

False Discovery Rate

A recent paradigm for multiple-testing correction known as the false discovery rate (FDR) was proposed by Benjamini and Hochberg (1995). This method controls the expected number of false discoveries in m tests by setting α at an appropriate level that is data-driven and makes no assumption about the relationship between the number of tests and π_0.

Using the notation from Table 6.1, the FDR is defined as

$$FDR = E\left(\frac{V}{R}\right),\tag{7}$$

in which E denotes expectation, V is the number of falsely rejected tests, and R is the total number of tests rejected. The possibility that $R = 0$ is problematic for the above definition, so that a mathematically more rigorous definition of the FDR is

$$FDR = E\left(\frac{V}{R}\,\bigg|\,R > 0\right)P(R > 0).\tag{8}$$

A closely related quantity, known as positive FDR, or pFDR, has been proposed (Storey 2003):

$$pFDR = E\left(\frac{V}{R}\,\bigg|\,R > 0\right).\tag{9}$$

The distinction between FDR and pFDR is rarely of practical importance, because these quantities are equivalent when $\Pr(R > 0) = 1$. This is usually the case when the number of tests is large. It is noteworthy that the FWER can be expressed as

$$FWER = P(V \geq 1).\tag{10}$$

The Bonferroni and Sidak procedures described previously control the FWER, which is conservative when a true association exists for more than one of the tests.

An intuitive justification of the FDR control strategy proposed by Benjamini and Hochberg (1995) is as follows. Among the m tests each with prior probability π_0 that H_0 is true, the expected

TABLE 6.1 Results of hypothesis testing for *m* tests

	Called not significant	Called significant	Total
H_0 true	U	V	m_0
H_0 false	T	S	m_1
	W	R	m

number of tests for which H_0 is true is $m\pi_0$. Assuming a uniform distribution of *p* values under H_0, a proportion α of these tests are expected to have *p* values less than α:

$$P(p \leq \alpha \mid H_0) = \alpha. \tag{11}$$

Therefore, the expected number of false positives = $\alpha m \pi_0$. Hence, out of the *R* tests declared significant, the expected proportion of false discoveries = $\alpha m \pi_0 / R$. If π_0 is close to 1, then this is approximately equal to $\alpha m / R$.

Because FDR $\approx \alpha m / R$, we should set

$$\alpha = \left(\frac{R}{m}\right) \text{FDR}. \tag{12}$$

The formal procedure proposed by Benjamini and Hochberg (1995) is as follows.

1. Set FDR to desired level (e.g., to 0.05).
2. Rank the tests in ascending order of *p* value, giving $p_1 \leq p_2 \leq \cdots \leq p_r \leq \cdots \leq p_m$.
3. Find the test with the highest rank, *r*, for which the *p* value, p_r, is less than or equal to $(r/m) \times$ FDR.
4. Declare the tests of rank 1, 2, ... , *r* as significant.

A minor modification is to replace *m* by $m\pi_0$. The above procedure essentially sets $\pi_0 = 1$.

q Value and Estimation of π_0

The *q* value is a simple extension of the FDR method (Storey 2003; Storey and Tibshirani 2003). Instead of setting a predefined threshold for FDR and rejecting tests accordingly, a *q* value can be assigned to every individual test performed. The *q* value for a particular test is defined as the expected proportion of false positives incurred when that test is called significant. Note that calling one test significant means that all other tests with stronger evidence against the null (i.e., tests with lower *p* values or more extreme test statistics) should also be called significant. In other words, the *q* value of a test is the expected proportion of false positives among all tests as extreme as or more extreme than the particular test.

It has been shown that when a large number of tests are performed and the *p* values follow "weak dependence," the estimated *q* values tend to be greater than or equal to the true *q* values. In other words, *q* value estimates across all tests are simultaneously conservative (Storey 2003; Storey and Tibshirani 2003). Weak dependence can be loosely defined as dependence that becomes negligible when the number of tests increases to infinity. This probably applies to genome-wide association data, as SNPs are correlated within finite linkage disequilibrium (LD) blocks.

The original FDR procedure sets $\pi_0 = 1$, which yields conservative results. To improve power, π_0 may be estimated from the data. One simple method for estimating π_0, proposed by Storey and Tibshirani (2003), assumes that *p* values of tests for which the alternative hypothesis is true will all tend to be smaller than a value set to be quite far from 0 (e.g., 0.5), whereas the *p* values for the

m_0 truly null tests should be uniformly distributed between 0 and 1. It follows that most p values near 1 belong to H_0. For instance, we may assume H_0 is true for all (or most) tests with $p \geq \lambda$, in which λ is a (relatively large) number between 0 and 1, say 0.5. The expected number of p values lying between between λ and 1 is therefore

$$E(N_{p > \lambda}) = (1 - \lambda)m_0 . \tag{13}$$

An estimate of m_0 is therefore $N_{p > \lambda}/(1 - \lambda)$. An estimate of π_0 is therefore

$$\hat{\pi}_0 = \frac{\#\{p \text{ values} > \lambda\}}{(1 - \lambda)m} . \tag{14}$$

An example of multiple-testing adjustments under the FDR framework is shown in Table 6.2. Both the fixed threshold and the q-value approach are illustrated.

FDR with Weights

It has been suggested that evidence from linkage scans could be used to weight p values in association studies to improve power (Roeder et al. 2006).

In a weighted FDR, each test is assigned a weight w_i according to its importance (Benjamini and Hochberg 1997). The weights should be normalized so that the average value of w over all tests is 1. The weighted FDR follows exactly the same procedure as the standard FDR, except that the r in $(r/m) \times$ FDR is replaced by W, which is defined as

$$W = w_1 + w_2 + \cdots + w_r . \tag{15}$$

The standard FDR is then a special case of the weighted FDR, with $w = 1$ for all tests.

FDR for Dependent Tests

The standard FDR procedure provides correct control of FDR even when the tests display "positive regression dependency" (Benjamini and Yekutieli 2001). When this condition is not met, a modified FDR procedure divides $(r/m) \times$ FDR by S, defined as

$$S = 1 + \frac{1}{2} + \frac{1}{3} + \cdots + \frac{1}{n} . \tag{16}$$

This modified FDR procedure provides conservative control of the FDR under *any* type of dependency between the tests.

TABLE 6.2 An example of multiple-testing correction using false discovery rate (FDR)

Rank	p value	(Rank/m) × FDR	Reject H_0?	q value
1	0.001	0.005	1	0.010
2	0.01	0.01	1	0.050
3	0.165	0.015	0	0.513
4	0.205	0.02	0	0.513
5	0.396	0.025	0	0.750
6	0.45	0.03	0	0.750
7	0.641	0.035	0	0.916
8	0.781	0.04	0	0.953
9	0.901	0.045	0	0.953
10	0.953	0.050	0	0.953

FDR threshold set at 0.05 for the third and fourth columns.

"Parametric FDR" Methods

Another approach to estimating the FDR is to fit a mixture model to the test statistics or *p* values such that some follow the null distribution, whereas others follow a *specified* alternative distribution. Some examples of the null and alternative distributions include central and noncentral chi-square distributions (Everitt and Bullmore 1999), central and noncentral normal distributions (Cox and Wong 2004), and uniform and β distributions (Allison et al. 2002).

One may then use method of moments or maximum likelihood estimation to estimate the proportion of tests that follow the alternative distribution and the parameters (e.g., mean and variance) of the alternative distribution. Finally, from parameter estimates, the posterior probability of each test belonging to the null distribution, if declared significant, can be calculated.

Parametric methods are more powerful than nonparametric approaches when the alternative distribution of the test statistic or *p* value can be correctly specified. They also allow the posterior probabilities of H_0 given the test statistic to be calculated directly for each of the hypotheses. However, they are computationally more complex and require more assumptions to be made. The nonparametric methods described above are much easier to implement and have been shown to control the FDR under weak dependence structures. Hence, for general purposes, the nonparametric approach may be preferred. Parametric methods may be considered when the purpose is to estimate posterior probability of H_0 for each hypothesis and when the form of alternative distributions can be estimated appropriately.

Potential Problems of False Discovery Rate Methods

There are a few pitfalls to note when using the FDR. First, the FDR was developed based on the assumption that the *p* values are distributed as uniform (0, 1) under H_0. However, such an assumption may not hold in many situations, for example in the presence of biased genotyping, population substructure, sample asymptotic tests, or correlated test statistics (Efron 2007). The FDR calculated may be misleading in these cases. In addition, it is noteworthy that FDR controls the *expected* proportion of false positives, rather than the true *exact* proportion. It is not easy to determine how "accurate" the calculated FDR really is for a real data set.

SUCCESSFUL AND EFFICIENT STUDY REQUIRES STATISTICAL POWER CALCULATION

There are two types of errors in hypothesis testing: type I errors refer to the rejection of H_0 when it is true, whereas type II errors refer to not rejecting the null hypothesis when it is false. Power is equal to one minus the probability of a type II error, that is, the probability of correctly rejecting the H_0 when a true association is present.

Power calculations are often required for the planning of studies and in writing project proposals for grant applications. In general, power calculations involve the following main steps.

1. Decide on all the necessary assumptions about the risk factor (e.g., its frequency and effect size), the disease (e.g., its prevalence), the study (e.g., sample size), and the statistical test (e.g., the form of the test statistic and the desired type I error rate).

2. Determine the frequency distributions of the test statistic under H_0 and H_1. Usually the distribution under H_0 will be determined by how the test statistic is constructed, and precisely known both in terms of its general form (e.g., normal, chi-square, etc.) and parameters (e.g., mean 0, variance 1). However, although the form of the distribution is also usually known under H_1, the parameters of the distribution will depend on the assumptions about the risk factor and the disease. Thus, step 2 is to calculate the values of the unknown parameters under H_1, based on the assumptions made in step 1.

3. From the inverse distribution function of the test statistic under H_0, we determine the critical value of the test statistic for rejecting H_0, to control the type I error rate at the desired level (e.g., 0.05).

4. From the distribution function of the test statistic under H_1 (determined in step 1), we determine the probability that the test statistic will exceed the critical value determined in step 2.

An Example Power Calculation

The principles of power calculations are illustrated here using a simple case–control association study as an example.

1. The necessary assumptions are
   - Biallelic risk locus with genotypes aa, Aa, and AA
   - Frequency of A = 0.1
   - Hardy–Weinberg equilibrium (HWE) in the population
   - Genotypic relative risks: aa = 1, Aa = AA = 2
   - Disease prevalence = 0.01
   - Statistical test is for H_0: odds ratio = 1
   - Significance level = 0.0000001 (10^{-7})
   - Sample size 1000 cases, 1000 controls

2. Under HWE, the expected genotype frequencies of AA, Aa, and aa are p^2, pq, and q^2, where p and q are the allele frequencies (i.e., 0.01, 0.18, and 0.81, respectively). Denoting the genotypes AA and Aa collectively as A* for convenience, the expected frequencies of A* and aa are 0.19 and 0.81, respectively. Let the penetrance (i.e., probability of disease given the genotype) of aa be r, then the penetrance of A* is $2r$. The disease prevalence is given by Prevalence = (0.81 × r) + (0.19 × $2r$) = 0.01. Solving this equation gives $r = 0.008403$. Using Bayes' rule, we can calculate the frequency of A* among cases as follows:

$$P(A^* \mid \text{Case}) = P(\text{Case} \mid A^*)P(A^*)/P(\text{Case})$$
$$= (2r)\,P(A^*)/P(\text{Case}) = 0.016807 \times 0.19/0.01 = 0.319328 . \quad (17)$$

Similarly, frequency of A* among controls is given by

$$P(A^* \mid \text{Control}) = P(\text{Control} \mid A^*)P(A^*)/P(\text{Control})$$
$$= (1 - 2r)\,P(A^*)/P(\text{Control}) = 0.983193 \times 0.19/0.99 = 0.188694 . \quad (18)$$

The expected genotype frequencies are summarized in Table 6.3. The OR of Table 6.3 is given by

$$OR = \frac{0.319328 \times 0.811306}{0.188694 \times 0.680672} = 2.017094 . \quad (19)$$

TABLE 6.3 Expected genotype frequencies for a case–control association study under hypothetical assumptions

	Cases	Controls
A*	0.319328	0.188694
aa	0.680672	0.811306

We note that this OR is slightly larger than the relative risk, as is expected for a moderately prevalent disease. A standard Z test statistic for the null hypothesis that the odds ratio is equal to 1 is given by

$$Z = \frac{\ln(OR)}{SE[\ln(OR)]},$$ (20)

where $SE[\ln(OR)]$ denotes the standard error of $\ln(OR)$. For the assumptions stated above, the numerator and the denominator of this statistic can be derived from the entries of Table 6.3:

$$\ln(OR) = \ln(2.017094) = 0.701658,$$ (21)

$$Var(\ln(OR)) = \frac{1}{319.328} + \frac{1}{188.694} + \frac{1}{680.672} + \frac{1}{811.306} = 0.011133,$$ (22)

$$\text{Critical value} = \Phi^{-1}(1 - 10^{-7}/2) = 5.326724.$$ (23)

Hence, under H_1, the test statistic Z^* is distributed as $N(6.64999, 1)$ (i.e., normal with mean 6.64999 and variance 1), whereas under H_0, the test statistic is distributed as $N(0, 1)$.

3. The critical value for a standard normal test, for a critical significance level of $\alpha = 10^{-7}$, is given by the inverse standard normal distribution function:

$$Z = \frac{0.701658}{\sqrt{0.011133}} = 6.649997.$$ (24)

4. Therefore, the power of the test under the above assumptions is

$$\begin{aligned} P(Z^* > 5.326724) &= P(Z^* - 6.64999 > 5.326724 - 6.64999) \\ &= P(Z > -1.32327) \\ &= 1 - 0.092872 \\ &= 0.907128. \end{aligned}$$ (25)

One can explore statistical power under the influence of the assumptions made by repeating the above calculations after changing some of the parameters, such as the relative risk, allele frequency, and sample size.

Relationship of Power to Sample Size

It is obvious that increasing the sample size will increase statistical power, other factors being equal. Indeed, there is a simple but nonlinear mathematical relationship that describes this dependence. For a two-by-two table with cell counts a, b in row 1 and c, d in row 2, the variance of $\ln(OR)$ is given by

$$Var(\ln(OR)) = \frac{1}{a} + \frac{1}{b} + \frac{1}{c} + \frac{1}{d}.$$ (26)

If the sample size is changed by a factor of k, then

$$\begin{aligned} Var(\ln(OR)) &= \frac{1}{ka} + \frac{1}{kb} + \frac{1}{kc} + \frac{1}{kd} \\ &= \frac{1}{k}\left(\frac{1}{a} + \frac{1}{b} + \frac{1}{c} + \frac{1}{d}\right). \end{aligned}$$ (27)

Therefore, when the sample size is changed by a factor of k, the mean of the distribution of Z^* is changed by a factor of $\sqrt{k}$. This means that, having performed a power calculation under a set of assumptions, for a certain sample size, it is not necessary to repeat the whole exercise if we wish to evaluate power under the same assumptions but for a different sample size. All we need to do is to obtain the new expected value of the normal test statistic by multiplying the original value by the square root of the ratio of sample sizes (new sample size/original sample size). We can then use statistical tables to look up the distribution function of the normal distribution with the new expected value.

The situation is even clearer if we consider the square of the Z test statistic, which would be a chi-square test statistic. If $Z \sim N(\mu, 1)$, then $Z^2 \sim \chi^2(\text{df} = 1, \text{NCP} = \mu^2)$, in which df is the degrees of freedom and NCP is the noncentrality parameter. Because μ is linearly related to the square root of sample size, NCP is linearly related to sample size. The NCP can be interpreted as the difference in the expected value of the chi-square statistic under H_0 and H_1.

In the above example NCP $= 6.49997^2 = 44.22246$. From the statistical tables for the inverse chi-square distribution, the critical chi-square value for $\alpha = 10^{-7}$ is 28.374 (note this is 5.326724^2). From the tables for the noncentral chi-square distribution with df $= 1$, and NCP $= 44.22246$, the probability that the chi-square statistic exceeds 28.374 is 0.907128. Halving the sample size halves the NCP to 22.11123, and the power reduces to 0.26616.

Power under Indirect Association

The above results have important implications for association studies using tag SNPs (i.e., SNPs that are in strong LD, usually $r^2 > 0.8$, with other SNPs and hence may serve as a proxy for them). If a direct association study of a causal SNP would provide an NCP of λ, then an indirect association study of a SNP in LD with the causal SNP has NCP of $R^2\lambda$. In other words, *NCP is linearly related to the magnitude of* R^2 *between the causal and the genotyped SNP.* As NCP is linearly related to sample size, the magnitude of R^2 also has a linear relationship with the sample size. Hence R^2 is the appropriate LD metric for the selection of tag SNPs, and division of the old sample size by R^2 gives the new sample size for the same power.

Practical Power Calculation for Genetic Studies

The power calculations shown previously may be performed using the genetic power calculator (Purcell et al. 2003; http://pngu.mgh.harvard.edu/~purcell/gpc/). The calculator also provides a variety of tools for power evaluation for other types of genetic studies, such as quantitative trait and family-based studies. The methods of power calculations for different types of studies can be found elsewhere (Risch and Merikangas 1996; Knapp 1999; Sham et al. 2000; Chen and Deng 2001; Lange et al. 2002; Lange and Laird 2002).

In performing power calculations for project proposals, investigators are often concerned about what assumptions to make regarding unknown quantities such as allele frequency and effect size. There are two common situations that require slightly different approaches—replication of a previous finding and identification of novel risk variants.

Replication of a Previous Finding

In this situation the previous report of a positive association would have provided information regarding the allele frequency and the effect size of the putative risk variant. In using the estimates of these quantities as assumptions in the power calculation, it is important to appreciate that effect size estimate is likely to be upwardly biased by the phenomenon of the "winner's curse," particularly if the original report was a screen of multiple variants. An intuitive explanation of this phenomenon is that chance is likely to have contributed positively to association signals of the most significant results from a screen of multiple variants. Methods for correcting this bias have been proposed (Zollner and Pritchard 2007; Ghosh et al. 2008; Zhong and Prentice 2008), and the corrected effect sizes should be used in power calculation of studies that attempt to replicate such findings.

Identification of Novel Risk Variants

For this scenario there may be multiple variants of different alleles and effect sizes. An ideal power calculation should consider a range of values of the parameters—for example, allele frequency ranging from 0.1 to 0.9, and odds ratio ranging from 1.2 to 4. The results can be plotted either as statistical power for a fixed sample size or as required sample size for fixed statistical power (e.g., 80%). In this way the investigator can judge the range of effects for which a study will have adequate power to detect.

ACKNOWLEDGMENTS

This work is supported by project grant EY-12562 for the National Eye Institute, USA, and by General Research Fund grants 757905, 766906, and 774707 from the Hong Kong Research Grants Council.

REFERENCES

Allison, D.B., Gadbury, G.L., Heo, M.S., Fernandez, J.R., Lee, C.K., Prolla, T.A., and Weindruch, R. 2002. A mixture model approach for the analysis of microarray gene expression data. *Computat. Statist. Data Anal.* **39:** 1–20.

Benjamini, Y. and Hochberg, Y. 1995. Controlling the false discovery rate: A practical and powerful approach to multiple testing. *J. R. Stat. Soc.* **57:** 289–300.

Benjamini, Y. and Hochberg, Y. 1997. Multiple hypotheses testing with weights. *Scand. J. Statist.* **24:** 407–418.

Benjamini, Y. and Yekutieli, D. 2001. The control of the false discovery rate in multiple testing under dependency. *Ann. Statist.* **29:** 1165–1188.

Chen, W.M. and Deng, H.W. 2001. A general and accurate approach for computing the statistical power of the transmission disequilibrium test for complex disease genes. *Genet. Epidemiol.* **21:** 53–67.

Cox, D.R. and Wong, M.Y. 2004. A simple procedure for the selection of significant effects. *J. R. Statist. Soc. Ser. B* **66:** 395–400.

Dudbridge, F. and Koeleman, B.P. 2004. Efficient computation of significance levels for multiple associations in large studies of correlated data, including genomewide association studies. *Am. J. Hum. Genet.* **75:** 424–435.

Efron, B. 2007. Correlation and large-scale simultaneous significance testing. *J. Am. Statist. Assoc.* **102:** 93–103.

Everitt, B.S. and Bullmore, E.T. 1999. Mixture model mapping of brain activation in functional magnetic resonance images. *Hum. Brain Mapp.* **7:** 1–14.

Ghosh, A., Zou, F., and Wright, F.A. 2008. Estimating odds ratios in genome scans: An approximate conditional likelihood approach. *Am. J. Hum. Genet.* **82:** 1064–1074.

Knapp, M. 1999. A note on power approximations for the transmission/disequilibrium test. *Am. J. Hum. Genet.* **64:** 1177–1185.

Lange, C. and Laird, N.M. 2002. Power calculations for a general class of family-based association tests: Dichotomous traits. *Am. J. Hum. Genet.* **71:** 575–584.

Lange, C., DeMeo, D.L., and Laird, N.M. 2002. Power and design considerations for a general class of family-based association tests: Quantitative traits. *Am. J. Hum. Genet.* **71:** 1330–1341.

Nyholt, D.R. 2004. A simple correction for multiple testing for single-nucleotide polymorphisms in linkage disequilibrium with each other. *Am. J. Hum. Genet.* **74:** 765–769.

Purcell, S., Cherny, S.S., and Sham, P.C. 2003. Genetic power calculator: Design of linkage and association genetic mapping studies of complex traits. *Bioinformatics* **19:** 149–150.

Risch, N. and Merikangas, K. 1996. The future of genetic studies of complex human diseases. *Science* **273:** 1516–1517.

Roeder, K., Bacanu, S.A., Wasserman, L., and Devlin, B. 2006. Using linkage genome scans to improve power of association in genome scans. *Am. J. Hum. Genet.* **78:** 243–252.

Salyakina, D., Seaman, S.R., Browning, B.L., Dudbridge, F., and Muller-Myhsok, B. 2005. Evaluation of Nyholt's procedure for multiple testing correction. *Hum. Hered.* **60:** 19–25; discussion 61–62.

Seaman, S.R. and Muller-Myhsok, B. 2005. Rapid simulation of P values for product methods and multiple-testing adjustment in association studies. *Am. J. Hum. Genet.* **76:** 399–408.

Sham, P.C., Cherny, S.S., Purcell, S., and Hewitt, J.K. 2000. Power of linkage versus association analysis of quantitative traits, by use of variance-components models, for sibship data. *Am. J. Hum. Genet.* **66:** 1616–1630.

Storey, J.D. 2003. The positive false discovery rate: A Bayesian interpretation and the q-value. *Ann. Statist.* **31:** 2013–2035.

Storey, J.D. and Tibshirani, R. 2003. Statistical significance for genomewide studies. *Proc. Natl. Acad. Sci.* **100:** 9440–9445.

Zhong, H. and Prentice, R.L. 2008. Bias-reduced estimators and confidence intervals for odds ratios in genome-wide association studies. *Biostatistics* **9:** 621–634.

Zollner, S. and Pritchard, J.K. 2007. Overcoming the winner's curse: Estimating penetrance parameters from case–control data. *Am. J. Hum. Genet.* **80:** 605–615.

WWW RESOURCES

http://pngu.mgh.harvard.edu/~purcell/gpc/ Purcell, S., Cherny, S.S., and Sham, P.C. 2003. Genetic power calculator. This calculator provides automated power analysis for variance components, quantitative trait locus linkage, and association tests in case–control, parent–child trios and sibships.

Introduction to Genetic Association Studies

Cathryn M. Lewis[1] and Jo Knight[2]

[1]*MRC Social, Genetic and Developmental Psychiatric Centre, Institute of Psychiatry, King's College London, London, United Kingdom and Department of Medical and Molecular Genetics, King's College London School of Medicine, London, United Kingdom;* [2]*MRC Social, Genetic and Developmental Psychiatric Centre, Institute of Psychiatry, King's College London, London, United Kingdom and Department of Medical and Molecular Genetics, King's College London School of Medicine, London, United Kingdom; and National Institute for Health Research, Biomedical Research Center, Guy's and St. Thomas NHS Foundation Trust and King's College London, London, United Kingdom*

INTRODUCTION

Genetic association studies identify candidate genes or genome regions that contribute to a specific disease by testing for a correlation between disease status and genetic variation. A higher frequency of a single-nucleotide polymorphism (SNP) allele or genotype in a series of individuals affected with a disease can be interpreted as meaning that the tested variant increases the risk of a specific disease (although several other interpretations are also valid; see the following sections). SNPs are the most widely tested markers in association studies (and this term will be used throughout), but microsatellite markers, insertion/deletions, variable-number tandem repeats (VNTRs), and copy-number variants (CNVs) are also used.

Association studies are a major tool for identifying genes conferring susceptibility to complex disorders. These traits and diseases are termed "complex" because both genetic and environmental factors contribute to the susceptibility risk. Extensive experience in genetic studies for many complex disorders (such as diabetes, heart disease, autoimmune diseases, and psychiatric traits) confirms that many different genetic variants control disease risk, with each variant having only a subtle effect.

Associations with polymorphisms in candidate genes have been confirmed in many different diseases (Lohmueller et al. 2003), and genome-wide association studies (GWAS) are identifying many novel associations in genes that had not been strong a priori candidates for the disease under test (Wellcome Trust Case Control Consortium 2007). However, the modest increase in risk implies that large well-designed and analyzed studies are required to detect and confirm signals for association.

This chapter provides a broad outline of the design and analysis of genetic association studies, but it focuses specifically on case–control studies in candidate genes or regions. Even in this era of genome-wide studies, case–control studies still form the majority of published reports. We illustrate the importance of quality control in performing these studies, describe basic analytical strategies for a SNP, and point the reader toward methods for analyzing haplotypes or multiple markers. We also highlight some of the pitfalls of performing powerful, accurate association studies and discuss how these challenges are reflected in the contradictory literature for many disease–gene investigations. Other approaches to genetic association studies are covered in the following chapters: Family-based association studies are considered separately in Chapter 12, quantitative trait locus studies are addressed in Chapter 5, and GWAS are discussed in detail in Chapter 8, and are not addressed in any detail here.

INTERPRETING SIGNIFICANT GENETIC ASSOCIATION

Significant genetic association may be interpreted as either (1) direct association, in which the genotyped SNP is the true causal variant conferring disease susceptibility; (2) indirect association, in which a SNP in linkage disequilibrium (LD) with the true causal variant is genotyped; or (3) a false-positive result, in which there is either chance or systematic confounding, such as population stratification.

Distinguishing between direct and indirect association is challenging and may require resequencing of the candidate region, dense genotyping of all available SNPs, or functional studies to confirm the role of a putative mutation in disease.

FINDING DIRECT ASSOCIATION

Case–Control Study

The simplest study design used to test for association is the case–control study, in which a series of cases affected with the disease of interest are collected together with a series of control individuals. The specific choice of phenotype for the cases may define the exact hypothesis to be tested, and applying strict clinical criteria for ascertainment is necessary to ensure a homogeneous set of cases. Two standard methods are used for collecting controls: the use of either a series of individuals who have been screened as negative for presence of the disease or of controls randomly ascertained from the population, whose disease status is unknown. Both control sets form a valid test for association, and they will have similar power for a rare disease. For a more common disease, a study with screened unaffected controls (often termed "supernormal" controls) will have higher power to detect association compared with a study using population-based controls, and the increase in power is notable for diseases with high prevalence. For some diseases, screening controls for the presence or absence of the disease may be difficult, and using a larger sample of unscreened controls may be more efficient.

Statistical Analysis of Case–Control Study

The genotypes of a single, biallelic SNP on a set of cases and controls can be summarized in a 2×3 contingency table of the genotype counts for each group, as shown in Figure 7.1. For a SNP with alleles G and T, we tabulate the number of cases and controls with each genotype GG, GT, and TT. Several different statistical analysis methods can be applied to this table. We will focus here on goodness-of-fit tests, rather than likelihood-based or regression methods. Pearson's chi-square test is used to assess departure from the null hypothesis that case and controls have the same the distribution of genotype counts. This test statistic has a chi-square distribution with two degrees of freedom on this 2×3 table.

This approach provides a valid statistical analysis of the data presented, but uses no genetic information. We have illustrated the data on a table with genotypes ordered as GG, GT, TT, with the inherent supposition that disease risk may increase (or decrease) as the number of T alleles increases. However, column order is not used in the test statistic, and reordering the table as GG, TT, GT gives the same value of the test statistic and p value. Other analysis methods that correspond to the under-

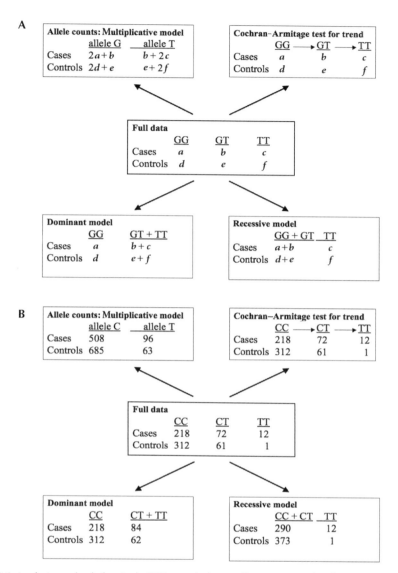

FIGURE 7.1 Analysis methods for single SNP association studies, testing under the assumption of specific genetic models for (A) arbitrary genotype counts and (B) the rheumatoid arthritis case–control study in Table 7.1.

lying genetic models we expect to be acting in complex diseases may be preferred. First, the table may be decomposed from genotypes into alleles, with cell counts of the number of G alleles, and the number of T alleles carried by cases and controls (regardless of the genotype combination in which these alleles were carried) (Fig. 7.1A, upper left). This test is valid under the null hypothesis of no association, or when the true model of association is multiplicative (or log additive), so the genotype relative risks for GG, GT, and TT genotypes can be modelled as 1, r, and r^2, with relative risk increasing by a factor r for each T allele carried (Sasieni 1997). An alternative test for this model is the Cochran–Armitage test for trend (CATT) (Fig. 7.1A, upper right), which, as its name implies, tests for a trend in differences in cases and controls across the ordered genotypes in the table. This test is asymptotically equivalent to the allele test described previously, although it is more robust to departures from Hardy–Weinberg equilibrium (HWE) (see the following) (Sasieni 1997). Further tests can be used to test specific genetic hypotheses—for example, that the SNP alleles increase disease risk under a dominant or a recessive model (Fig. 7.1A, bottom). Assuming T is a high-risk allele, these tests compare GG genotypes to CT + TT genotypes (dominant model), or CC + CT to TT genotypes (recessive model).

Although the tests described previously are all valid methods for analysis of an association study, any such study should have a prespecified analysis plan because applying all tests will increase the probability of a false-positive result. Candidate gene association studies most commonly test for a difference in allele frequency. The allele frequencies in cases and controls provide useful, direct summary statistics for the data. The CATT has become popular in GWAS (e.g., O'Donovan et al. 2008). Other test statistics, such as analyzing under a dominant or recessive model, may also be applied to ensure that interesting findings are not missed because of the specific analysis method used. These tests are rarely a primary analysis tool for complex genetic disorders, but may be used as secondary analyses to explore the potential mode of inheritance of an associated SNP, or to test a prespecified hypothesis. When several analysis methods are used, a correction for multiple testing should be applied. This is not straightforward owing to the correlation between test statistics, and simulation studies may be required.

Example of a Statistical Analysis

PTPN22 is associated with several autoimmune phenotypes, with the strongest association seen at R602W. Table 7.1 shows the genotypes at this variant (SNP rs2476601, C1858T) in a study of London rheumatoid arthritis (RA) cases and randomly ascertained controls (Steer et al. 2005). Genotypes for both cases and controls were in HWE, with p values of 0.060 and 0.267, respectively. The genotype counts show that cases have a higher frequency of both CT and TT genotypes compared with controls. In RA cases, the frequency of allele T (15.9%) is higher than in controls (8.4%). Analyzing the allele counts contingency table shows strong evidence for association at this SNP ($p = 0.00003$) (Fig. 7.1B). Significant evidence for association is also found using the CATT ($p = 0.00004$) (Fig. 7.1B).

A summary measure of the effect of this SNP on risk for RA can be obtained through calculating odds ratios (ORs). These can be calculated separately for CT and TT genotypes by comparing each to the baseline CC genotype, which is most common in the population. For the CT genotype, the OR is the odds of the CT genotype compared with the CC genotype in cases, divided by the same quantity in controls: $(72/218) / (61/312) = 1.69$. Confidence intervals on the OR can be calculated using the Woolf method: the standard error of ln(OR) is approximately

$$\sqrt{\frac{1}{a} + \frac{1}{b} + \frac{1}{c} + \frac{1}{d}},$$

in which a, b, c, and d are the entries in the relevant genotype subtable. The OR, or genotype relative risk, for CT and TT genotypes compared with CC genotypes confirm that both these genotypes have an elevated risk of RA (because neither confidence interval contains 1), although the confidence interval for TT genotypes is very wide because only a single TT control individual is observed (Table 7.1). Examination of ORs (1.69 and 17.2) suggests that a gene dosage model is acting, with much

TABLE 7.1 *PTPN22* C1858T genotypes for rheumatoid arthritis (RA) case–control study

Cohort	No. of individuals	Genotypes			Frequency of allele T
		CC	CT	TT	
RA cases	302	218 (72.2%)	72 (23.8%)	12 (4.0%)	15.9%
Controls	374	312 (83.4%)	61 (16.3%)	1 (0.3%)	8.4%
OR (95% CI)		1	1.69 (1.15–2.48)	17.17 (2.22–133.06)	

Data from Steer et al. (2005).

higher risk in mutation homozygotes (TT) than the heterozygotes (CT). The increase in risk from CT to TT is higher than would be expected under a multiplicative model, in which estimates from the allele count table give an OR of 2.06 for each T allele carried, implying an approximately fourfold increase risk for TT individuals. However, for an association of this strength, analyzing the genotype counts assuming a multiplicative model still results in highly significant evidence of association.

Using Quantitative Measures

Some complex phenotypes, such as high blood pressure, height, and obesity, are better characterized by quantitative rather than qualitative measures. Several options are available for the analysis of such data. The quantitative measure can be tested for association in a linear regression framework, assessing whether the genotypes (as an explanatory variable) predict trait value. Similar to the analysis options described in case–control studies previously discussed, genotypes may be coded as a three-level factor, or as a count of T alleles carried (0, 1, 2), or as a dominant or recessive model. Quantitative measures may be analyzed in a case–control framework by dichotomizing the sample. However, this method may result in a loss of power because all information on the distance of an individual's observed phenotype from the dichotomizing threshold is lost. The power of a quantitative trait association study may be increased by ascertaining individuals only from the extremes of the distribution (Slatkin 1999).

FINDING INDIRECT ASSOCIATION

Linkage Disequilibrium

LD measures the correlation between SNP alleles at sites in the same region of the genome. Dependence between SNPs arises because any novel SNP (e.g., a change from base pair C → A at a genomic site) occurs on a background of fixed alleles at other SNPs in the region. For example, for five flanking SNPs in each direction, the existing chromosomal haplotype may have been ACTCG-**C**-GGATC, which becomes ACTCG-**A**-GGATC. The A allele at this SNP is fully correlated with these flanking haplotypes, so that initially all copies of allele A have allele G at the neighboring SNP. As this chromosome is transmitted through the generations, the length of this haplotype is diminished by recombination, and different copies of the original *A* allele will have different recombination patterns, and be flanked by different lengths of DNA from the original chromosome. Figure 7.2 illustrates this process, in which a disease mutation D occurs in meiosis from generation 1 to generation 2, on a specific background chromosome. This mutation (if it is not lost to the population through nontransmission) is transmitted through the generations, with recombinations reducing the length of the original ancestral chromosome. However, all copies of this mutation D arising from the same mutation event will harbor some portion of the ancestral chromosome, with the length of retained chromosome depending on the pattern of recombination events through the generations.

LD is an important phenomenon in association testing because it induces correlation in short regions of the genome. In Figure 7.2, mutation D occurs close to a polymorphic marker bearing the M allele. In the current generation, most chromosomes carrying mutation D also carry allele M. Thus, we have two opportunities to detect association with the disease, by genotyping either M or D. Genotyping the true disease mutation D (direct association) should have higher power to detect association, but where M and D are in strong LD, and sample sizes are adequate, significant association should be detectable by genotyping M (indirect association).

Intuitively, LD measures the correlation between SNP alleles. Given a chromosome with a specific SNP allele, how does this influence the probability distribution of alleles carried at other SNPs within the same genetic region? Many different statistical measures to quantify LD between two

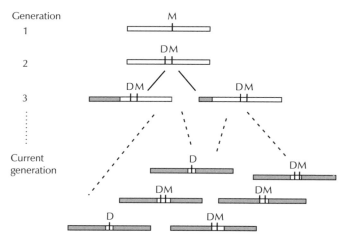

FIGURE 7.2 Association with disease through direct association (D) and indirect association (M). Disease susceptibility mutation D arises on an ancestral chromosome (white) close to a SNP marker, M. The ancestral chromosome flanking D is lost through recombination through the generations. Observing chromosomes in the current generations shows that all copies of D carry some region of the ancestral (white) chromosome and many of these will also carry the marker allele M.

SNPs have been proposed (Devlin and Risch 1995), with D' and r^2 being most widely used. The International HapMap Project is a valuable resource for study design, allowing researchers to investigate LD in a region and to select an informative subset of available SNPs to be genotyped in an association study (http://www.hapmap.org).

Analysis of Multiple Markers and Haplotypes

Although high-throughput genotyping has increased the number of SNPs it is feasible to genotype, studies still consider only a subset of available SNPs and test for indirect association using such a subset. In such circumstances, statistical analysis of individual SNPs (as described previously) may not be the most effective strategy and may lack the ability to detect association at an ungenotyped SNP. The pattern of alleles at multiple markers is usually better able to predict the allele at an untyped locus; hence, simultaneous analysis of multiple markers can improve the power of association studies (de Bakker et al. 2005).

Perhaps the most obvious method for analyzing multiple markers simultaneously is multiple logistic regression. Logistic regression is an adaptation of linear regression in which a logit transformation is used to allow for analysis of a binary outcome (i.e., case–control status). In Equation 1, p is the probability of having disease, β_0 represents the intercept, β_1 and β_2 represent the main effect of each marker on the trait, and β_3 represents the interaction term. The variables x_1 and x_2 contain information about the genotype at the two markers and can be coded in a number of different ways— for example, $-1, 0, 1$. The interaction term ($x_1 * x_2$) can also be coded in a number of different ways:

$$\text{logit}(p) = \ln\frac{p}{1-p} = \beta_0 + \beta_1 x_1 + \beta_2 x_2 + \beta_3(x_1 * x_2) \ . \tag{1}$$

Coefficients β_i can be estimated for each SNP as well as for interactions between them. Stepwise regression can be used systematically to compare different genetic models and to investigate whether multiple markers have independent effects on the trait or are simply in LD with each other, with either marker capturing evidence for association and no improvement in model fit when both markers are included (Cordell and Clayton 2002).

An alternate analysis approach is to phase genotypes into haplotypes and use these as the unit of analysis. This method is attractive because the haplotype is the functional unit of the gene. It is often impossible to be certain about the combination of haplotypes carried by any one individual. However, it is straightforward to determine all possible combinations, and techniques like the E-M algorithm can be used to assign a probability to each haplotype pair (Excoffier and Slatkin 1995). Haplotype effects can be estimated using regression techniques adapted to handle phase uncertainty—for example, a weighted regression technique in which the likelihood function of a finite mixture regression is a weighted sum over all possible haplotypes for each individual (Sham et al. 2004).

Interaction between Genes in Disease Risk

The previous discussion of analysis of multiple markers is focused on markers within a short genetic region (and potentially in LD), but analysis of multiple markers across the genome is also important to identify interaction between genes in disease risk. Interaction is most simply defined as the interdependence of effects at two loci. If the disease risk conferred by the presence of risk alleles at two markers can be inferred from the marginal effects of the presence of each risk allele individually, then no interaction is present. When the joint effect of risk alleles at both markers is much larger (or smaller) than implied by the marker-specific effects, then interaction exists. Statistical interaction (as defined previously) may differ from biological interaction between two genes (Cordell 2002).

The presence of interaction between loci may make each locus more difficult to identify in single SNP tests. Despite the increase in numbers of tests, regression techniques with interaction terms are both computationally feasible and powerful for GWAS (Marchini et al. 2005). Another method for analysis of interaction is multifactor dimensional reduction (MDR). This is a nonparametric approach with a focus on overcoming low numbers of observations in high-order data sets (Ritchie et al. 2003). This technique has been applied to several different traits but none of the results has yet been replicated independently (Milne et al. 2008).

ADDRESSING PROBLEMS IN ANALYSIS

Quality Control

One disadvantage of a case–control study design compared with family-based association studies is the lack of an internal check on genotyping quality. Standard laboratory practice of assigning both cases and controls to each plate, checking for differences in genotype frequency across plates, and genotyping duplicate samples can help eliminate systematic errors. Testing for HWE in controls can also identify problems with genotyping quality.

Hardy–Weinberg Equilibrium

Under HWE, alleles segregate randomly in the population, allowing expected genotype frequencies to be calculated from allele frequencies. A comparison of the expected and observed genotype frequencies provides a test of HWE (e.g., using a chi-square statistic). For alleles G and T, in which the frequency of allele G is p and the frequency of allele T is $q = (1 - p)$, the expected frequencies of genotypes GG, GT, and TT are p^2, $2pq$, and q^2. Allele frequencies (p, q) are usually estimated from the genotype sample under test, rather than obtained from external genotyping data.

Departure from HWE is generally tested for by using the Pearson chi-square test to assess goodness of fit (of the observed genotype counts to their expectation under HWE). Table 7.2 shows the step-by-step calculation with observed counts for genotypes GG, GT, and TT of *a, b, c,* and an application to a data set of 100 control genotypes (GG: 60, GT: 30, TT: 10). The estimated frequency

TABLE 7.2 Testing for departure from Hardy–Weinberg equilibrium

| | Genotype counts | | | | Estimated frequency |
	GG	GT	TT	Total	of G allele
General					
Observed (*O*)	*A*	*b*	*c*	$N = a + b + c$	$p = (2a + b)/(2N)$
Expected (*E*)	Np^2	$2Np(1-p)$	$N(1-p)^2$		
Test data set					
Observed (*O*)	60	30	10	100	$p = (2 \times 60 + 30)/200 = 0.75$
Expected (*E*)	56.25	37.5	6.25		

of allele G is 0.75 (= [2 × 60 + 30]/200), noting the division by the number of alleles (2*N*) here, not genotypes (*N*). The chi-square goodness-of-fit test statistic is then calculated from summing $(O - E)/E^2$ across genotypes, giving chi-square = 4.0. Under the null hypothesis of no departure from HWE, the test statistic has one degree of freedom (not two degrees of freedom, as implied by the table dimensions), because the allele frequency *p* has been estimated from the observed data. In this test data set, a *p* value of 0.046 is obtained, giving slight evidence of departure from HWE, with a deficit in the number of observed heterozygotes.

Departures from HWE in control samples may be caused by the following.

1. Genotyping error. In many genotyping platforms, calling heterozygotic individuals is more challenging than homozygotic individuals, and a higher rate of missing individuals for this genotype can distort HWE.

2. Assortative mating. HWE requires random mating for the SNP under test, which is reasonable for a random SNP across the genome, but may be violated for SNPs that affect mate choice, such as height.

3. Selection. Any genotype increasing the risk of fetal loss or early death is likely to be underrepresented.

4. Population stratification. Control samples that arise from a combination of genetically distinct subpopulations may not be in HWE.

5. Chance. HWE *p* values for studies of more than one SNP should be corrected appropriately for multiple testing.

Departures from HWE may be caused by any of these factors, but also by the genotyped SNP playing a role in disease susceptibility. Case genotypes for a disease mutation will only be in HWE if the genetic model is multiplicative, with genotype relative risks of 1, *r*, r^2. However, for modest effect sizes, the power to detect departures from HWE may be low in cases.

No standard guidelines for rejecting SNPs that depart from HWE have been developed. In practice, all SNPs for which HWE *p* values decrease below a predetermined threshold should be checked manually for genotyping quality. Investigators should also be aware of SNPs showing significant association in which HWE *p* values are close to this threshold and unsupported by neighboring SNPs in LD.

Missing Genotypes

Another indication of poor genotyping quality is low call rates, with many missing genotypes for each SNP or each individual. This is a major issue in GWAS, but it is also applicable to candidate gene association studies. Genotypes that are missing at random will not bias a test, but poor genotype call rates may indicate nonrandom missingness, with one specific genotype (often heterozygotes) having a lower call rate. This may bias tests of association. Differential rates of missingness between cases and controls (for example, because of differences in DNA extraction and storage) may also be a problem (Clayton et al. 2005).

Population Stratification

Population stratification arises in case–control studies when the two study groups are poorly matched for genetic ancestry. Confounding then occurs between disease state (case, control) and genetic ancestry, with a subsequent increase in false-positive associations. For population stratification to occur, the underlying populations must differ in SNP allele frequency and be represented at different frequencies in the case and control groups. Detecting and controlling for population stratification is important, particularly in GWAS, in which even subtle differences between cases and controls can have major effects on the analysis. Several methods are available to detect and correct for population stratification, including genomic control, the Cochran/Mantel–Haenszel test, and the transmission disequilibrium test.

Genomic control (GC) assumes that population stratification inflates the association test statistics by a constant factor λ, which can be estimated from the median or mean test statistic from a series of unlinked SNPs genotyped in both cases and controls (Devlin and Roeder 1999). Test statistics are then divided by λ and compared with a chi-square distribution or an F distribution) to test for association (Devlin et al. 2004). Genotypes at SNPs uncorrelated with disease status can also be used to infer population ancestry, assigning the samples to distinct population groups, which can then be controlled for in the analysis (Pritchard et al. 2000). In GWAS, population substructure can be identified through a principal components analysis, which models ancestral genetic differences between cases and controls and then corrects for this in the analysis (Price et al. 2006).

Where individuals can be classified into known subgroups (e.g., by birthplace), analysis can be performed within each subgroup and combined using a Cochran/Mantel–Haenszel test (Clayton et al. 2005). The issue of population stratification can be avoided by using family-based studies (see Chapter 12). The most widely used method is the transmission disequilibrium test (TDT) (Spielman et al. 1993), which tests for non-Mendelian transmission of SNP alleles from heterozygous parents to affected offspring; overtransmission suggests that the SNP allele increases risk of disease.

PITFALLS AND PROBLEMS OF ASSOCIATION STUDIES

A major challenge in association studies of candidate genes has been nonreplication of significant findings. For many diseases and genes, the literature contains papers with little consistent pattern in the results obtained. Typically, this comprises an initial report showing significant association, with follow-up studies showing little or no evidence of association. We discuss here reasons for these discrepancies between studies.

False-Positive Finding

The initial report of association may have been a false-positive finding that arose by chance or systematic bias in the study. The "Quality Control" section discussed several problems that can lead to such results, and each of these should be checked (population stratification and genotyping errors). False-positive results may arise through a failure to correct for multiple testing across the number of genes, SNPs, statistical analysis methods, or phenotypic subgroups tested, although this can be difficult to determine from a publication. For independent tests (e.g., multiple genes that are not in LD), a Bonferroni correction may be applied to the *p* values. Where tests are correlated, an appropriate correction may be difficult to determine, but permutation tests can be used to determine empirical levels of significance. A noted phenomenon is that the first published study tends to overestimate the effect size, with subsequent studies detecting more moderate contribution of the genotyped variant to disease risk (Ioannidis et al. 2001).

Replication Study Lacks Power

Alternatively, replication studies may lack power to detect the true association. Most genes contributing to complex disorders confer only a very modest increase in disease risk, and to detect these with high power requires large sample sizes. For example, for a SNP of 10% frequency, under a multiplicative model with heterozygote relative risk of 1.3, at least 1146 cases and controls are required to obtain 80% power at a significance level of 5% with no correction for multiple testing (Purcell et al. 2003). Including the multiple testing correction greatly increases the numbers needed. Many association studies have used samples of hundreds, not thousands, of cases and controls, and therefore lack the ability to detect such associations. Meta-analysis of published data provides a possible solution, and such studies have confirmed many associations that were unclear from individual study reports (Altshuler et al. 2000; Ioannidis et al. 2001; Lohmueller et al. 2003).

Heterogeneity between Studies

Another problem is that heterogeneity between studies may validly lead to different conclusions about the role of a SNP in disease risk. Sources of heterogeneity include the precise clinical criteria used in case definition for each study, differences in disease severity, disease subtype, age of diagnosis, or duration of disease. If a genetic variant contributes predominantly to a specific subphenotype of disease, then the mix of cases ascertained in different studies will substantially affect the power of each study to detect association. Information from family or twin studies on heritability of different components of disease definition can help refine the hypothesis to be tested, with some studies choosing to ascertain cases likely to be more heavily genetically predisposed, for example, those with a family history of disease, or early onset (Antoniou and Easton 2003).

Heterogeneity across Studies

Population heterogeneity across studies may also lead to differences in study outcomes. Variations in SNP frequencies are seen across the major population groups because of random drift, novel mutations, and (less commonly) selection. However, meta-analyses of replicated genetic association studies suggest that even when the SNP frequency differs across populations, the effect size of mutations remains approximately constant (Ioannidis et al. 2004). Some mutations may be absent in specific population groups; for example, NOD2 mutations, which are present in >30% of Crohn's disease patients in European populations are absent in Asian populations (Mathew and Lewis 2004).

CONCLUSION

This chapter has given a broad outline of the design and analysis of genetic association studies, as well as the pitfalls of performing powerful, accurate association studies. These challenges are reflected in the contradictory literature for many disease–gene investigations. However, consistent findings of disease–gene associations have been detected, and the realization that most mutations confer only modest increases in risk has led to an improvement in study design. Larger studies are now being performed and internal replication of significant findings is becoming standard practice.

REFERENCES

Altshuler, D., Hirschhorn, J.N., Klannemark, M., Lindgren, C.M., Vohl, M.C., Nemesh, J., Lane, C.R., Schaffner, S.F., Bolk, S., Brewer, C., et al. 2000. The common PPARγ Pro12Ala polymorphism is associated with decreased risk of type 2 diabetes. *Nat. Genet.* **26:** 76–80.

Antoniou, A.C. and Easton, D.F. 2003. Polygenic inheritance of breast cancer: Implications for design of association studies. *Genet. Epidemiol.* **25:** 190–202.

Clayton, D.G., Walker, N.M., Smyth, D.J., Pask, R., Cooper, J.D., Maier, L.M., Smink, L.J., Lam, A.C., Ovington, N.R., Stevens, H.E., et al. 2005. Population structure, differential bias and genomic control in a large-scale, case-control association study. *Nat. Genet.* **37:** 1243–1246.

Cordell, H.J. 2002. Epistasis: What it means, what it doesn't mean, and statistical methods to detect it in humans. *Hum. Mol. Genet.* **11:** 2463–2468.

Cordell, H.J. and Clayton, D.G. 2002. A unified stepwise regression procedure for evaluating the relative effects of polymorphisms within a gene using case/control or family data: Application to HLA in type 1 diabetes. *Am. J. Hum. Genet.* **70:** 124–141.

de Bakker, P.I., Yelensky, R., Pe'er, I., Gabriel, S.B., Daly, M.J., and Altshuler, D. 2005. Efficiency and power in genetic association studies. *Nat. Genet.* **37:** 1217–1223.

Devlin, B. and Risch, N. 1995. A comparison of linkage disequilibrium measures for fine-scale mapping. *Genomics* **29:** 311–322.

Devlin, B. and Roeder, K. 1999. Genomic control for association studies. *Biometrics* **55:** 997–1004.

Devlin, B., Bacanu, S.A., and Roeder, K. 2004. Genomic control to the extreme. *Nat. Genet.* **36:** 1129–1130; author reply 1131.

Excoffier, L. and Slatkin, M. 1995. Maximum-likelihood estimation of molecular haplotype frequencies in a diploid population. *Mol. Biol. Evol.* **12:** 921–927.

Ioannidis, J.P., Ntzani, E.E., Trikalinos, T.A., and Contopoulos-Ioannidis, D.G. 2001. Replication validity of genetic association studies. *Nat. Genet.* **29:** 306–309.

Ioannidis, J.P., Ntzani, E.E., and Trikalinos, T.A. 2004. "Racial" differences in genetic effects for complex diseases. *Nat. Genet.* **36:** 1312–1318.

Lohmueller, K.E., Pearce, C.L., Pike, M., Lander, E.S., and Hirschhorn, J.N. 2003. Meta-analysis of genetic association studies supports a contribution of common variants to susceptibility to common disease. *Nat. Genet.* **33:** 177–182.

Marchini, J., Donnelly, P., and Cardon, L.R. 2005. Genome-wide strategies for detecting multiple loci that influence complex diseases. *Nat. Genet.* **37:** 413–417.

Mathew, C.G. and Lewis, C.M. 2004. Genetics of inflammatory bowel disease: Progress and prospects. *Hum. Mol. Genet.* **13:** R161–R168.

Milne, R.L., Fagerholm, R., Nevanlinna, H., and Benitez, J. 2008. The importance of replication in gene-gene interaction studies: Multifactor dimensionality reduction applied to a two-stage breast cancer case-control study. *Carcinogenesis* **29:** 1215–1218.

O'Donovan, M.C., Craddock, N., Norton, N., Williams, H., Peirce, T., Moskvina, V., Nikolov, I., Hamshere, M., Carroll, L., Georgieva, L., et al. 2008. Identification of loci associated with schizophrenia by genome-wide association and follow-up. *Nat. Genet.* **40:** 1050–1055.

Price, A.L., Patterson, N.J., Plenge, R.M., Weinblatt, M.E., Shadick, N.A., and Reich, D. 2006. Principal components analysis corrects for stratification in genome-wide association studies. *Nat. Genet.* **38:** 904–909.

Pritchard, J.K., Stephens, M., and Donnelly, P. 2000. Inference of population structure using multilocus genotype data. *Genetics* **155:** 945–959.

Purcell, S., Cherny, S.S., and Sham, P.C. 2003. Genetic Power Calculator: Design of linkage and association genetic mapping studies of complex traits. *Bioinformatics* **19:** 149–150.

Ritchie, M.D., Hahn, L.W., and Moore, J.H. 2003. Power of multifactor dimensionality reduction for detecting gene–gene interactions in the presence of genotyping error, missing data, phenocopy, and genetic heterogeneity. *Genet. Epidemiol.* **24:** 150–157.

Sasieni, P.D. 1997. From genotypes to genes: Doubling the sample size. *Biometrics* **53:** 1253–1261.

Sham, P.C., Rijsdijk, F.V., Knight, J., Makoff, A., North, B., and Curtis, D. 2004. Haplotype association analysis of discrete and continuous traits using mixture of regression models. *Behav. Genet.* **34:** 207–214.

Slatkin, M. 1999. Disequilibrium mapping of a quantitative-trait locus in an expanding population. *Am. J. Hum. Genet.* **64:** 1764–1772.

Spielman, R.S., McGinnis, R.E., and Ewens, W.J. 1993. Transmission test for linkage disequilibrium: The insulin gene region and insulin-dependent diabetes mellitus (IDDM). *Am. J. Hum. Genet.* **52:** 506–516.

Steer, S., Lad, B., Grumley, J.A., Kingsley, G.H., and Fisher, S.A. 2005. Association of R602W in a protein tyrosine phosphatase gene with a high risk of rheumatoid arthritis in a British population: Evidence for an early onset/disease severity effect. *Arthritis Rheum.* **52:** 358–360.

Wellcome Trust Case Control Consortium (WTCCC). 2007. Genome-wide association study of 14,000 cases of seven common diseases and 3,000 shared controls. *Nature* **447:** 661–678.

8 | Genome-Wide Association Studies

Ammar Al-Chalabi

MRC Centre for Neurodegeneration Research, Institute of Psychiatry, King's College London, London SE5 8AF, United Kingdom

INTRODUCTION

The goal of association studies is to discover genetic variation that differs in frequency between cases and controls or between individuals with different phenotypic values. Until a few years ago, the only method available for such studies was low-throughput analysis in which a single gene was selected and either genotyped for known genetic variants or sequenced to identify such variants. The well known association between APOE alleles and Alzheimer's disease was found this way (Corder et al. 1993). Candidate gene studies still have an important part to play in association studies, and for rare non-Mendelian diseases this is the only possible approach.

With the completion of the Human Genome Mapping Project (http://genome.ucsc.edu) (Lander et al. 2001; Venter et al. 2001), we have learned that single-nucleotide polymorphisms (SNPs) are frequent in the genome and that variants in physical proximity tend to correlate in genotype. Therefore, a major international effort was started to map this correlation in the form of the International HapMap Project (http://www.hapmap.org) (2003). The concurrent advances in genetic laboratory techniques, statistical methods, and computing power, coupled with the information from the HapMap, have allowed large-scale microchip-based technologies to be used to assay large numbers of SNPs quickly and easily. Thus, truly genome-wide association studies (GWAS) can now be performed, analogous to linkage studies of Mendelian diseases in having no prior hypothesis of the chromosomal location responsible for disease. Although GWAS offer huge potential for revealing genetic contributions to disease, their large scale results in potential problems with false-positive results because of multiple testing, false-negative results because of the much more stringent p values, and, therefore, large sample sizes required, and the need for strict quality control to avoid multiplying up genotyping and other possible sources of error (NCI-NHGRI Working Group on Replication in Association Studies 2007).

In this chapter we will only discuss case–control studies in which family members are not analyzed, but the principles apply to large-scale family-based association studies as well (see Chapter 11).

GWAS TEST THE COMMON DISEASE/COMMON VARIANT HYPOTHESIS

Although there is no hypothesis for the disease gene location, the use of microchip technologies to perform a GWAS is still testing a specific hypothesis: The common disease/common variant hypothesis. This is the idea that polymorphic variation in the population of more than about 5% frequency might increase susceptibility to common disease (Lander 1996; Cargill et al. 1999; Chakravarti 1999; Reich and Lander 2001). Such variants would be in an "evolutionary shadow" either because the effect is only a little deleterious or because the diseases impact only individuals who have reached old age, something that was unusual until relatively recently in human history. Even though the relative risk imparted by such a variant might be very small, because it is common, such variation would have a significant effect on public health and disease frequency. The combination of many such variants and the effect of environmental factors would combine to cause disease. We now know that at least for several common diseases, this hypothesis appears to be true. The alternative hypothesis—the rare variant hypothesis—is not disproved even in these cases, however. This hypothesis states that low-penetrance rare variants (<1% population frequency) impart a moderately large relative risk and are responsible for disease (Pritchard 2001).

TAG SNPS AND LINKAGE DISEQUILIBRIUM ARE THE BASIS OF GWAS

Tag SNPs

To use the common disease/common variant hypothesis to test for association still means that every common variant needs to be typed, which is no small undertaking given that the HapMap currently records more than 3.1 million common SNPs. One solution is to use the highly correlated structure of the genome to type a subset of SNPs that capture the variation in the untyped SNPs, thus greatly reducing the amount of work needed (Johnson et al. 2001). This is the reason for the HapMap.

A typed SNP with a genotype correlating with that of other, untyped SNPs is said to "tag" the information in the untyped SNPs. For example, take two loci, A and B, at each of which there is a SNP, and that the SNPs at these loci have strongly correlating genotypes. The genotype at SNP A predicts the genotype at SNP B and therefore the genotype at SNP B can be estimated with a high degree of certainty by genotyping SNP A. Thus, SNP A is a tag SNP for SNP B. This relationship between SNPs is known as linkage disequilibrium (LD), and it can be estimated by comparison between the allele frequencies observed at the two loci and the haplotype frequencies observed. If there were no correlation between the genotypes at the two SNPs, the haplotype frequency would be the same as the product of the respective allele frequencies. The degree to which the haplotype frequency deviates from the product of the allele frequencies is a measure of LD.

Linkage Disequilibrium

For example, denoting alleles 1 and 2 at each locus as a subscript and signifying the haplotype frequencies with H, we can write the following table (Table 8.1):

TABLE 8.1

Haplotype	Observed frequency
A_1B_1	H_{11}
A_1B_2	H_{12}
A_2B_1	H_{21}
A_2B_2	H_{22}

The allele frequencies observed are denoted as follows (Table 8.2):

TABLE 8.2

Allele	Observed frequency
A_1	p_1
A_2	p_2
B_1	q_1
B_2	q_2

But we could work out what Table 8.1 should be, assuming that the alleles at each SNP are independent using the observed allele frequencies in Table 8.2 (Table 8.3):

TABLE 8.3

Haplotype	Calculated frequency
A_1B_1	$p_1 q_1$
A_1B_2	$p_1 q_2$
A_2B_1	$p_2 q_1$
A_2B_2	$p_2 q_2$

The difference between the calculated haplotype frequency, assuming independence, and that actually observed is a measure of the independence between the two loci and is usually denoted with a capital D as follows (Table 8.4):

TABLE 8.4

Haplotype	Observed frequency	Calculated frequency	D
A_1B_1	H_{11}	$p_1 q_1$	$H_{11} - p_1 q_1$
A_1B_2	H_{12}	$p_1 q_2$	$p_1 q_2 - H_{12}$
A_2B_1	H_{21}	$p_2 q_1$	$p_2 q_1 - H_{21}$
A_2B_2	H_{22}	$p_2 q_2$	$H_{22} - p_2 q_2$

or alternatively (Table 8.5):

TABLE 8.5

	A_1	A_2	Total
B_1	$H_{11} = p_1 q_1 + D$	$H_{21} = p_2 q_1 - D$	q_1
B_2	$H_{12} = p_1 q_2 - D$	$H_{22} = p_2 q_2 + D$	q_2
Total	p_1	p_2	1

Although it is easy to calculate D, it is not a very intuitive measure of LD because it depends on the underlying allele frequencies and is maximal when the allele frequencies are each 0.5. Therefore, it is commonly standardized by dividing by the theoretical maximum when $D \geq 0$ to give $D' = D/D_{max}$, and the minimum when $D \leq 0$ to give $D' = D/D_{min}$. D_{max} is given by the smaller of the products $p_1 q_2$ and $p_2 q_1$, and D_{min} by the larger of the two products $p_1 q_1$ and $p_2 q_2$. The advantage of D' is that it is easily understood, with large absolute values implying strong LD between the two loci. D' is therefore a measure of the association of the two loci. This is not quite the same as the ability of one locus to predict the genotype at the other, however, as this is maximal when the allele frequencies at the two loci are the same. This is derived by taking the square of D' and standardizing by all allele frequencies as follows: $r^2 = D^2/p_1 q_1 p_2 q_2$. Measuring LD using r^2 is also intuitive, with large values indicating high LD. The advantage of r^2 over D' is that it has useful statistical properties. For example, the change in sample size for a given power resulting from genotyping a tag SNP rather than the relevant causal SNP can be easily calculated by dividing the sample size by r^2. LD relationships between SNPs can be visualized using a matrix as shown in Figure 8.1.

Ancestral Mutation and Haplotypes

We can understand how LD is generated by considering how SNPs arise in the population. Imagine two loci, A and B, with no genetic variation. If a point mutation arises at locus A, then there are two types of haplotypes in the population: A_1-B and A_2-B. Now imagine there is a second point mutation at site B. This must occur either in an A_1-B haplotype or an A_2-B haplotype. Let us imagine

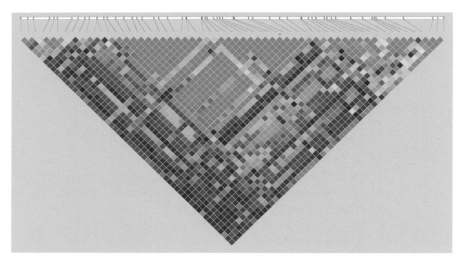

FIGURE 8.1 A triangle plot showing the LD relationships of SNPs as visualized using Haploview software. The strength of LD between any pair of SNPs is represented by coloring, with red being strong LD and blue weak. Plots can show any measure of LD, including r^2, D', LOD, and χ^2.

it has arisen on an A_2-B haplotype, generating a B_2 allele and designating the original allele as B_1. This means there are only three haplotypes in existence: A_1-B_1, A_2-B_1, and A_2-B_2. There is no A_1-B_2 at first because the B_2 allele arose on an A_2 background, and the identification of a B_2 reliably predicts an A_2 at the first locus, whereas an A_1 at the first locus reliably predicts a B_1 at the second. On the other hand, B_1 does not reliably predict the allele at A, and A_2 does not reliably predict the allele at B. This is a situation in which D' is 1 but r^2 is not. This complete LD gradually decays with time because of recombination between loci A and B, so that at some point it is likely that A_1-B_2 will be generated. It can also decay through mutation back to the original allele.

The HapMap records the genotypes at each locus in a selection of individuals of different ancestries representing large swathes of the worldwide population and allows the genotype correlations between SNPs to be analyzed. This information is used by chip manufacturers to generate DNA microarrays that probe for tag SNPs preferentially, allowing, for example, a chip with just 300,000 SNPs to capture about 90% of the common variation in the genome. To put this in context, one needs to remember that for this chip there are 10,000 ungenotyped base pairs of genome for every SNP tested.

CHIPS AND PLATFORMS USED FOR GWAS

The two commonest DNA microarray genotyping platforms currently in use are those manufactured by Affymetrix and Illumina, but several other companies use similar or identical technology, including NimbleGen and Perlegen. Each uses the concept of a chip containing probes that hybridize to the SNP of interest. The hybridization results in a fluorescent signal that indicates which allele is present and may include probes for specificity of binding as an error check. The Affymetrix chips use SNPs defined according to their proximity to various restriction enzyme cutting sites. The Illumina chips use SNPs preferentially selected in exons. The latest chips for both platforms also include SNPs in regions with variation in copy number and can assay more than 500,000 SNPs.

The fluorescence at each probe is analyzed using a clustering algorithm, and the most likely genotype is reported by the software. Such an algorithm can be confused by some resulting patterns of fluorescence and for this reason it is wise to look at the actual raw output of the scanner for SNPs of interest, although this may not be possible if the work was outsourced or is public genotype data. In addition, in some cases, the unusual patterns of fluorescence arise because of copy-number variation at the locus, and direct analysis of the raw data allows an estimate of the copy number. Similarly, a rare minor allele (one with less than about 20–30 copies in the analysis) may not yield enough data points for the clustering algorithm to reliably call the genotype. SNPs this rare should probably be discarded from the analysis.

HOW TO PREPARE FOR GWAS ANALYSIS

Data Handling

See Figure 8.2 for a summary of data handling and quality control. The main issues of study design have been discussed in Chapter 4. The standard GWAS design is a case–control study in which controls are matched for ancestry and, if relevant, age and sex. Power calculations need to take into account the multiple testing inherent in analyzing millions of SNPs. Because the SNPs used are mainly tag SNPs, they theoretically do not correlate with each other, only with the untyped SNPs, and a strict multiple testing correction needs to be applied (e.g., Bonferroni correction) (Bonferroni 1936). As a result, unless the signal to be detected is large and there is little allelic or disease heterogeneity, thousands or tens of thousands of samples need to be analyzed to detect an effect. This means that there may be, for example, a million genotypes for 10,000 people, making two billion data points. This is a significant data handling problem in terms of manipulation of the data for analysis and tracking of the results. In addition, quality control is important, and tracking 10,000 DNA samples and their associated clinical and genotype data is not trivial. Several software packages exist to handle these large data sets successfully, the most popular of which is PLINK (Purcell et al. 2007). Others include an R package, snpMatrix (Clayton and Leung 2007), PBAT (Lange et al. 2004; Van Steen and Lange 2005), SNPTEST (Marchini et al. 2007), and EIGENSTRAT/EIGENSOFT (Patterson et al. 2006; Price et al. 2006). Haploview 4.0 (Barrett et al. 2005), which allows visualization of haplotype blocks in the genome and analysis of local regions for tagging and association, also allows the import of GWAS results from PLINK for reordering and plotting.

Quality Control

The first steps in analysis are quality control steps. Perhaps the most important quality control step is ensuring that the phenotype data are robust, because a problem here will make the subsequent analyses meaningless. DNA samples should ideally be tracked using an automated tracking system from donation to analysis to reduce the risk of clerical errors. The use of bar codes is a common method. Good-quality DNA should be used where possible, and cases and controls should be drawn, tracked, and extracted in the same way, ideally at the same time and in the same center. These procedures reduce the risk of a systematic error affecting cases differently from controls because of handling. Plates used for genotyping should contain a random mixture of cases and controls in the same plate, again reducing the chance of a systematic error.

Once genotyping has occurred, chip call quality measures generated by the chip scanning software should be used to filter out SNPs that are likely to be unreliable, and simple tests used to reduce the possibility of a genotype assigned to the wrong person. For example, the sex of each person should match with the genetic sex predicted by examining homozygosity on the X chromosome. Individuals for whom large numbers of SNPs could not be genotyped should be eliminated, because this suggests a problem with the DNA sample. Similarly, SNPs that failed to be genotyped in large numbers of individuals should be eliminated because this suggests that the SNP discovery probe was not reliable.

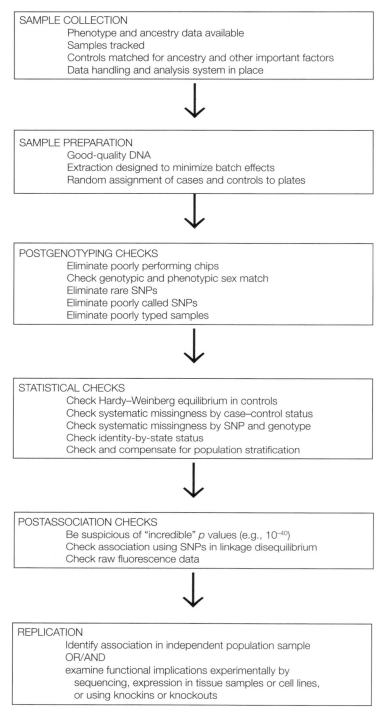

FIGURE 8.2 A summary of data handling and quality control considerations. Because of the large scale of GWAS, data considerations that do not apply to single candidate gene studies are important.

Genotypes can be analyzed to flag likely genotyping problems. Hardy–Weinberg equilibrium (HWE) is expected in controls (Hardy 1908; Weinberg 1908). It is common practice to use deviation from HWE as a measure of genotyping error, because the assumption is that heterozygotes are difficult to identify and homozygotes are therefore overrepresented (Hosking et al. 2004). Although theoretically deviation from HWE is not a good measure of genotyping error (Zou and Donner 2006), it has proven itself as a reasonable quality control metric in practice. However, because loss of HWE can occur for reasons other than genotyping error, such as population stratification or true association, SNPs that otherwise pass quality control should probably be reevaluated. Alternatively, a very stringent HWE threshold can be used, such as $p = 10^{-6}$. A lack of HWE in cases is not a problem and may even suggest association, unless the cases and controls were ascertained or genotyped on separate occasions, in which case it is possible there is a systematic problem arising from the laboratory work on one day.

Such systematic "missingness," in which a genotype is missing in cases more frequently than controls or vice versa, can be looked for using, for example, PLINK. SNPs for which this is the case should be eliminated. Another more subtle form of systematic missingness is when one genotype is more difficult to call than others. For example, if heterozygotes are more likely to be missing, this may not show up on a Hardy–Weinberg test, but can be detected using flanking SNPs in LD to predict the missing genotype. If, for example, every time the flanking SNPs predict heterozygosity in the test SNP the genotype at the test SNP is missing, this suggests a systematic problem with calling heterozygotes and those SNP results need to be treated with caution. Again, this is implemented in PLINK.

Finally, identity-by-state (IBS) measures should be used to flag duplicate samples (or identical twins) or related individuals using genome-wide SNP data. It is possible to prune out SNPs in LD without affecting the reliability of the results, thus reducing the data set and the time taken to perform the IBS analysis, which otherwise can take hours to days. (This data pruning is purely for generating an IBS file, and the SNPs should be restored for further analysis.) Genetic relatedness is about 0.5 for nonidentical siblings and parent–child relationships; 0.25 for half-siblings, avuncular relationships, and grandparents; and 0.125 for first cousins. The statistical measure of relatedness is symbolized by $\hat{\pi}$. The analysis should reveal individuals with a high value for $\hat{\pi}$ easily, and it is often clear that there is a set of samples with $\hat{\pi}$ almost exactly 1, 0.5, 0.25, and 0.125, representing duplicate samples and the differing degrees of relatedness. The remaining samples may still show high degrees of relatedness, however, and a reasonable background relatedness can be taken as $\hat{\pi} = 0.05$ (although in founder populations this would obviously be higher).

Population Stratification

An extremely important element of quality control is ensuring the cases are well matched to the controls for ancestry. This is because hidden variables in an association analysis can cause false-positive results or even reverse the direction of the association, a problem known as Simpson's paradox (Simpson 1951). This can be illustrated with the following example: Imagine 1750 cases and 1800 controls. We genotype for a SNP and the allele counts look like this (Table 8.6):

TABLE 8.6

	Allele 1	Allele 2
Cases	3000	500
Controls	2700	900

A quick calculation shows an odds ratio of 2.00 (i.e., [3000 × 900]/[500 × 2700]), with $p < 10^{-6}$. In other words we find that allele 1 doubles the odds of being a case. Now we discover that there are really two populations here, one in which allele 1 is relatively common, and the other in which it is less common. If we now split the table to reflect these two populations we find the following (Tables 8.7 and 8.8):

TABLE 8.7

	Allele 1	Allele 2
Cases	2800	300
Controls	1100	50

TABLE 8.8

	Allele 1	Allele 2
Cases	200	200
Controls	1600	850

The odds ratio is now 0.42, $p < 10^{-6}$ for Table 8.7 and 0.53, $p < 10^{-6}$ for Table 8.8 (you can check for yourself that the counts in Tables 8.7 and 8.8 add up to the same as those in Table 8.6). In other words, the effect in *both* underlying populations is actually that allele 1 halves the odds of being a case (i.e., allele 2 is associated with being a case rather than allele 1). This example, in which association in one direction in the underlying populations is converted to association in the other direction when they are combined, is of course extreme, but less extreme versions simply convert lack of association in the underlying populations into association in the combined population or destroy a true association. Even if we have collected our samples as diligently as possible, it is still possible to have an underlying population structure capable of distorting the results. The problem of population stratification is not confined to case–control designs and can also affect quantitative trait GWAS—for example, if the populations in Tables 8.7 and 8.8, instead of having different frequencies of cases, had different mean trait levels. Fortunately there are methods that can deal with this.

Correction for Population Stratification

A relatively simple correction method is to use "genomic control" (Devlin and Roeder 1999). With this method, an inflation factor is calculated from loci not relevant to the disease to correct the chi-square statistics globally. When we test the hypothesis that there is no underlying difference between the cases and controls in a GWAS, for nearly all our tests this will be true. In other words it will be as if we are testing association in two sample sets randomly drawn from the same population. In this situation, any test statistic value is as likely as any other. This means that if we were to rank our test statistics by size and plot the expected values against the observed values (a so-called quantile–quantile, or Q-Q, plot), we should have a straight line from the origin to (1, 1). If there is a systematic problem in our data, such as an underlying population structure, this has the effect of inflating (usually) the test statistic and the Q-Q plot deviates upward from the (0, 0)(1, 1) line. The correction factor λ_{GC} to revert the line back to the correct position is calculated using the ratio between the Cochran–Armitage trend test statistic and the chi-square statistic for an allelic test of association as follows. The counts are denoted as in Table 8.9:

TABLE 8.9

Alleles	aa	Aa	AA	Total
Case	r_0	r_1	r_2	R
Control	s_0	s_1	s_2	S
Total	n_0	n_1	n_2	N

The trend test is

$$Y^2 = \frac{N(N(r_1 + 2r_2) - R(n_1 + 2n_2))^2}{N(n_1 + 4n_2) - (n_1 + 2n_2)^2} , \qquad (1)$$

and the chi-square statistic for allelic association is

$$\chi^2 = \frac{2N(2N(r_1 + 2r_2) - R(n_1 + 2n_2))^2}{4R(N-R)(2N(n_1 + 2n_2) - (n_1 + 2n_2)^2)} , \qquad (2)$$

with the two statistics being approximately equal in a population in HWE. The trend test statistic is inflated by stratification by a factor λ_{GC} so that $Y^2 \sim \lambda_{GC} \chi^2$. Another way to estimate λ_{GC} is to use the median or the mean of the trend statistics, which should be 1.0 if there is no inflation.

An alternative method is to use genome-wide data, or a subset of ancestry informative SNPs, to analyze the population structure. There are various methods to do this implemented in programs such as *structure* (Pritchard et al. 2000), EIGENSTRAT (Price et al. 2006), and PLINK (Purcell et al. 2007). The first step is usually to dissect out the ancestry of each individual using some sort of data reduction method such as principal components analysis. It is then relatively straightforward to identify groups of individuals with the same ancestry. If there are only a few individuals with a different ancestry, these can be eliminated from the analysis. If there are many, the analysis can be stratified by the two or more groups. Alternatively, the first two or more components of the ancestry information can be used as covariates in an analysis, thus effectively accounting for the population structure.

DATA ANALYSIS METHODS USED IN GWAS

The most basic test of association for a discrete trait, such as disease affection, is a chi-square (χ^2) test. There are six genetic models that could be tested at each SNP. These are allelic (or multiplicative) in which allele counts are used to make a 2 × 2 table; additive in which genotypes are counted in a 3 × 2 table and the null hypothesis is that no genotype is associated with disease making a two degrees of freedom (df) test; a trend to test AA > Aa > aa (usually the Cochran–Armitage trend test is used) in which the genotypes are counted in a 3 × 2 table but the test is a 1-df test for a dose effect; a dominant model (AA and Aa vs. aa); a recessive model (AA vs. Aa and aa); and an overdominant model (also called heterozygous advantage) in which Aa is compared with AA and aa. If the samples are from different centers or there is some other reason for the data to be analyzed in strata, a Cochran/Mantel–Haenszel χ^2 test can be used in which a 2 × 2 × k table is tested, in which k is the number of strata. This will correctly handle confounding by a categorical variable such as ancestral group.

Regression methods can also be used. For example, discrete traits can be analyzed by logistic regression, the advantage being that other variables may be used as covariates. Situations in which there is no control category because quantitative traits such as blood pressure, cholesterol level, or IQ are being investigated can be tested by linear regression. Again, covariates may be included in the model. Similarly, association testing can be used in case-only studies to examine disease modifier genes. For example, age of onset for diabetes could be studied by regressing the age of onset against the genotype.

Complex disease genes under the common disease common variant hypothesis generally have small effect sizes, with odds ratios of the order of 1.3 or smaller (Bodmer and Bonilla 2008). Because the number of tests is large, p values of the order of 10^{-7} are needed to guard against large numbers of false positive associations from the statistical noise generated by multiple testing. This, in turn, requires large numbers of people to be studied to have sufficient power. There are various statistical strategies to help overcome this multiple testing problem as described in Chapter 6, but there are also some genetic methods that can help to increase confidence in a finding as real. The simplest is to use a gene-based test of association. This reduces the number of independent tests from hundreds of thousands to tens of thousands. One such test is Hotelling's T^2 test in which all the SNPs in a gene are considered jointly for analysis (Hotelling 1931). A second strategy is to investigate a putative association by examining association using surrounding SNPs. This is implemented in the program PLINK as the proxy-haplotype test. In this procedure, three SNPs on either side of the associated variant are used to form a seven-SNP haplotype. Association is then tested using the haplotype without the associated variant. If the association is not caused by some genotyping error at the associated SNP, it should hold. In addition, nearby SNPs in LD with the associated SNP should also show association, thus further increasing confidence that the association is technically real. Finally, if the association with the surrounding haplotype is stronger than that with the associated SNP, there may be a causal SNP that is untyped on the associated haplotype. A third strategy is to use permutation testing. Permutation of case–control status can be used to show that the association at a SNP is not something that is likely to have occurred by chance. This is the most robust method of dealing with multiple testing as the probability distribution is derived empirically from the data, but it has the disadvantage of being computationally intensive. More recently, methods in which pathways or networks of genes are analyzed as a unit have been developed as a means of pulling out a likely signal (Wang et al. 2007). Even though the multiple testing burden remains high, the finding that more genes in a particular pathway or network tend to be associated than would be expected by chance, increases the confidence with which it can be said that the pathway is relevant. Finally, replication of the association in a second independent sample is the ideal confirmation. This is most powerfully analyzed jointly with the original data (Skol et al. 2006).

HAPLOTYPE ANALYSES CAN LOCATE A FUNCTIONAL VARIANT

Identifying Haplotypes

When an association is identified, it can be useful to examine haplotype associations that include the SNP of interest as a means to locate a functional variant. As described in the previous section on "Ancestral Mutation and Haplotypes," the association between a mutant allele and the haplotype on which it arose can only be destroyed through recombination or (less commonly) mutation. Thus, the implication of a strong haplotypic association is that the true causal variant lies on the associated haplotype background and may not have been typed. This is similar to the outcome from imputation, which also uses local LD patterns to attempt to locate a causal variant, the main difference being that imputation uses a reference set of LD data such as the HapMap to impute missing data and therefore the genotype at the untyped SNP. In a family-based study, a haplotype is a segment of chromosome between two crossover points. In a population-based study, the crossover points are unknown because previous generations are not genotyped, so a likely haplotype has to be estimated. Haplotypes can be assigned to an individual using a systematic method such as the expectation–maximization (E-M) algorithm, which uses existing data to estimate missing data (Fig. 8.3) (Dempster et al. 1977). Some haplotypes can be identified with certainty. For example, if we take two loci, A and B, with alleles A, a, B, and b, and denote separate chromosomes with a slash, a doubly homozygous individual A/A and B/B must have two AB haplotypes. Similarly, someone who is A/A and B/b must have one AB and one Ab haplotype. In the case of a double heterozygote A/a and B/b there are two equally likely possibilities, AB/ab and Ab/aB. Given a series of genotypes, we can estimate the likely phase of the uncertain haplotypes using information from the known haplotypes.

The E-M Algorithm

We start by assigning weighted counts to the haplotypes (Fig. 8.3). A haplotype that can be identified with certainty is given a weight of 1. A haplotype that is of uncertain phase is assigned an estimated

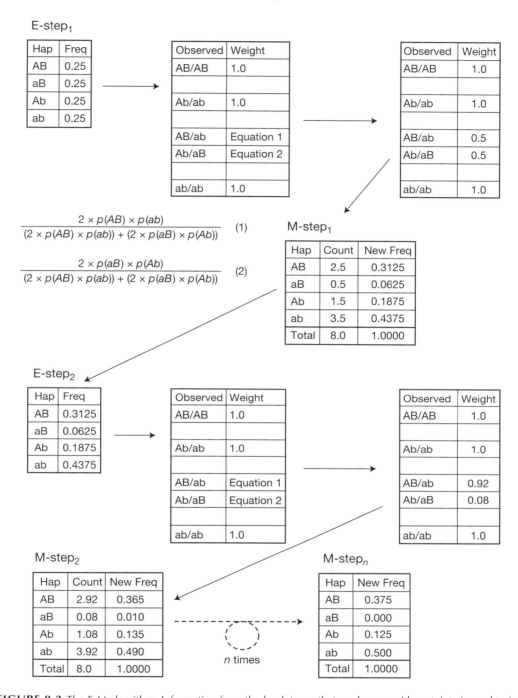

$$\frac{2 \times p(AB) \times p(ab)}{(2 \times p(AB) \times p(ab)) + (2 \times p(aB) \times p(Ab))} \quad (1)$$

$$\frac{2 \times p(aB) \times p(Ab)}{(2 \times p(AB) \times p(ab)) + (2 \times p(aB) \times p(Ab))} \quad (2)$$

FIGURE 8.3 The E-M algorithm. Information from the haplotypes that are known with certainty is used to inform the likely haplotypes with ambiguous phase. The expectation step is used to calculate likely weightings (using Equations 1 and 2, which generate the probabilities over all possibilities) and these are used to count the likely haplotypes in the maximization step. This generates new frequencies that are used to calculate the new weights and so on. In this case, the AB/ab haplotype is chosen as the correct phase and the aB haplotype does not exist. (Example courtesy of Shaun Purcell.)

weight based on, for example, the prior probability that the haplotype could exist. This is the "expectation" step. The number of people with each haplotype can now be counted (including fractions for partial weights), generating the haplotype frequencies. This is the "maximization" step. The new haplotype frequencies can now be used to weight the counts again, and the counts can again be used to modify the haplotype frequencies. This process continues iteratively until there is no change in the result. Thus, a likely haplotype phase can be assigned to an individual using the information from individuals with known phase.

Haplotype Blocks

Increasing numbers of markers can be included in the haplotype. The point at which it is reasonable to say one haplotype ends and another begins is largely a matter of opinion, but there are various techniques to try to define haplotype blocks objectively. For example, one method is to use measures of D' between SNP alleles to define where ancestral recombinations have occurred and therefore where a haplotype block ends (Gabriel et al. 2002). Another method is to search for haplotypes in which the fourth haplotype (the A_1-B_2 haplotype in the example from the previous section on "Ancestral Mutation and Haplotypes") is missing, because this implies that no recombination has occurred since the mutation arose (Wang et al. 2002). For analysis purposes, another strategy is to use a sliding window of say three SNPs. Because of the very large number of ways that haplotypes can be made and tested, permutation testing is particularly important in deciding if a haplotypic association is significant.

GWAS HAVE PRODUCED SOME KEY RESULTS

Age-Related Macular Degeneration

The first association reported from a GWAS was for age-related macular degeneration and complement factor H. Three independent studies found an association with, what is in retrospect, a surprisingly large effect size (Edwards et al. 2005; Haines et al. 2005; Klein et al. 2005). In one of the studies, 96 cases and 50 controls were sufficient to find the associated SNP with a p value $< 10^{-7}$. The 95% confidence limits for the odds ratio for the causal variant in one study was between about 2.45 and 5.57, which is extremely high for a complex disease and is the reason such a small study could detect the signal.

Wellcome Trust Case Control Consortium

The Wellcome Trust Case Control Consortium (WTCCC) paper is a landmark because the large numbers involved, the validation of the GWAS approach, and the public health interest of the several diseases studied have provided many lessons (Wellcome Trust Case Control Consortium 2007). Despite the very large numbers of individuals studied, no convincing signal was seen for hypertension, probably because of disease heterogeneity. The diseases in which an association was found had SNPs with odds ratios of the order of 1.3 or smaller, confirming the findings of other studies that complex diseases need large sample sizes to have the power to detect the effect sizes seen under the common disease/common variant hypothesis. Imputation was used to estimate the likely genotypes at the untyped SNPs present in the HapMap, resulting in 2.5 million genotypes per person (Marchini et al. 2007). The study design also showed that a "universal control" strategy, in which one well-characterized set of controls can be used for several studies, could be successful. The generation of so many genotypes from the U.K. population also helped confirm that if people with non-European ancestry are excluded from the sample, there is only modest population stratification in the British population.

Type 1 Diabetes

In the original WTCCC study of 5000 people and a partly overlapping study of nonsynonymous SNPs in 7000 people, six chromosomal regions were associated with type 1 diabetes (T1D). Those six and a further six top hits were followed up in 18,000 more people in a second study (Todd et al. 2007). In total, four of these loci were convincingly associated with T1D, possible only because of the huge numbers of people studied, enabling small effect sizes to be seen above the noise of multiple testing. Eight were regarded as a small effect or false positive. These gigantic sample sets were also used to explore the proportion of T1D that is genetic and environmental and to shed light on the T1D metabolic pathway. They were also used to study geographical changes in allele frequency and to look for gene–gene interaction by searching for deviation from a multiplicative model.

MY STUDY IS NEGATIVE—HELP!

The difficulties nongeneticists have with leaving behind the notion of one gene, one disease and Mendelian genetics are compounded when GWAS are involved, as the misunderstanding of many people is that "gene chips" effectively assay the genome for mutations. The interpretation of a GWAS in which there is no significant association is that the disease does not have a significant genetic component, rather than the far more likely interpretations that the study was underpowered, rare variants were involved, copy-number variants were involved, other difficult-to-tag variation such as microsatellites were responsible, or gene–gene interactions were important. Thus, it is vital to reflect on the hypothesis that is being tested with the typical GWAS, which is the common disease/common variant hypothesis, and to reinforce that even a well-powered negative study simply means that this hypothesis is likely to be false, but says nothing about rare variants or the other variation described because these cannot be tested with tag SNPs.

OTHER STRATEGIES ARE IMPORTANT

GWAS currently consist of a search for common variants predisposing to disease. At least two other strategies are becoming more important—a search for structural variants (copy-number variants) in the genome that might be commoner in cases than controls and a search for rare variants that might be less common but have a larger effect size.

Copy-Number Variants

Structural variation in the genome is now regarded as an important cause of phenotypic variation (Iafrate et al. 2004; Sebat et al. 2004). This is defined as a region of duplication or deletion greater than 1 kb, and such loci are regarded as variant in copy number. There are several methods for analysis of copy-number variants and these are discussed in detail in Chapter 12. Recently, an integrated approach has been developed using a software package, Birdsuite, which can test for association of copy-number variants using the PLINK software package (Korn et al. 2008). The best methods for interpreting the contribution of copy-number variants to disease are still being debated but are likely to be resolved soon.

Rare Variants

Genetic sequence variations that are of less than 1% frequency are likely to be neutral or deleterious, but they may take some time to be eliminated from the population if only mildly deleterious. In fact, statistical theory suggests that rare variants are likely to be a common component of the genetic contribution to disease (Bodmer and Bonilla 2008). Whereas the odds ratio for common variants is typically around 1.3 or less, the odds ratio for rare variants based on an examination of the current literature is likely to be 2 or more, with a current average of 3.84. Rare variants do not need to cause familial clustering. Because the variants we are interested in do not cause Mendelian disease by definition (we are interested in complex diseases), any one variant must have reduced penetrance. We can use the binomial distribution to estimate the penetrance that would produce a sufficiently low familial rate. For example, given a dominant model (because the variant is rare a homozygote is unlikely) and a sibship size of 4, a penetrance of 0.2 means there is only a 0.05 probability of more than one affected in the sibship. Even a penetrance of 0.5 means that only one-quarter of such sibships would have more than one affected. If the sibship is size 3, then the probability drops to about 1 in 6, and in today's smaller families, only 1 in 16 sibships of size 2 would have both affected. There are several striking consequences of this observation. The first is that rare variants with even quite high penetrance could be responsible for complex diseases. Second, as family sizes reduce, more diseases will appear sporadic, even if the variant is highly penetrant. For example, assuming a dominant mode of inheritance and 90% penetrance, 80% of sibships of size 2 would have only one affected (although of course in this example one parent would be very likely affected). Third, many diseases have a familial and sporadic form. For example, in many neurological diseases, about 10% of cases are familial. This could be accounted for by a gene with about 30% penetrance because many of the gene carriers would not be manifesting.

Genetic technology is now moving toward whole genome sequencing at a rapid rate. The current generation of sequencers is capable of sequencing the entire coding sequence (55 Mb) in about a week, and it is likely that in the next decade we will have access to huge quantities of sequence data that can be mined for rare variants. A chi-square test for association can be performed by collapsing all variants in a single gene into one class, but the interpretation of which variants are causal and which are neutral will require bioinformatics and functional analyses on a large scale.

CONCLUSION

Genome-wide association studies are relatively new and have already taught us many lessons in the design and interpretation of such studies and in the causes of complex traits. Larger data sets, cheaper technologies, and greater international cooperation will mean that smaller and smaller effect sizes will be dissected out over the next few years, greatly increasing our understanding of the causes of complex diseases.

REFERENCES

Barrett, J.C., Fry, B., Maller, J., and Daly, M.J. 2005. Haploview: Analysis and visualization of LD and haplotype maps. *Bioinformatics* **21:** 263–265.

Bodmer, W. and Bonilla, C. 2008. Common and rare variants in multifactorial susceptibility to common diseases. *Nat. Genet.* **40:** 695–701.

Bonferroni, C.E. 1936. Teoria statistica delle classi e calcolo delle probabilità. *Pubblicazioni del R Istituto Superiore di Scienze Economiche e Commerciali di Firenze* **8:** 3–62.

Cargill, M., Altshuler, D., Ireland, J., Sklar, P., Ardlie, K., Patil, N., Shaw, N., Lane, C.R., Lim, E.P., Kalyanaraman, N., et al. 1999. Characterization of single-nucleotide polymorphisms in coding regions of human genes. *Nat. Genet.* **22:** 231–238.

Chakravarti, A. 1999. Population genetics—making sense out of sequence. *Nat. Genet.* (suppl. 1) **21:** 56–60.

Clayton, D. and Leung, H.T. 2007. An R package for analysis of whole-genome association studies. *Hum. Hered.* **64:** 45–51.

Corder, E.H., Saunders, A.M., Strittmatter, W.J., Schmechel, D.E., Gaskell, P.C., Small, G.W., Roses, A.D., Haines, J.L., and Pericak-Vance, M.A. 1993. Gene dose of apolipoprotein E type 4 allele and the risk of Alzheimer's disease in late onset families. *Science* **261:** 921–923.

Dempster, A.P., Laird, N.M., and Rubin, D.B. 1977. Maximum likelihood from incomplete data via the EM algorithm. *J. Roy. Statist. Soc. Ser. B* **39:** 1–38.

Devlin, B. and Roeder, K. 1999. Genomic control for association studies. *Biometrics* **55:** 997–1004.

Edwards, A.O., Ritter 3rd, R., Abel, K.J., Manning, A., Panhuysen, C., and Farrer, L.A. 2005. Complement factor H polymorphism and age-related macular degeneration. *Science* **308:** 421–424.

Gabriel, S.B., Schaffner, S.F., Nguyen, H., Moore, J.M., Roy, J., Blumenstiel, B., Higgins, J., DeFelice, M., Lochner, A., Faggart, M., et al. 2002. The structure of haplotype blocks in the human genome. *Sci-*

ence **296:** 2225–2229.

Haines, J.L., Hauser, M.A., Schmidt, S., Scott, W.K., Olson, L.M., Gallins, P., Spencer, K.L., Kwan, S.Y., Noureddine, M., Gilbert, J.R., et al. 2005. Complement factor H variant increases the risk of age-related macular degeneration. *Science* **308:** 419–421.

Hardy, G.H. 1908. Mendelian proportions in a mixed population. *Science* **28:** 49–50.

Hosking, L., Lumsden, S., Lewis, K., Yeo, A., McCarthy, L., Bansal, A., Riley, J., Purvis, I., and Xu, C.F. 2004. Detection of genotyping errors by Hardy–Weinberg equilibrium testing. *Eur. J. Hum. Genet.* **12:** 395–399.

Hotelling, H. 1931. The generalization of Student's ratio. *Ann. Math. Statist.* **2:** 360–378.

Iafrate, A.J., Feuk, L., Rivera, M.N., Listewnik, M.L., Donahoe, P.K., Qi, Y., Scherer, S.W., and Lee, C. 2004. Detection of large-scale variation in the human genome. *Nat. Genet.* **36:** 949–951.

The International HapMap Consortium. 2003. The International HapMap Project. *Nature* **426:** 789–796.

Johnson, G.C., Esposito, L., Barratt, B.J., Smith, A.N., Heward, J., Di Genova, G., Ueda, H., Cordell, H.J., Eaves, I.A., Dudbridge, F., et al. 2001. Haplotype tagging for the identification of common disease genes. *Nat. Genet.* **29:** 233–237.

Klein, R.J., Zeiss, C., Chew, E.Y., Tsai, J.Y., Sackler, R.S., Haynes, C., Henning, A.K., SanGiovanni, J.P., Mane, S.M., Mayne, S.T., et al. 2005. Complement factor H polymorphism in age-related macular degeneration. *Science* **308:** 385-389.

Korn, J.M., Kuruvilla, F.G., McCarroll, S.A., Wysoker, A., Nemesh, J., Cawley, S., Hubbell, E., Veitch, J., Collins, P.J., Darvishi, K., et al. 2008. Integrated genotype calling and association analysis of SNPs, common copy number polymorphisms and rare CNVs. *Nat. Genet.* **40:** 1253–1260.

Lander, E.S. 1996. The new genomics: Global views of biology. *Science* **274:** 536–539.

Lander, E.S., Linton, L.M., Birren, B., Nusbaum, C., Zody, M.C., Baldwin, J., Devon, K., Dewar, K., Doyle, M., FitzHugh, W., et al. 2001. Initial sequencing and analysis of the human genome. *Nature* **409:** 860–921.

Lange, C., DeMeo, D., Silverman, E.K., Weiss, S.T., and Laird, N.M. 2004. PBAT: Tools for family-based association studies. *Am. J. Hum. Genet.* **74:** 367–369.

Marchini, J., Howie, B., Myers, S., McVean, G., and Donnelly, P. 2007. A new multipoint method for genome-wide association studies by imputation of genotypes. *Nat. Genet.* **39:** 906–913.

NCI-NHGRI Working Group on Replication in Association Studies. 2007. Replicating genotype–phenotype associations. What constitutes replication of a genotype–phenotype association, and how best can it be achieved? *Nature* **447:** 655–660.

Patterson, N., Price, A.L., and Reich, D. 2006. Population structure and eigenanalysis. *PLoS Genet.* **2:** e190.

Price, A.L., Patterson, N.J., Plenge, R.M., Weinblatt, M.E., Shadick, N.A., and Reich, D. 2006. Principal components analysis corrects for stratification in genome-wide association studies. *Nat. Genet.* **38:** 904–909.

Pritchard, J.K. 2001. Are rare variants responsible for susceptibility to complex diseases? *Am. J. Hum. Genet.* **69:** 124–137.

Pritchard, J.K., Stephens, M., and Donnelly, P. 2000. Inference of population structure using multilocus genotype data. *Genetics* **155:** 945–959.

Purcell, S., Neale, B., Todd-Brown, K., Thomas, L., Ferreira, M.A., Bender, D., Maller, J., Sklar, P., de Bakker, P.I., Daly, M.J., et al. 2007. PLINK: A tool set for whole-genome association and population-based linkage analyses. *Am. J. Hum. Genet.* **81:** 559–575.

Reich, D.E. and Lander, E.S. 2001. On the allelic spectrum of human disease. *Trends Genet.* **17:** 502–510.

Sebat, J., Lakshmi, B., Troge, J., Alexander, J., Young, J., Lundin, P., Maner, S., Massa, H., Walker, M., Chi, M., et al. 2004. Large-scale copy number polymorphism in the human genome. *Science* **305:** 525–528.

Simpson, E.H. 1951. The interpretation of interaction in contingency tables. *J. Roy. Statist. Soc., Ser. B* **13:** 238–241.

Skol, A.D., Scott, L.J., Abecasis, G.R., and Boehnke, M. 2006. Joint analysis is more efficient than replication-based analysis for two-stage genome-wide association studies. *Nat. Genet.* **38:** 209–213.

Todd, J.A., Walker, N.M., Cooper, J.D., Smyth, D.J., Downes, K., Plagnol, V., Bailey, R., Nejentsev, S., Field, S.F., Payne, F., et al. 2007. Robust associations of four new chromosome regions from genome-wide analyses of type 1 diabetes. *Nat. Genet.* **39:** 857–864.

Van Steen, K. and Lange, C. 2005. PBAT: A comprehensive software package for genome-wide association analysis of complex family-based studies. *Hum. Genomics* **2:** 67–69.

Venter, J.C., Adams, M.D., Myers, E.W., Li, P.W., Mural, R.J., Sutton, G.G., Smith, H.O., Yandell, M., Evans, C.A., Holt, R.A., et al. 2001. The sequence of the human genome. *Science* **291:** 1304–1351.

Wang, K., Li, M., and Bucan, M. 2007. Pathway-based approaches for analysis of genomewide association studies. *Am. J. Hum. Genet.* **81:** 1287–1283.

Wang, N., Akey, J.M., Zhang, K., Chakraborty, R., and Jin, L. 2002. Distribution of recombination crossovers and the origin of haplotype blocks: The interplay of population history, recombination, and mutation. *Am. J. Hum. Genet.* **71:** 1227–1234.

Weinberg, W. 1908. Ber den Nachweis der Vererbung Beim Menchen. *Jahreshefte Verein* **64:** 368–382.

The Wellcome Trust Case Control Consortium. 2007. Genome-wide association study of 14,000 cases of seven common diseases and 3,000 shared controls. *Nature* **447:** 661–678.

Zou, G.Y. and Donner, A. 2006. The merits of testing Hardy–Weinberg equilibrium in the analysis of unmatched case-control data: A cautionary note. *Ann. Hum. Genet.* **70:** 923–933.

WWW RESOURCES

http://www.bioconductor.org/packages/2.3/bioc/html/snpMatrix.html   Clayton and Leung 2007. snpMatrix.

http://www.biostat.harvard.edu/~clange/degault.htm   Lange et al. 2004; Van Steen and Lange 2005. PBAT.

http://www.broad.mit.edu/mpg/birdsuite/   Korn et al. 2008. Birdsuite.

http://www.broad.mit.edu/mpg/haploview   Barrett et al. 2005. Haploview 4.0.

http://genepath.med.harvard.edu/~reich/EIGENSTRAT.htm   Patterson et al. 2006; Price et al. 2006. EIGENSTRAT/EIGENSOFT.

http://www.hapmap.org   International HapMap Project.

http://www.ncbi.nlm.nih.gov/dbgap   Database of genotypes and phenotypes.

http://www.ncbi.nlm.nih.gov/projects/projects/SNP/   Single nucleotide polymorphism database.

http://pngu.mgh.harvard.edu/~purcell/plink/download.shtml   Purcell et al. 2007. PLINK.

http://pritch.bsd.uchicago.edu/software.html   Pritchard et al. 2000. *structure*.

http://www.stats.ox.ac.uk/~marchini/software/gwas/snptest.html   Marchini et al. 2007. SNPTEST.

9 | Introduction to Linkage Disequilibrium, the HapMap, and Imputation

Benjamin M. Neale

Center for Human Genetic Research, Massachusetts General Hospital,
Boston, Massachusetts 02114 and Broad Institute, Cambridge, Massachusetts 02142

INTRODUCTION

Throughout the human genome, a correlation structure exists across genetic variation of different loci. Such a correlational structure means that knowing the genotype at one locus might provide information about the genotype at a second locus. This correlation between variation at different loci is termed linkage disequilibrium (LD). (See Chapters 7 and 8 for more extensive discussions of LD.) LD has implications in numerous avenues of genetic research. In this chapter, we discuss the importance of LD in genetics, touching on both population genetics and association studies. We then introduce the seminal collaborative scientific endeavor to map LD in the human genome— the International HapMap Project—and its relevance for imputation. The concepts in this chapter lead directly to the issues discussed in Chapter 10.

FIRST, SOME BASIC LD STATISTICS

LD has a long and storied history in human genetics. The initial formalization of independence between two loci in the population was introduced by Hilda Geiringer in 1944 in a paper entitled "On the probability theory of linkage in Mendelian heredity" (Geiringer 1944). Specifically, Geiringer noted that if the probability of recombination between two loci (e.g., two biallelic variants A_1/A_2 and B_1/B_2) exceeds zero, then the probability of each haplotype (A_1B_1, A_1B_2, A_2B_1, and A_2B_2) is equal to the product of the allele frequencies of the two variants in question (e.g., $P[A_1B_1] = P[A_1]*P[B_1]$). Geiringer did not propose statistics to test for deviation from this expectation, but this formulation provided the core for future LD statistics.

The LD Statistic D'

Lewontin and Kojima (1960) proposed the first major LD statistic D in 1960. D aims to capture the deviation from linkage equilibrium, which Geiringer had defined. D is rarely reported currently, because the scale of D is frequency-dependent, i.e., the maximum value of D is a function of p_{A1} and p_{B1}. For example, if $p_{A1} = 0.9$ and $p_{B1} = 0.9$, then the maximum value of D is when $p_{A1B1} = 0.9$. This yields a D of 0.09. When $p_{A1} = 0.5$ and $p_{B1} = 0.5$, then the maximum value of D is when $p_{A1B1} = 0.5$. This yields a D of 0.25, which is the largest value D can take. To resolve this frequency dependency, Lewontin standardized D to D' by dividing by D_{max}, which is the maximum value that D can take as a function of the allele frequencies (Lewontin 1964).

$$D' = D/D_{max}.$$ (1)

D_{max} is defined as min($p_{A1}*p_{B1}$, $p_{A2}*p_{B2}$) if $D < 0$ and min($p_{A1}*p_{B2}$, $p_{A2}*p_{B1}$) if $D > 0$. Generally, the absolute value of D' is reported because the sign of D' is a function of which allele is defined as the major or minor allele at locus A and locus B. In the event that p_{A1B1}, p_{A1B2}, p_{A2B1}, or p_{A2B2} is equal to 0, then D' will equal 1.

Conceptually, this provides a convenient way to think about the relationship between D' and LD. Because one of the four possible haplotypes was not observed, one implication is that the chance of recombination having occurred between these two loci is greatly diminished. We cannot rule out recombination because of issues of power or "hidden recombination."

In the case of power, if the minor alleles at the two loci are rare, then the probability of observing the haplotype composed of the two rare variants is the product of those minor allele frequencies. For example, if the allele frequency for both rare alleles is 1%, then the frequency of that haplotype under a model of no LD is 0.01%.

As for hidden recombination, consider the three following haplotypes out of the four possible (A_1B_1, A_1B_2, A_2B_1). If a recombination event occurred between A_1B_1 and A_2B_1 or between A_1B_1 and A_1B_2, then no new haplotype will be observed in the population. Thus, the recombination is hidden by the lack of novelty in the generated haplotypes.

The LD Statistic r^2

In contrast to D', the other major LD statistic is r^2. Defined by Hill and Roberstson in 1968, r^2 is simply the squared correlation coefficient for a 2 × 2 table and takes the form of (Hill and Robertson 1968):

$$r^2 = D^2/(p_{A1}*p_{A2}*p_{B1}*p_{B2}) .$$ (2)

The value r^2 is more directly relevant to association testing and analysis because the magnitude of the r^2 between a pair of loci is directly proportional to the percentage of noncentrality parameter shared. The noncentrality parameter is a measure of how different the true distribution is from the null. This is related to the power, which is a measure of the probability that the null hypothesis is not rejected, given the true distribution. Thus, for association tagging, r^2 is much more useful because it is directly related to the power, rather than tracking the behavior of recombination in the genome. For a description of other proposed LD statistics, see Devlin and Risch (1995) and Evans and Cardon (2005). For multiallelic extensions of these statistics, see the review by Zhao et al. (2005).

USING THE INTERNATIONAL HAPMAP PROJECT TO MAP LD

At the turn of the 21st century, a great debate in human genetics raged about the extent to which LD persists in the human genome. In 1999, Leonid Kruglyak (1999) had made a range of predictions about the distribution of LD in the human genome, which pressaged the genome-wide association era. In this landmark paper, Kruglyak argued that ~500,000 single-nucleotide polymorphisms (SNPs) ought to be sufficient to obtain coverage of common variation in the human genome. From that point, a number of important data-driven papers were published about LD patterns in the human genome. *Nature Genetics* published the first major examinations of LD patterns (Daly et al. 2001; Johnson et al. 2001). Both of these papers, and a paper in *Science* (Patil et al. 2001), demonstrated much higher than expected regions of LD punctuated by apparent recombination hot spots. This block-like nature of the genome speaks to a lack of randomness in the nature of recombination across human history. From these data sets, Reich and colleagues (2002) and Wall and Pritchard (2003) both demonstrated that the empirical SNP data did not follow a neutral recombination model (Reich et al. 2002; Wall and Pritchard 2003). However, apparent "holes" in the LD structure of the genome existed, where a "lonely SNP" showed very low LD with close

SNPs. At the same time as these early observations about LD in the genome became demonstrated empirically, other research demonstrated that recombination in male sperm was "punctate," or non-randomly distributed across a region (Jeffreys et al. 2001). Such an apparent block-like nature provided the impetus for the International HapMap Project, a global initiative to map the LD patterns in human populations. As the HapMap website states:

> The International HapMap Project is a multi-country effort to identify and catalog genetic similarities and differences in human beings. Using the information in the HapMap, researchers will be able to find genes that affect health, disease, and individual responses to medications and environmental factors. The Project is a collaboration among scientists and funding agencies from Japan, the United Kingdom, Canada, China, Nigeria, and the United States. All of the information generated by the Project will be released into the public domain.

The first two phases of HapMap focused on genotyping 30 parent–offspring trios from the Centre d'Etude du Polymorphisme Humain (CEPH) collection of European ancestry in Utah, 30 parent–offspring trios from Yoruban individuals of African ancestry from Nigeria, 45 Han Chinese individuals from Beijing, and 45 Japanese individuals from Tokyo. The genotyping for these two phases included two major SNP genotyping efforts as well as deep sequencing of a handful of random regions in a subset of samples (ENCODE projects, http://genome.ucsc.edu/ENCODE; International Hapmap Consortium 2005). A sequence of papers has described this project in great detail. This data set records basic patterns of recombination across the genome in a range of human populations and provides a guide to SNP discovery efforts and how well that reflects coverage in the genome. Most importantly, the need for different population annotations is demonstrated. Generally speaking, African populations show lower levels of LD than European or Asian populations, which is consistent with the "out of Africa" hypothesis for human evolution (International HapMap Consortium 2003; Frazer et al. 2007), because the African population, being older, will have had more chance for recombination to break down patterns of LD.

The HapMap project also completed a third phase, in which the Illumina and Affymetrix million SNP products were genotyped. In addition to the previously described three samples, seven additional sample sets have been added: an African American set from the southwestern United States; a Chinese set from Denver, Colorado; Gujarati Indians from Houston, Texas; Luhya tribe members from Webuye, Kenya; Mexican Americans from Los Angeles, California; Maasai tribe members from Kinyawa, Kenya; and Tuscans from Italy. These sets of data particularly facilitated imputation analysis.

MOVING FROM TAGGING TO IMPUTATION

Observations about LD progressed from discussions of population genetic models to relevance to association testing. As noted above, r^2 provides a much better estimate of power to detect association at the disease locus than D'. Carlson and colleagues outlined a simple "greedy" algorithm that identifies tag SNPs to act as proxies for the other SNPs in a region (Carlson et al. 2004). This approach considers the pair-wise r^2 for all SNPs in a region and selects the SNP with the highest number of tagged SNPs at a given r^2 threshold. For the remaining SNPs, the same process is repeated until either the maximum number of SNPs capable of being genotyped is selected or all SNPs are captured (Table 9.1). de Bakker and colleagues extended this model by adding in two and three marker haplotypes (de Bakker et al. 2005). Thus, the algorithm is the same as before, but when two tag SNPs are selected, two and three marker haplotypes are formed to determine if they predict any SNPs that were not already captured. It is this approach of taking two and three marker haplotypes that sets the stage for imputation approaches.

Imputation aims to use the LD patterns in a considerably more systematic fashion to obtain information about untyped loci (Nicolae 2006; Marchini et al. 2007). Numerous approaches have

TABLE 9.1 The Carlson Greedy algorithm for tagSNP selection

1. Define MAF threshold $= M$.
2. Identify all SNPs with MAF $> M$.
3. Calculate pairwise r^2 values for all included SNPs.
4. Define r^2 threshold $= R$.
5. Identify all pairwise associations where $r^2 > R$.
6. Identify SNP with highest number of associations with $r^2 > R$.
7. Extract the SNP from Step 6 and all additional SNPs captured for evaluation.
8. Examine pairwise r^2 for the subset of extracted SNPs.
9. All SNPs with $r^2 > R$ with all other extracted SNPs are potential tagSNPs.
10. Select a single tag SNP from the set of potential tagSNPs (potentially including information about biology, ease of genotyping, or other ranking criteria).
11. Repeat from Step 5 restricting considerations to untagged SNPs.
12. Any SNPs not captured by other SNPs are defined as singletons.

been suggested, but two main categories exist: Hidden Markov Models (HMM) and extended tagging. In brief, HMM methods treat the problem like any other Markov model, such that the observed data (genotyped SNPs) are reflective of the latent structure (the "true haplotype" for each individual). These models tend to use each chromosome to generate the model for determining the latent state. The extended tagging approaches logically extend the de Bakker et al. model, such that multiple loci are used to predict each untyped locus.

Imputation approaches tend to use probabilistic genotype counts for association analysis. A range of possible approaches has been suggested to accommodate this uncertainty in analysis. Logistic regression for a single study is unbiased, as long as the errors affect cases and controls equally. A simple chi-squared test on the dosages will yield a deflated test statistic. To inflate this test statistic back to appropriate levels, the use of the empirical variance rather than the theoretical binomial variance is appropriate. In the context of meta-analysis, studies should be weighted not only on their sample size, but also on the quality of imputation at untyped loci (de Bakker et al. 2008). Finally, a joint likelihood that incorporates both the information about the imputation and the case–control status has been implemented by a number of groups (Lin et al. 2008). Such approaches may prove more statistically powerful while still being less analytically accurate. By far, the largest contributions to the quality of imputation across the genome are the size of the reference population and the starting coverage of the chip. Imputation is discussed further in Chapter 10.

CONCLUSION

LD is an essential component of disease mapping in the context of human genetics. The correlational structure between loci in the genome facilitates much of the current work by dramatically reducing the cost of thoroughly assaying the human genome. Many statistical approaches have been built around the selection of variation for typing and analysis of this variation through the use of LD.

REFERENCES

Carlson, C.S., Eberle, M.A., Rieder, M.J., Yi, Q., Kruglyak, L., and Nickerson, D.A. 2004. Selecting a maximally informative set of single-nucleotide polymorphisms for association analyses using linkage disequilibrium. *Am. J. Hum. Genet.* **74:** 106–120.

Daly, M.J., Rioux, J.D., Schaffner, S.F., Hudson, T.J., and Lander, E.S. 2001. High-resolution haplotype structure in the human genome. *Nat. Genet.* **29:** 229–232.
de Bakker, P.I., Yelensky, R., Pe'er, I., Gabriel, S.B., Daly, M.J., and

Altshuler, D. 2005. Efficiency and power in genetic association studies. *Nat. Genet.* **37:** 1217–1223.

de Bakker, P.I., Ferreira, M.A., Jia, X., Neale, B.M., Raychaudhuri, S., and Voight, B.F. 2008. Practical aspects of imputation-driven meta-analysis of genome-wide association studies. *Hum. Mol. Genet.* **17:** R122–R128.

Devlin, B. and Risch, N. 1995. A comparison of linkage disequilibrium measures for fine-scale mapping. *Genomics* **29:** 311–322.

Evans, D.M. and Cardon, L.R. 2005. A comparison of linkage disequilibrium patterns and estimated population recombination rates across multiple populations. *Am. J. Hum. Genet.* **76:** 681–687.

Frazer, K.A., Ballinger, D.G., Cox, D.R., Hinds, D.A., Stuve, L.L., Gibbs, R.A., Belmont, J.W., Boudreau, A., Hardenbol, P., Leal, et al. 2007. A second generation human haplotype map of over 3.1 million SNPs. *Nature* **449:** 851–861.

Geiringer, H. 1944. On the probability theory of linkage in Mendelian heredity. *Ann. Math. Statist.* **15:** 25–57.

Hill, W.G. and Robertson, A. 1968. Linkage disequilibrium in finite populations. *Theor. Appl. Genet.* **38:** 226–231.

International HapMap Consortium. 2003. The International HapMap Project. *Nature* **426:** 789–796.

International HapMap Consortium. 2005. A haplotype map of the human genome. *Nature* **437:** 1299–1320.

Jeffreys, A.J., Kauppi, L., and Neumann, R. 2001. Intensely punctate meiotic recombination in the class II region of the major histocompatibility complex. *Nat. Genet.* **29:** 217–222.

Johnson, G.C., Esposito, L., Barratt, B.J., Smith, A.N., Heward, J., Di Genova, G., Ueda, H., Cordell, H.J., Eaves, I.A., Dudbridge, F., et al. 2001. Haplotype tagging for the identification of common disease genes. *Nat. Genet.* **29:** 233–237.

Kruglyak, L. 1999. Prospects for whole-genome linkage disequilibrium mapping of common disease genes. *Nat. Genet.* **22:** 139–144.

Lewontin, R. 1964. The interaction of selection and linkage I. General considerations; heterotic models. *Genetics* **49:** 49–67.

Lewontin, R. and Kojima, K. 1960. The evolutionary dynamics of complex polymorphisms. *Evolution* **14:** 458–472.

Lin, D.Y., Hu, Y., and Huang, B.E. 2008. Simple and efficient analysis of disease association with missing genotype data. *Am. J. Hum. Genet.* **82:** 444–452.

Marchini, J., Howie, B., Myers, S., McVean, G., and Donnelly, P. 2007. A new multipoint method for genome-wide association studies by imputation of genotypes. *Nat. Genet.* **39:** 906–918.

Nicolae, D.L. 2006. Testing untyped alleles (TUNA)-applications to genome-wide association studies. *Genet. Epidemiol.* **30:** 718–727.

Patil, N., Berno, A.J., Hinds, D.A., Barrett, W.A., Doshi, J.M., Hacker, C.R., Kautzer, C.R., Lee, D.H., Marjoribanks, C., McDonough, D.P., et al. 2001. Blocks of limited haplotype diversity revealed by high-resolution scanning of human chromosome 21. *Science* **294:** 1719–1723.

Reich, D.E., Schaffner, S.F., Daly, M.J., McVean, G., Mullikin, J.C., Higgins, J.M., Richter, D.J., Lander, E.S., and Altshuler, D. 2002. Human genome sequence variation and the influence of gene history, mutation and recombination. *Nat. Genet.* **32:** 135–142.

Wall, J.D. and Pritchard, J.K. 2003. Assessing the performance of the haplotype block model of linkage disequilibrium. *Am. J. Hum. Genet.* **73:** 502–515.

Zhao, H., Nettleton, D., Soller, M., and Dekkers, J.C. 2005. Evaluation of linkage disequilibrium measures between multi-allelic markers as predictors of linkage disequilibrium between markers and QTL. *Genet. Res.* **86:** 77–87.

WEB RESOURCES

http://genome.ucsc.edu/ENCODE ENCODE.

http://www.hapmap.org International HapMap Project.

10 Meta-Analysis of Genome-Wide Association Studies

Paul I.W. de Bakker,[1,2] Benjamin M. Neale,[2,3] and Mark J. Daly[2,3]

[1]*Division of Genetics, Department of Medicine, Brigham and Women's Hospital, Harvard Medical School, Boston, Massachusetts 02115;* [2]*Program in Medical and Population Genetics, Broad Institute of MIT and Harvard, Cambridge, Massachusetts 02142;* [3]*Center for Human Genetic Research, Massachusetts General Hospital, Boston, Massachusetts 02114*

INTRODUCTION

Individual genome-wide association studies (GWAS) have only limited power to find novel loci underlying complex traits and common diseases. With relatively modest sample and effect sizes, a true association between genotype and phenotype may never meet genome-wide statistical significance ($p < 5 \times 10^{-8}$) in a single study. Through meta-analysis, novel susceptibility loci can be discovered by effectively summing the statistical evidence of individually underpowered studies. Most genetic discoveries for complex traits are now made through meta-analysis collaborations (Barrett et al. 2008; Ferreira et al. 2008; Zeggini et al. 2008; Kathiresan et al. 2009; Newton-Cheh et al. 2009). These efforts so far have been restricted to single-locus analyses, testing for main effects at a single polymorphism at a time. A key benefit of this approach is that individual-level genotype (and phenotype) data do not need to be exchanged between research groups (which in practice can be a genuine obstacle). In this chapter, we focus only on meta-analysis at single single-nucleotide polymorphisms (SNPs), paying particular attention to how imputation uncertainty can be incorporated into the association analysis and subsequent meta-analysis.

Probably the most important aspect of genome-wide association meta-analysis is harmonization of the study results (de Bakker et al. 2008). Not surprisingly, studies differ in design, sample collection, genotyping platforms, association analysis methods, and so forth. The goal for meta-analysis is that the association results (per SNP) of each study can be formatted, exchanged, and analyzed in such a way that the statistical evidence can be combined appropriately and that no valuable information is lost. Without minimizing the importance of having a clear phenotype definition (and corresponding measurements), we will assume for the sake of brevity that investigators representing the various studies have made sensible agreements about phenotype definitions, necessary sample exclusions, and appropriate covariate modeling.

IMPUTATION FILLS IN MISSING GENOTYPE DATA

Genotyping platforms differ in terms of their SNP content (and more recently, also in terms of copy-number polymorphism [CNP] content). Thus, combining disparate data sets from different studies would seem to be a difficult task. Fortunately, linkage disequilibrium (LD) among nearby variants makes it possible to predict (or "impute") polymorphisms that are not directly genotyped. As described elsewhere, the International HapMap Project (http://www.hapmap.org) has genotyped >3 million SNPs across the human genome in 270 DNA samples from four populations, providing 120 phased haplotypes for Utah residents with ancestry from northern and western Europe (CEU), 120 for Yoruba in Ibadan, Nigeria (YRI), and another 180 for Han Chinese from Beijing, China (CHB) and Japanese from Tokyo, Japan (JPT) (International HapMap Consortium 2007). A number of software tools are now available for imputation, including IMPUTE (Marchini et al. 2007), MACH (Li and Abecasis 2006), PLINK (Purcell et al. 2007), BIMBAM (Servin and Stephens 2007), and BEAGLE (Browning and Browning 2009). These programs require as input the genotype data collected in a sample and the HapMap genotypes (or haplotypes) as reference data and generate genotypes for all SNPs present on HapMap as output. Thus, the imputation procedure fills in the missing genotype data, effectively allowing data sets to be analyzed for a common set of SNPs.

Imputation Accuracy and Quality

Imputation accuracy is limited by two factors. First, the genotyping platform affects the imputation accuracy, because the SNP content directly determines the effective genome-wide coverage of variation. Second, imputation accuracy is limited by the SNP density and sample size of the reference panel (i.e., HapMap). By design, HapMap is strongly biased toward common variation (with good coverage of alleles with frequency >5%). Consequently, SNP genotyping arrays based on HapMap provide good coverage of common SNPs (Pe'er et al. 2006). In contrast, representation of rare variants (alleles with frequency <5%) is much less complete, because only a few haplotypes are observed per minor allele for those rare SNPs covered in HapMap. Therefore, the expectation is that the prediction accuracy for rare alleles will be worse than for common variation when using HapMap as the reference data set. Importantly, the accuracy of the (imputed) genotypes for a given SNP is likely to vary between studies, given that some will have directly genotyped it while others imputed it with varying degrees of success.

To illustrate some of these effects on imputation quality, we explore here data from the Diabetes Genetics Initiative (http://www.broad.mit.edu/diabetes). In that study, 1464 cases (affected with type 2 diabetes) and 1467 controls (matched for age, gender, BMI, and geographic locale) were genotyped on the Affymetrix 500K platform (Saxena et al. 2007). Figure 10.1 shows a pie chart of all HapMap SNPs split into four bins according to the pairwise r^2 between each SNP and any of the genotyped SNPs on the 500K array. Approximately 46% are perfectly captured by either genotyping or perfect LD. About one quarter of all HapMap SNPs have weaker LD (pairwise $r^2 < 0.5$) and are thus more difficult to impute. Table 10.1 shows the number of SNPs in these four bins split by their minor allele frequency. Of the low-frequency SNPs, most (58%) are poorly captured by the genotyped SNPs. Generally, high-frequency SNPs are better captured than low-frequency SNPs,

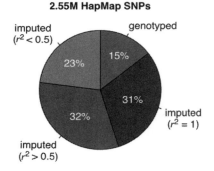

FIGURE 10.1 Breakdown of 2.5 million polymorphic SNPs in HapMap-CEU (release 21) by the pairwise correlation (r^2) to any of the genotyped SNPs on the Affymetrix 500K array. Almost two thirds of all HapMap SNPs can be imputed relatively straightforwardly as pairwise LD is strong: 31% of all HapMap SNPs are perfectly correlated ($r^2=1$), and another 32% have $r^2>0.5$ to a genotyped SNP; 23% of HapMap SNPs are more difficult to capture and likely will require haplotype information to be imputed accurately. (Based on data produced by the Diabetes Genetics Initiative [Saxena et al. 2007].)

TABLE 10.1 Number of SNPs by minor allele frequency split by pairwise r^2 to any of the genotyped SNPs on the Affymetrix platform

	Minor allele frequency (MAF)		
	Low frequency (MAF <5%)	Intermediate frequency (MAF 5%–20%)	High frequency (MAF 20%–50%)
r^2	[MAF<5%]	[MAF 5–20%]	[MAF 20–50%]
$r^2 = 1$	112,153 (33%)	291,171 (40%)	379,247 (34%)
$r^2 > 0.5$	33,724 (10%)	249,031 (34%)	530,915 (48%)
$r^2 < 0.5$	198,368 (58%)	194,498 (26%)	195,753 (18%)
Total	344,245 (100%)	734,700 (100%)	1,105,915 (100%)

consistent with previous observations that common SNPs in HapMap CEU tend to have more proxies (International HapMap Consortium 2005).

Starting with all quality control (QC)-passing SNPs from the 500K array (~380K SNPs) as input genotypes, we used the MACH program to impute all SNPs on HapMap CEU (release 21). As output, MACH computes the dosage (the estimated number of minor alleles per individual, ranging between 0 and 2) for every SNP in each individual (output in mldose files). The dosage is based on the posterior probabilities for each of the three genotype possibilities (AA, AB, BB) in each individual (output in mlprob files):

$$dosage = 1 \times p(AB) + 2 \times p(BB) , \qquad (1)$$

where p(AB) and p(BB) are the posterior probabilities of the heterozygote (AB) and minor homozygote (BB), respectively.

After the imputations are done, MACH computes the average maximal posterior probability (called "Quality") for each SNP averaged over the entire sample and estimates the correlation (called "Rsq") between the imputed genotype and the actual genotype (which can be interpreted as a measure of the imputation uncertainty). This Rsq metric is equivalent to the ratio between the observed variance of the dosage and the expected (binomial) variance of the dosage. A low observed/expected dosage variance ratio would indicate poor imputation accuracy, whereas accurate genotypes should approach unity (fluctuation around 1 due to Hardy–Weinberg deviations). In our example, we find that common SNPs are generally imputed well, with a clear decrease in imputation accuracy for low-frequency SNPs (as reflected by the heavy tail toward low observed/expected dosage variance ratios in Fig. 10.2A).

For the analyses presented here, we excluded 326 discordant sibships and kept only unrelated cases and matched controls, thus allowing us to use the 1-degree-of-freedom chi-square test for association:

$$\chi^2 = \frac{\left(p_{case} - p_{control}\right)^2}{\left(\dfrac{1}{n_{case}} + \dfrac{1}{n_{control}}\right)\left(p(1-p)\right)} , \qquad (2)$$

where p_{case} and $p_{control}$ are the minor allele frequency of a given SNP in cases and controls, respectively, p is the combined allele frequency, and n_{case} and $n_{control}$ are the number of chromosomes in cases and controls, respectively.

Chi-Squared Correction

Although the genomic inflation factor λ (described elsewhere) for the genotyped SNPs in our analysis was modest (1.04), we observed that the test statistic was conservative, resulting in a substantial

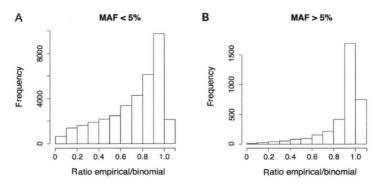

FIGURE 10.2 Histogram of the observed/expected dosage variance ratios for (*A*) rare SNPs (MAF <5%) and (*B*) common SNPs (MAF >5%). (MAF) Minor allele frequency.

deflation of the distribution for poorly imputed SNPs (Fig. 10.3). This illustrates that the χ^2 test statistic based on raw dosages (when uncertain) does not behave properly under the null hypothesis. We propose here a simple correction for this conservative behavior. With increasing imputation uncertainty, the variance of the dosage decreases due to lack of information. To correct for the uncertainty, we need to appropriately reduce the variance term in the denominator of the χ^2 formulation above. We can achieve this by replacing the binomial variance term $p(1 - p)$ with the empirically observed variance. (To compute the empirically observed variance, we calculate the mean dosage for the total sample and sum the square deviations.)

We validate this statistical correction for a subset of poorly imputed SNPs in our example to mimic a worst-case scenario. Figure 10.3 shows the quantile–quantile (Q-Q) plot for 198,368 SNPs of frequency <5% and pairwise $r^2 < 0.5$ to any of the genotyped SNPs. Before correction, we observe a strongly deflated distribution of the test statistic. After correcting for the variance deflation, we recover a test statistic distribution consistent with the null distribution (along the diagonal of the Q-Q plot).

To further explore the impact of imputation accuracy on the association analysis, a subset of common (~8000) and rare (~36,000) genotyped SNPs was hidden from the imputation. For these SNP sets, we compared the test statistic based on the "true" genotype to the corrected test statistic based on the imputed dosage. Figure 10.4 shows the distribution of the respective Z scores (we converted the chi-square into a Z score for the sake of visualization). There is a clear positive trend

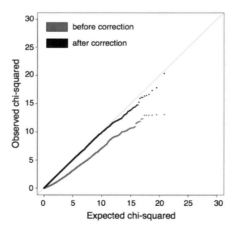

FIGURE 10.3 Quantile–quantile plot for the test statistic distribution of the chi-square test for SNPs of frequency <5% and pairwise r^2 to any of the genotyped SNPs before and after correction for the imputation uncertainty. The uncorrected distribution is deflated relative to the expected null distribution. The proposed correction is able to recover a proper null distribution along the diagonal.

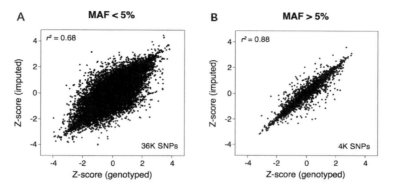

FIGURE 10.4 Comparison of the test statistic (as a *Z* score) between imputed genotypes and experimentally observed genotypes for randomly selected SNPs: (*A*) Minor allele frequency (MAF) < 0.05 (*n* = 36,000) and (*B*) MAF > 5% (*n* = 8,000). The lower correlation coefficient indicates overall lower imputation accuracy for the low-frequency SNPs. (Based on data produced by the Diabetes Genetics Initiative [Saxena et al. 2007].)

between the correlation of the *Z* scores and the allele frequency. The Pearson correlation between the *Z* scores is 0.9 for high-frequency SNPs and 0.7 for low-frequency SNPs. The correlation drops progressively with lower frequency, with a $r^2 = 0.3$ for SNPs with frequency <1%. So far, we have demonstrated that it is possible to obtain proper test statistics from imputation dosages by using a simple correction to the chi-squared test and that, as expected, the performance for rare variant imputation is worse than for common SNPs.

Incorporating Imputation Uncertainty into Meta-Analysis

The depression in the variance of the test statistic not only underestimates the true significance, but also causes further problems in the context of meta-analysis. Combining uncorrected test statistics causes an initial loss in power, but the weighting of a study is also typically proportional to its sample size or the inverse of the variance (of the effect). By not correcting the sample size, we allow for too great a contribution of an uncertain genotype. Therefore, an important consideration is how imputation uncertainty is incorporated into the meta-analysis (de Bakker et al. 2008). No single study should be able to distort (or disproportionately contribute to) the meta-analysis. Only when these conditions are met, can the association information be safely combined across multiple studies in the meta-analysis.

PERFORMING THE META-ANALYSIS

Having ensured that the test statistic distribution looks good, we are ready to do the actual meta-analysis. Probably the simplest approach is to sum the association evidence across studies where every study is weighted by the effective sample size. First, we convert *p* values to *Z* scores as follows (in R code):

```
#
# this is the routine to convert a pvalue into a zscore.
#
# the direction of the odds ratio (or) determines the sign of
# the resulting z-score
#
convert.pvalue <- function(pval, or) {
```

```
if ( or > 1 ) {
z <- qnorm( pval / 2 );
} else {
z <- -(qnorm( pval / 2 ));
}

return(z);
}
```

The resulting Z scores can then be summed weighted by the sample size, as follows:

$$z_{meta} = \sum z_i \times w_i ,\tag{3}$$

where $w_i = \sqrt{N_i/N_{total}}$ and N_i and N_{total} are the study sample size and the total sample size, respectively.

Often, studies have many more controls than cases, and power will be primarily limited by the number of cases. (The power improvement for including additional controls saturates after adding three to four times the number of cases.) In these cases, we need to compute an "effective" sample size (as weighting a study by its total sample size would overestimate its true contribution). One possible way to do this is to compute the noncentrality parameter (NCP) for a given disease model (including case and control numbers) and use the resulting NCP as the N_i (see Appendix A).

To take into account imputation uncertainty, we can scale the sample size by the observed/expected variance ratio (or a similar information metric output by the imputation program). As imputation quality varies per SNP, this must be done on a per SNP basis. Because there is random fluctuation of the observed/expected variance ratio (even for accurate genotype calls), we should only use the test statistic correction (if using the χ^2 test for a binary trait above) and sample size scaling when the average maximal posterior probability is below some threshold (e.g., 0.99).

The resulting meta-analysis Z scores can be converted back to p values as follows (R code):

```
#
# to convert the meta-analytic z-scores into p-values
#
pmeta <- pnorm(-(abs(zmeta))) * 2;
```

Instead of meta-analysis by sample size weighting, we can perform meta-analysis by weighting by the inverse variance for linear and logistic regression association results. This is relatively straightforward. First, we compute the weighted β and standard error (SE) terms based on the estimated β coefficient and corresponding SE that have been calculated for each study:

$$\langle\beta\rangle = \frac{\sum_i\left[\beta_i/(SE_i)^2\right]}{\sum_i\left[1/(SE_i)^2\right]}\tag{4}$$

and

$$\langle SE\rangle = \sqrt{1/\sum_i\left[1/(SE)^2\right]}.\tag{5}$$

From these two weighted terms, we can compute a meta-analysis Z score:

$$z = \langle\beta\rangle/\langle SE\rangle .\tag{6}$$

Using linear or logistic regression as the primary association analysis approach has a key advantage. Although count-based test statistics (such as the χ^2 test) require explicit corrections (as we proposed above), linear and logistic regression models can implicitly deal with imputation uncertainty as the variance of the allele (or dosage) frequency becomes deflated with growing uncertainty. This variance deflation is automatically absorbed by regression modeling by generating a large(r) standard error of the estimated β coefficient. This means that, in principle, the genome-wide distribution of the test statistic by linear or logistic regression modeling is overall well behaved. For other methods that more explicitly deal with imputation uncertainty, see, for example, Guan and Stephens (2008).

Conventional meta-analysis literature advocates the preferred use of random-effects models (rather than fixed-effects models). For genetic discovery, the main focus is on finding novel associated loci, not on accurate estimation of the effect size. A random-effects meta-analysis effectively penalizes the test statistic for observed heterogeneity between studies, thus lowering power for discovery. This is why all genome-wide meta-analysis efforts to date are based on a fixed-effects (and not random-effects) model. However, we note that there are certainly situations where heterogeneity can be informative (especially when the number of contributing studies is large).

The 1000 Genomes Project (http://www.1000genomes.org) is an ongoing effort to build a genome-wide inventory of all segregating sequence variation down to a frequency of 1% (0.1% in genic regions) through sequencing in large population samples and to enable imputation-based approaches for rare variants. We expect that imputation-based meta-analysis will continue to have an important role in genetic discovery as more complete resources such as 1000 Genomes are being developed.

Finally, in support of this chapter, we have made available a meta-analysis example based on the genome-wide association results of the type 2 diabetes scan performed by the Wellcome Trust Case Control Consortium and the Diabetes Genetics Initiative (http://www.broad.mit.edu/~debakker/meta_t2d.html). R scripts and Perl code that were developed for and used for a meta-analysis of electrocardiographic QT interval duration (Newton-Cheh et al. 2009) can be found on the same web page.

APPENDIX A

The routine below (in R code) computes the noncentrality parameter (NCP) for a given disease model, assuming risk allele frequency, relative risks for the heterozygote and major homozygote, and sample size.

```
#
# this routine computes NCP given the following parameters:
#
# fA = risk allele frequency
# k = population prevalence of trait
# rAa = relative risk of genotype Aa
# rAA = relative risk of genotype AA
# n_case = number of cases
# n_control = number of controls
#
cc_gpc <- function( fA, k, rAa, rAA, n_case, n_control ) {
    # frequency of non-risk allele
    fa <- 1 - fA

    # calculate genotype frequencies
    fAA <- fA * fA
    fAa <- 2 * fA * fa
    faa <- fa * fa

    # baseline risk of genotype aa
```

```
raa <- k / (fAA*rAA + fAa*rAa + faa)

# risk of genotypes
rrAA <- rAA * raa
rrAa <- rAa * raa
rraa <- raa

nrrAA <- 1 - rrAA
nrrAa <- 1 - rrAa
nrraa <- 1 - rraa

# compute odds ratio
orAa <- ( rrAa / nrrAa ) / ( rraa / nrraa )
orAA <- ( rrAA / nrrAA ) / ( rraa / nrraa )

# genotype frequencies in cases
case_AA <- fAA * rrAA
case_Aa <- fAa * rrAa
case_aa <- faa * rraa
case_sum <- case_AA + case_Aa + case_aa

case_AA <- case_AA / case_sum
case_Aa <- case_Aa / case_sum
case_aa <- case_aa / case_sum

# genotype frequencies in controls
control_AA <- fAA * nrrAA
control_Aa <- fAa * nrrAa
control_aa <- faa * nrraa
control_sum <- control_AA + control_Aa + control_aa

control_AA <- control_AA / control_sum
control_Aa <- control_Aa / control_sum
control_aa <- control_aa / control_sum

# allele frequencies in cases and controls
case_A <- case_AA + case_Aa / 2
control_A <- control_AA + control_Aa / 2

# turn into case-control counts
n_case_A <- 2 * n_case * case_A
n_case_a <- 2 * n_case * (1 - case_A)
n_control_A <- 2 * n_control * control_A
n_control_a <- 2 * n_control * (1 - control_A)

# compute 2x2 chi-square test for association
x2 <- chisq.test(matrix(c(n_case_A, n_control_A,
                          n_case_a, n_control_a),
                        nrow=2, ncol=2),
                 correct=F)
# NCP is the chi-square statistic
ncp <- x2$statistic

# return frequency of A allele in cases and controls, and NCP
c(case_A, control_A, ncp)
}
```

REFERENCES

Barrett, J.C., Hansoul, S., Nicolae, D.L., Cho, J.H., Duerr, R.H., et al. 2008. Genome-wide association defines more than 30 distinct susceptibility loci for Crohn's disease. *Nat. Genet.* **40:** 955–962.

Browning, B.L. and Browning, S.R. 2009. A unified approach to genotype imputation and haplotype-phase inference for large data sets of trios and unrelated individuals. *Am. J. Hum. Genet.* **84:** 210–223.

de Bakker, P.I., Ferreira, M.A., Jai, X., Neale, B.M., Raychauduri, S., and Voight, B.F. 2008. Practical aspects of imputation-driven meta-analysis of genome-wide association studies. *Hum. Mol. Genet.* **17:** R122–R128.

Ferreira, M.A., O'Donovan, M.C., Meng, Y.A., Jones, I.R,. Ruderfer, D.M., et al. 2008. Collaborative genome-wide association analysis supports a role for ANK3 and CACNA1C in bipolar disorder. *Nat. Genet.* **40:** 1056–1058.

Guan, Y. and Stephens, M. 2008. Practical issues in imputation-based association mapping. *PLoS Genet.* **4:** e1000279.

International HapMap Consortium. 2005. A haplotype map of the human genome. *Nature* **437:** 1299–1320.

International HapMap Consortium. 2007. A second generation human haplotype map of over 3.1 million SNPs. *Nature* **449:** 851–861.

Kathiresan, S.,Willer, C.J., Peloso, G.M., Demissie, S., Musunuru, K., et al. 2009. Common variants at 30 loci contribute to polygenic dyslipidemia. *Nat. Genet.* **41:** 56–65.

Li, Y. and Abecasis, G.R. 2006. MACH 1.0: Rapid haplotype reconstruction and missing genotype inference. *Am. J. Hum. Genet.* **S79:** 2290.

Marchini, J., Howie, B., Myers, S., McVean, G., and Donnelly, P. 2007. A new multipoint method for genome-wide association studies by imputation of genotypes. *Nat. Genet.* **39:** 906–913.

Newton-Cheh, C., Eijgelsheim, M., Rice, K.M., de Bakker, P.I., Yin, X., et al. 2009. Common variants at ten loci influence QT interval duration in the QTGEN Study. *Nat. Genet.* **41:** 399–406.

Pe'er, I., de Bakker, P.I.W., Maller, J., Yelensjy, R., Altschuler, D., and Daly, M.J. 2006. Evaluating and improving power in whole-genome association studies using fixed marker sets. *Nat. Genet.* **38:** 663–667.

Purcell, S., Neale, B., Todd-Brown, K., Thomas, L., Ferreira, M.A., et al. 2007. PLINK: A tool set for whole-genome association and population-based linkage analyses. *Am. J. Hum. Genet.* **81:** 559–575.

Saxena, R., Voight, B.F., Lyssenko, V., Burtt, N.P., de Bakker, P.I., et al. 2007. Genome-wide association analysis identifies loci for type 2 diabetes and triglyceride levels. *Science* **316:** 1331–1336.

Servin, B. and Stephens, M. 2007. Imputation-based analysis of association studies: Candidate regions and quantitative traits. *PLoS Genet.* **3:** e114.

Zeggini, E., Scott, L.J., Saxena, R., Voight, B.F., Marchini, J.L., et al. 2008. Meta-analysis of genome-wide association data and large-scale replication identifies additional susceptibility loci for type 2 diabetes. *Nat. Genet.* **40:** 638–645.

WEB RESOURCES

http://www.1000genomes.org 1000 Genomes Project.

www.broad.mit.edu/diabetes Diabetes Genetics Initiative.

http://debakker.med.harvard.edu/ de Bakker lab page.

http://www.hapmap.org HapMap Project.

http://pngu.mgh.harvard.edu/~purcell/plink/gplink.shtml Purcell et al. 2007. PLINK.

http://quartus.uchicago.edu/~yguan/bimbam/index. html

Servin and Stephens 2007. BIMBAM.

http://www.sph.umich.edu/csg/abecasis/MACH/download/ Li and Abecasis 2006. MACH.

http://www.stat.auckland.ac.nz/~bbrowning/beagle/beagle.html Browning and Browning 2009. BEAGLE.

http://www.stats.ox.ac.uk/~marchini/software/gwas/impute. html Marchini et al. 2007. IMPUTE.

11 | Gene–Environment Interaction and Common Disease

Ruth J.F. Loos and Nicholas J. Wareham

Medical Research Council Epidemiology Unit, Institute of Metabolic Science, Addenbrooke's Hospital, CB2 0QQ Cambridge, United Kingdom

INTRODUCTION

Over the past 3 years, genome-wide association studies (GWAS) have led to a rapid increase in the number of genetic variants that have been associated incontrovertibly with disease (Manolio et al. 2008). Although these recently discovered loci are set to improve fundamentally our insights into the pathophysiology of these diseases and traits, their contribution to the variation in risk at the population level is small and their predictive value is generally low. Most common diseases arise from a combination of genetic and environmental influences. The lifestyle changes that have taken place over the last three decades, rather than our genome, which has remained unchanged over the past few generations, are a more important contributor to the increasing prevalence of common diseases, such as type 2 diabetes, obesity, hypertension, hypercholesterolemia, and cancer (Mokdad et al. 2004). Yet, not every individual responds in a similar way to the same environment; the response to the environment depends on the genetic susceptibility of an individual. Indeed, environmental and genetic factors do not act strictly independently, but instead interact with each other in leading to disease. It has been said that "genes load the gun, but the environment pulls the trigger," and this metaphor accurately describes the intricate interplay between genes and environment, an interplay clearly shown by the recent obesity epidemic (Box 11.1).

Research into the existence of and biological explanations for gene–environment interactions is growing, but it has been hampered not only by challenges common to all epidemiological studies (von Elm and Egger 2004; Ioannidis et al. 2006), but also by issues specific to the field of gene–environment interaction. These include the use of small, underpowered sample sizes, irreproducible results, poor reporting on the statistical significance of interactions, and selective reporting of positive results. In addition, the typically large measurement error in the assessment of lifestyle factors, such as physical activity and food intake, make testing for the presence of gene–environment interaction even harder than the detection of main effects. An appropriate study design, accurate measurement of environmental factors, and large sample size are the first steps toward overcoming some of these problems.

This chapter describes the evidence for gene–lifestyle interaction in common disease with a particular focus on obesity and type 2 diabetes. We outline the epidemiological principles and study designs for identifying such interactions and address the challenges and potential pitfalls that are inherent to gene–environment interaction studies.

BOX 11.1 How Genetic Susceptibility and Lifestyle Interact in Causing the Obesity Epidemic

Obesity is a common, multifactorial condition that arises through the joint actions of multiple genetic and environmental factors. Family and twin studies have shown that genetic factors account for 40%–70% of the variation in obesity risk (Maes et al. 1997). Because our genome has not substantially changed over the last few decades, genetic variation alone cannot explain the rapid increase in obesity prevalence that we have witnessed over the past three decades. A changing environment that promotes excessive calorie intake and that discourages physical activity appears to be the main culprit for the dramatic increase in obesity prevalence worldwide. Yet, this obesogenic environment does not affect every person to the same extent, as illustrated in Box 11.1, Figure 1 (Ravussin and Bouchard 2000). Individuals with a high genetic susceptibility to obesity who live in an obesogenic environment will gain the most weight. In an environment that does not favor obesity, these susceptible individuals would be of normal weight or slightly overweight. Individuals with a low genetic susceptibility will be of normal weight in a restrictive environment, whereas in an obesogenic environment these individuals will remain at normal weight or become slightly overweight.

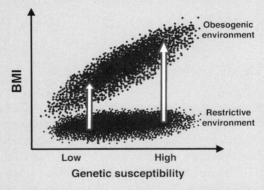

FIGURE 1 Gene–environment interaction in relation to body mass index (BMI). In a restrictive environment, BMI in genetically susceptible individuals is similar to that of those who are genetically resistant. In an obesogenic environment, those who are genetically prone to become obese will have a substantially higher BMI than those who are genetically resistant. (Adapted, with permission from Elsevier, from Ravussin and Bouchard 2000.)

WHAT GENE–ENVIRONMENT INTERACTION MEANS IN GENETIC EPIDEMIOLOGY

The concept of "gene–environment interaction" has different meanings to different people, and the interpretation depends largely on the type of research with which one is involved. A biologist or physiologist might interpret gene–environment interaction at the molecular or cellular level as the direct (such as nutrients) or indirect (such as physical activity-induced release of nitric oxide) effect of an environmental factor on the genome. As such, environmental factors could affect gene transcription, expression, and function, which may, in turn, influence risk of disease.

A public health epidemiologist interprets gene–environment interaction at the population level as the differential response to an environmental factor (e.g., diet, physical activity, and smoking) on disease risk dependent on the individual genotype. A genetic epidemiologist will see it as the difference in genetic susceptibility to disease depending on the environment in which one lives. This chapter focuses on the epidemiological definition of gene–environment interaction from both a public health as well as a genetic perspective.

EVIDENCE OF GENE–ENVIRONMENT INTERACTION IN COMMON DISEASE

Suggestive evidence of gene–lifestyle interaction in the development of common diseases, such as obesity and type 2 diabetes, was first provided by descriptive epidemiological studies—for example, migration studies that compare the risk of disease among genetically related populations that live different lifestyles. A classical example is the comparison of the risk of obesity and type 2 diabetes between Pima Indians living in the "obesogenic" environment of Arizona (69% of whom are obese and 55% have type 2 diabetes) and those living in the "restrictive" environment of the remote Mexican Sierra Madre mountains (13% of whom are obese with only 6% having type 2 diabetes) (Ravussin et al. 1994; Esparza et al. 2000). These findings illustrate that, despite similar genetic predisposition, different lifestyles result in strikingly different prevalences of obesity and type 2 diabetes. White Americans living in a similar obesogenic environment, but who have a different genetic background, are much less susceptible to becoming obese (~32%) or to developing type 2 diabetes (~8%) compared with the Pima Indians living in Arizona.

Several well-controlled twin studies have provided additional evidence concerning the role of genetic factors in the response to diet and exercise intervention. In an overfeeding study, 12 monozygotic twin pairs of young adult men were overfed by 1000 kcal/d surplus (6 d/wk) for 100 d (Bouchard et al. 1990). The mean weight gain was 8.1 kg, but there was considerable variation in response with a threefold difference between the lowest and highest gainers (range ~4–12 kg). This heterogeneity in response was not random but was dependent on the participants' genetic backgrounds. The variation in weight gain between twin pairs (genetically different) was more than three times (F ratio = 3.4, $p < 0.02$) larger than the variation within pairs (genetically identical).

Along the same lines, evidence is available from two well-controlled weight loss intervention studies in monozygotic twins. In the first study, the negative energy balance was induced through calorie restriction (Hainer et al. 2000) and in the other through exercise training (Bouchard et al. 1994). Large individual differences in weight loss were observed, whereas genetically identical individuals responded more similarly. The ratio of the between-pair to the within-pair variance in weight loss was even more pronounced with calorie restriction (F ratios were 12.8, $p < 0.001$) or with exercise training (F ratio 6.8, $p < 0.01$) than for weight gain (F ratio 3.4, $p < 0.02$). These experiments confirm that there are considerable differences in the way individuals respond to alterations in energy balance and that the magnitude of response to changes in diet by either overfeeding or energy restriction is influenced by genetic predisposition.

SEEING THE GENE–ENVIRONMENT INTERACTION FROM DIFFERENT PERSPECTIVES

Figure 11.1 illustrates the concept of gene–environment by categorizing the environmental exposure. However, gene–environment interaction can also be tested with an environmental factor measured as a continuous trait. Likewise, the outcome can be categorical (e.g., case vs. control) or can be a continuous trait (e.g., body mass index [BMI], glucose concentration, and lipid levels).

Genetic Perspective

The genetic perspective starts from a main-effect hypothesis that tests for association between a genetic variant and a disease or trait (Fig. 11.1A, Step 1) (e.g., whether a genotype is associated with BMI). Next, the interaction hypothesis tests whether the genotype–disease association is different across different levels of environmental exposure (Fig. 11.1A, Step 2) (e.g., whether the genotype–BMI association is different in individuals on a low-fat diet compared with individuals on a high-fat diet). The example illustrated in Figure 11.1A would suggest that the A allele of the genetic variant increases the susceptibility for increased BMI (Step 1), but that susceptibility can be overcome by maintaining a low-fat diet (Step 2). The gene–environment interaction would be statistically significant if the slopes (dotted lines) of the two associations (low-fat diet vs. high-fat diet) differ significantly from each other.

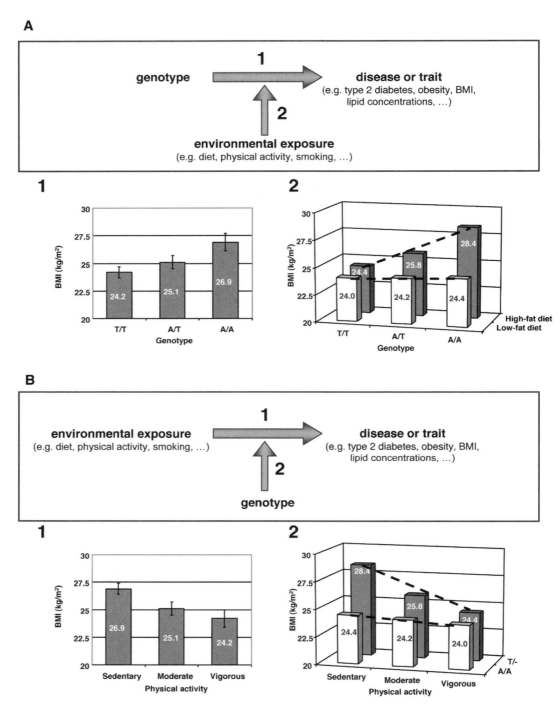

FIGURE 11.1 (A) Gene–environment interaction from a genetic perspective; the association between a genetic variant and an outcome is dependent on the environmental factor. (B) Gene–environment interaction from a public health perspective; the association between an environmental exposure and an outcome is dependent on the individual's genotype.

Public Health Perspective

The public health perspective first questions the association between an environmental factor and disease or trait (Fig. 11.1B, Step 1) (e.g., whether physical inactivity is associated with BMI). Subsequently, the interaction hypothesis tests whether carriers of a certain genotype are more susceptible for the influence that the environment has on disease than noncarriers (Fig. 11.1B, Step 2) (e.g., whether the detrimental effects of physical inactivity on BMI is more pronounced in T-allele carriers than A/A homozygotes). The fictitious example illustrated in Figure 11.1B would suggest that a sedentary lifestyle is associated with higher BMI, in particular when the individual is a T-allele carrier, whereas A/A homozygotes are resistant to physical activity. Again, the gene–environment interaction will be statistically significant if the slopes of the two associations (A/A vs. T/–) differ significantly from each other.

To use the examples in Figure 11.1, we assume that the association is present in one stratum while absent in the other. However, gene–environment interaction can also be present when both strata show association with the same direction of effect but are statistically more pronounced in one stratum than in the other or when the association in the strata show opposite direction of effect. It is possible for gene–environment interaction to be present even in the absence of a main effect. The exact mathematics behind the statistical testing of gene–environment are beyond the scope of this chapter and are described in detail by Thomas (2004).

WHY STUDY GENE–ENVIRONMENT INTERACTION?

Unidentified gene–environment interaction may mask the presence of a genetic effect in a certain environment or it may mask the presence of an environmental effect for a given genotype. For example, recent GWAS (Zeggini et al. 2008; Loos et al. 2008) as well as large-scale meta-analyses of candidate genes (Ludovico et al. 2007; Young et al. 2007; Kurokawa et al. 2008) have shown that the overall effect size of a genetic variant on the risk of a disease or on variation of a trait is generally very small. Often these meta-analyses show significant heterogeneity of effect sizes across studies, raising the possibility that the genetic contribution may be modified by unobserved environmental factors. Because of the expected small genetic effect and potential heterogeneity, main-effect association studies require data from tens of thousands of individuals to identify convincing and robust associations. Likewise, establishing the role of publicly accepted healthy lifestyle factors on the risk or prevention of disease has proved to be challenging. In a systematic review, we found that the overall protective effect of habitual physical activity on weight gain in adults was small, with mixed results across studies (Wareham et al. 2005). One of the reasons for the limited beneficial effect could be that the response to physical activity is dependent on individual genotype.

Gene–environment interaction studies allow the identification of lifestyles for which genetic susceptibility to disease is more pronounced and, conversely, the identification of genetically susceptible individuals for whom the effect of environmental factors on disease is bigger. Restricting analyses to more homogeneous groups of individuals in terms of lifestyle or environment should increase the genetic contribution to a disease or trait. This approach is elegantly shown in two intervention studies showing that the effect of the *TCF7L2* risk allele on the progression toward type 2 diabetes was abolished in the lifestyle intervention group but apparent in the placebo control group. This result suggests an interaction between common variants in the *TCF7L2* gene and lifestyle in the risk of progression to type 2 diabetes (Box 11.2) (Florez et al. 2006; Wang et al. 2007).

By restricting analyses to genetically susceptible individuals, the influence of environmental factors on the risk of disease might be stronger compared with less susceptible individuals. This is illustrated by an analysis of the fat mass and obesity-related gene *FTO* and physical activity in 704

Box 11. 2 *TCF7L2*–Lifestyle Interaction in the Diabetes Prevention Program

The Diabetes Prevention Program is a multicenter, randomized clinical trial designed to test whether lifestyle intervention (physical activity and diet) or pharmacologic treatment (metformin) can prevent or delay the development of type 2 diabetes (Diabetes Prevention Program Research Group 2002). The trial enrolled 3234 individuals who were already at increased risk for type 2 diabetes, were overweight, and had elevated fasting and postload glucose concentrations. Volunteers were randomly assigned to a placebo group, a lifestyle-intervention group, or a metformin-treated group. A fourth group was treated with troglitazone, but this treatment was discontinued because of its toxic effects on the liver. Over an average follow-up period of 3 yr, lifestyle intervention and metformin treatment reduced the incidence of type 2 diabetes by 58% (95% confidence interval [CI], 48%–66%) and 31% (17%–43%), respectively, compared with the placebo group (Diabetes Prevention Program Research Group 2002).

Besides the importance of environmental factors, such as lifestyle and pharmacological treatment, it has been estimated that 30%–70% of the risk of type 2 diabetes can be attributed to genetic factors (Poulsen et al. 1999). Common variants in the gene that encodes the transcription factor 7-like 2 gene (*TCF7L2*) have been strongly and reproducibly associated with the risk of type 2 diabetes in various ethnic populations, each risk allele conferring 40%–50% risk (Grant et al. 2006; Cauchi et al. 2007).

Florez et al. (2006) examined whether the association between *TCF7L2* variants and the risk of type 2 diabetes is attenuated by lifestyle or metformin treatment in the Diabetes Prevention Program. The overall analysis confirmed the strong influence of the *TCF7L2* gene as homozygotes for the T allele at rs7903146 were more likely to progress to diabetes than those homozygous for the C allele (hazard ratio [HR] 1.55 [1.2–2.01], $p < 0.001$) (Box 11.2, Fig. 1D). However, lifestyle and metformin treatment seemed to attenuate the genetic susceptibility. Although the influence of the *TCF7L2* variant on progression to type 2 diabetes was most pronounced in individuals of the placebo group (HR 1.81 [1.21–2.7], $p = 0.004$) (Box 11.2, Fig. 1A), no difference in risk according *TCF7L2* genotype was observed in individuals of the lifestyle intervention group (HR 1.15 [0.68–1.94], $p = 0.60$) (Box 11.2, Fig. 1C). In the metformin treatment group (Box 11.2, Fig. 1B), the effect of the *TCF7L2* genotype was attenuated compared with the placebo group. Although the *TCF7L2*-intervention interaction in the Diabetes Prevention Program was not statistically significant, corroborating results from the Finnish Diabetes Prevention Study (Wang et al. 2007) provide further evidence of attenuation of the *TCF7L2* susceptibility on diabetes risk by diet and exercise.

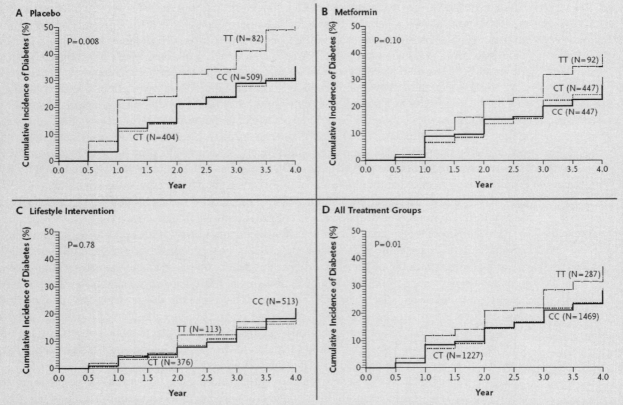

FIGURE 1 Incidence of type 2 diabetes according to treatment group and genotype at variant rs7903146. (Adapted, with permission, from Florez et al. 2006, ©Massachusetts Medical Society. All rights reserved.)

Old Order Amish individuals, in whom physical activity was associated with BMI but only in individuals who were homozygous for the *FTO* risk allele (Box 11.3) (Rampersaud et al. 2008).

Thus, results from gene–environment interaction studies may allow the unmasking of associations that are hidden because of heterogeneity in the population and advance our understanding of physiological pathways that underlie disease. Another important implication of this type of study is that it can produce important public health messages. Individuals might be genetically susceptible to develop disease, but this does not mean that they are destined to become diseased. Changes in lifestyle can overcome genetic susceptibility, as illustrated by the *FTO*–physical activity example (Andreasen et al. 2008; Rampersaud et al. 2008) or the *TCF7L2*–lifestyle intervention studies (Florez et al. 2006; Wang et al. 2007) (see Boxes 11.2 and 11.3).

STUDY DESIGN

Various classical epidemiological study designs have been applied to identifying gene–environment interactions, including cross-sectional and prospective designs, cohort and case–control design, and randomized intervention designs.

Box 11.3 *FTO*–Physical Activity Interaction in the Old Order Amish

This study examined the *FTO*–physical activity interaction in relation to BMI in the population-based Heredity and Phenotype Intervention (HAPI) Heart Study of 704 men and women who were members of the Old Order Amish Community in Lancaster County, Pennsylvania (Rampersaud et al. 2008). Physical activity was measured objectively for 7 consecutive days using an accelerometer.

GWAS identified common variants in the *FTO* gene to be highly significantly associated with BMI, obesity, and related traits in adults and children (Frayling et al. 2007; Scuteri et al. 2007). Subsequently, the association was replicated unequivocally in populations of different ethnicity. Each *FTO* risk allele increased BMI by ~0.40–0.66 kg/m² (equivalent to 1–1.5 kg body weight) and the risk of obesity by ~30%.

Rampersaud et al. (2008) showed that increased physical activity is associated with a lower BMI, but only in individuals homozygous for the *FTO* risk allele (e.g., rs1861868); no such association was observed in the carriers of the protective allele (Box 11.3, Fig. 1). These data suggest that the responsiveness to physical activity is attenuated by one's genetic susceptibility to obesity. Similar findings were observed in a large Danish population-based study with self-reported physical activity data (Andreasen et al. 2008).

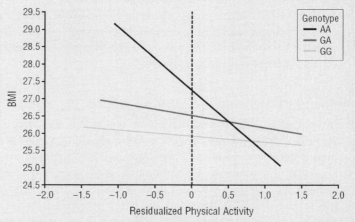

FIGURE 1 Predicted BMI as a function of residualized physical activity accelerometer counts according to *FTO* rs1861868 genotypes. Increased physical activity is associated with lower BMI; however, this association is significantly more pronounced in those who carry the A/A genotype. (Adapted, with permission, from Rampersaud et al. 2008.)

Cross-Sectional Case–Control Studies

The traditional cross-sectional case–control design is one of the most widely used study designs and, despite its well-known weaknesses and caveats (Sackett 1979), this design has proved to be effective for studying the main effects of genetic and environmental exposures on the risk of disease. Case–control studies compare the prevalence of an exposure (genetic or environmental) in a representative sample of cases with a representative sample of controls ascertained from the same population as the cases. This study design tends to be of relatively low cost, works quickly, and requires a relatively small sample size. It has important advantages when the disease is rare and the exposure is frequent.

The cross-sectional nature of this study design is not a limitation when investigating the influence of a genetic variant; however, it is a major limitation for gene–environment interaction studies. In cross-sectional studies, the data on the environmental exposures are collected retrospectively, after the diagnosis of disease in cases, which makes this design particularly susceptible to recall bias. Particularly for diseases such as type 2 diabetes, recall bias can have a substantial impact on the outcome of gene–environment analyses because the initial treatment to manage the condition usually involves education and advice about physical activity and dietary intake, which may be the very lifestyle exposure we want to study. Therefore, cases may recall these behaviors differently from controls who have not been exposed to lifestyle education. In the case of obesity, the situation might be even more complicated because not only recall bias, but potentially other biases affect the assessment of the environmental exposure. It is well documented that reported dietary intake of total energy is systematically different according to the degree of obesity (Lissner 2002).

Case-Only Studies

The use of case-only designs or prospective study design can at least in part avoid biases that hamper the cross-sectional case–control design. An important challenge for case–control studies is the recruitment of adequate controls to avoid selection bias. The case-only (or case–case) design has therefore been proposed as an alternative and efficient approach for studying gene–environment interaction. When well-designed disease registries are available, an appropriate set of cases is generally easier to recruit than controls and, because this design only requires cases, it is relatively inexpensive and quick. The key parameter is the interaction term between a genetic marker and an environmental risk factor (Khoury and Flanders 1996). Although this design allows one to determine whether or not such an interaction is present, its major weakness is that it is unable to quantify the individual and combined effects of these factors on risk (Clayton and McKeigue 2001).

The validity of the case-only design is sensitive to the critical assumption that the genetic and environmental factors are independent (Piegorsch et al. 1994), an assumption that may easily be violated for environmental factors, such as dietary intake or habitual physical activity that are themselves influenced by genetic factors.

Prospective Cohort Designs

In prospective cohort studies, data on the genetic and environmental exposures of a representative sample of the population are collected at baseline before the onset of disease. The cohort sample is then systematically followed over time until the manifestation of a specified disease outcome. Cases are then compared with a selection of controls who had an equal likelihood to develop the disease. This produces a case–control study "nested" within the original cohort. Information on the environmental exposure is collected on the entire cohort at baseline, but it is only processed and analyzed at follow-up on those individuals who have developed a disease and on a selection of controls.

The real value of this design lies in its longitudinal nature, which minimizes recall bias—the important weakness that hampers the retrospective case–control design. The likelihood that assessment of the lifestyle exposure differs according to disease outcome is low. This design is efficient if follow-up rates are high and independent of the disease status such that cases and controls are representative for the population. Ideally, the cohort is surveyed on a regular basis as repeated measurements will provide a more accurate reflection of the pattern of the environmental exposure and of the timing of disease onset.

However, when the disease of interest is rare, this study design might be inefficient because an entire cohort needs to be studied prospectively for relatively few cases to be informative as incident cases. Even with common diseases, large sample sizes are needed to produce a sufficient number of cases for an adequately powered case–control analysis. Prospective studies are typically of long duration, and thus it generally takes several years, even decades, to accrue cases. The systematic follow-up and the large sample size required make the prospective cohort design time-intensive and costly.

POWER AND SAMPLE SIZE IN GENE–ENVIRONMENT INTERACTION STUDIES

The sample size required to detect associations with sufficient power largely depends on three factors: (1) the prevalence of the exposure (genetic or environmental), (2) the expected effect size, and (3) the measurement precision of exposures and outcome. The lower the minor allele frequency of the genetic variant, the lower the prevalence of the environmental exposure; the smaller the expected effect size, the larger the sample size needs to be. As a rule of thumb, the sample size for studies of the interaction between two exposures needs to be four times the sample size required to test for the main effect of each of the exposures separately (Smith and Day 1984). With a case–control design, for example, a sample size of at least 7500 is required to achieve 80% power to detect a moderate interaction effect ($\theta = [OR_{E|G=1}]/[OR_{E|G=0}] = 1.5$) with a common gene variant (allele frequency of 25%) and significance at 5% (Luan et al. 2001). Thus, even for relatively common diseases such as type 2 diabetes, a cohort of 500,000 people would need to be studied for 5 yr to yield sufficient incident cases. Several such large-scale prospective studies to investigate the influence of genes and environment on common diseases are in progress (Box 11.4). Often such sizable studies have to compromise on the measurement precision of the environmental factors such as dietary intake and physical activity because of their size.

Yet, measurement precision is a key factor that impedes the power to identify associations and interactions. The better the precision of the measurements, the more power a study will have. Environmental factors, such as physical activity and dietary intake, are not easily assessed in epidemiological studies because they are complex, multidimensional behaviors. Often their data collection is performed by basic questionnaires that reduce the complexity of these behaviors to a global self-reported index, which is generally a relatively imprecise measure. In an analysis of the power to detect gene–lifestyle interaction in studies of continuously distributed outcomes, a study of fewer than 10,000 individuals using reasonably accurate measurement of exposure and outcome would be equally well powered to detect an interaction as one of more than 150,000 individuals that used less precise measures (Wong et al. 2003) (Box 11.5). In addition to the measurement precision of the environmental factor and the outcome, the genotyping quality contributes to the power of gene–environment interaction studies. Although, in principle, genetic variation can be measured more precisely and objectively than outcomes and environmental factors, it is important to recognize that genotyping is not completely free of measurement error, particularly high-throughput genotyping, which is still prone to misclassification of individuals despite major progress in technology. Quality control measures such as genotyping success rate, concordance rate of duplicates, and tests for Hardy–Weinberg equilibrium need to be undertaken before analyses. (See Chapters 8 and 16 for more on quality control and error checking for genotype data.)

Box 11.4 Large-Scale Gene–Environment Interaction Studies

The Diogenes Project (http://www.diogenes-eu.org/). An integrated project of the European Union (EU) sixth framework to examine the interaction between genetic variation and diet (glycemic index and protein content) in relation to weight gain. The study is designed as a prospective case–cohort study including 6000 cases (weight gainers) and 6000 cohort individuals ascertained from EPIC (European Prospective Investigation into Cancer and Nutrition) cohorts from five European countries.

InterAct (http://www.inter-act.eu/). An integrated project of the EU sixth framework to examine the interaction between genetic variation and lifestyle (diet and physical activity) in relation to the incidence of type 2 diabetes. The study is designed as a prospective case–control study including 12,000 incident type 2 diabetes cases and 12,000 control individuals ascertained from the EPIC cohorts from ten European countries.

GENEVA Gene Environment Association Studies (http://www.genevastudy.org/). This study is part of the Genetics Program of Genes, Environment, and Health Initiative (GEI) and will be using GWAS to identify genetic variants for common conditions, such as tooth decay, heart disease, cancer, and diabetes, and will also assess their interplay with nongenetic risk factors.

Box 11.5 The Impact of Measurement Precision on Sample Size in Gene–Environment Interaction Studies

Environmental factors such as habitual physical activity and dietary intake are not easily assessed. Particularly in large-scale studies, such as those required for gene–environment interaction analyses, information on environment is often restricted to questionnaire data because objective measurements tend to be expensive and time-consuming in sizable cohorts. However, this turns out to be a vicious circle because measurement precision is critical toward the power of a study. Thus, with more precise measurements, smaller sample sizes are required to obtain the same power as with large studies with less precise measurements.

Box 11.5, Table 1 shows how different levels of measurement precision of environment and outcome affect the sample size required to significantly ($\rho = 10^{-4}$) identify a reasonable interaction ($\beta_1/\beta_2 = 2$) at 95% power for a variant with a minor allele frequency of 20% (Wong et al. 2003). The critical terms are the correlation between the observed and the "true" measure of exposure (ρ_{TY}) and outcome (ρ_{TX}). The "true" measure can be assessed by a calibration study using a gold standard technique.

For example, for studies with poor assessment of both environmental exposure and outcome ($\rho_{TX} = 0.3$ and $\rho_{TY} = 0.4$), a sample size of at least 150,000 individuals would be required to detect the given interaction. However, the same interaction can be identified with less than 15,000 people in studies using slightly more accurate measurements ($\rho_{TX} = 0.6$ and $\rho_{TY} = 0.6$). For comparison, the correlation between physical activity energy expenditure assessed by questionnaires and the gold standard assessment method is only 0.30 (Wareham et al. 2002; Friedenreich et al. 2006).

TABLE 1 Sample size required to identify a given interaction ($\beta_1/\beta_2 = 2$) for different degrees of precision (ρ_{TY}, ρ_{TX}) of a continuously distributed exposure and outcome, with 95% power and a significance level of 10^{-4}

| | | ρ_{TX} | | | | | | |
		0.3	0.4	0.5	0.6	0.7	0.8	0.9
ρ_{TY}	0.4	150,989	84,787	54,116	37,501	27,464	20,950	16,484
	0.5	87,705	49,191	31,364	21,680	15,841	12,051	9,453
	0.6	53,329	29,854	18,988	13,086	9,527	7,217	5,633
	0.7	32,602	18,195	11,526	7,904	5,720	4,302	3,330
	0.8	19,149	10,627	6,683	4,541	3,249	2,410	1,836

Adapted, with permission of Oxford University Press, from Wong et al. (2003).
The parameters fixed in this calculation are power of 95%, α-level 10^{-4}, the minor allele frequency $p = 0.2$, the gene misclassification $P_A = P_a = 0.025$, the interaction $\beta_1/\beta_2 = 2$ (i.e., the association in one-risk allele carriers is two times larger as compared with the association in noncarriers). (P_A and P_a) Probabilities of misclassification of either allele; (α-level) significance level for the intraction.

Taken together, these sample size calculations clearly show the benefit of precisely measured outcomes, environmental factors, and genetic markers. Because objective measurement is generally more precise than self-report questionnaires but less feasible on a mass scale, the question of the appropriate size for studies to examine the combined genetic and lifestyle determinants of disorders such as obesity and type 2 diabetes becomes critical.

REPLICATION OF RESULTS AND META-ANALYSES

Replication of findings in two or more studies is important as a stage toward the acceptance of new observations. The reproducibility of main-effect association findings is generally rather modest or even low; however, replication of gene–environment interaction is likely to be even lower. One of the main reasons for the lack of replication is that the expected effect sizes of association and interaction are generally small in the case of common diseases, such as type 2 diabetes and obesity. Thus, only studies with large sample sizes will have sufficient power to detect these small effects. New large-scale studies specifically designed to examine gene–environment interaction are in progress (see Box 11.4), but the majority of the currently available studies tend to be underpowered for identifying gene–environment interaction and only relatively few examples are available that show robust replication (Hunter et al. 2007).

Ideally, one would combine the available studies in a meta-analysis. However, this proves to be particularly difficult in the case of gene–environment interaction studies. Meta-analysis requires at least some degree of standardization of the three components (genes, environment, and outcome) among studies. This is fairly straightforward for outcome and genetic markers, but it is much harder for the environmental factor. Even when the same genetic marker is studied in relation to the same disease, the environmental factor often differs across studies. Even if the same environmental factor is studied, it is not unusual for it to be measured differently using different measurement tools. The lack of uniformity in the measurement of environmental factors makes it very difficult to perform meta-analyses.

CANDIDATE GENE VERSUS GENOME-WIDE STUDIES

The candidate gene approach is a hypothesis-driven approach that is based on the current understanding of the biology and pathophysiology of the disease of interest. Until recently, this was the main approach for testing associations between a genetic variant and an outcome of interest. Genes that have shown evidence of association with traits in animal models, cellular systems, or extreme/monogenic forms of the disease are tested at the population level. In the context of gene–environment interaction, the candidate gene approach has been the leading approach. For example, variation in the *PPARγ* gene, particularly the Pro12Ala variant, has been a well-studied candidate gene in gene–environment interaction studies in relation to the risk of type 2 diabetes not only because this gene plays a key role in adipocyte differentiation and forms a target of thiazoladinediones, but also because it is responsive to dietary fat at the molecular level (Nakano et al. 2006). A clear a priori interaction hypothesis that is plausible at the biological level reduces the risk of false-positive results. Yet, although the a priori hypothesis can be formulated at the gene level, it often cannot be narrowed down to a single genetic variant. Therefore, one should always remain cautious regarding false-positive findings, especially for genes with a high degree of variation.

Recently, genome-wide association has become available as an approach for identifying genetic variants associated with disease. Genome-wide association is a hypothesis-generating approach that aims to identify new, unanticipated genetic variants associated with disease or traits through screening of the whole genome. This approach has proved to be very successful and continues to identify new genetic variants that were not previously linked to common disease (Manolio et al. 2008). These promising findings have raised the question of whether genome-wide association can be used to identify gene–environment interaction without prior biological knowledge. This may not be that straightforward. Main-effect GWAS for common diseases or traits, such as type 2 diabetes or obesity, already require tens of thousands of individuals combined in meta-analyses to discover new genetic variants that, as expected, have small effects (Loos et al. 2008; Zeggini et al. 2008). Given that gene–environment interaction studies require larger sample sizes than main-effect studies and that combining this type of study in meta-analyses would require standardization, especially of the environmental exposure, the use of the genome-wide association approach to identify gene–environment interaction is a difficult proposition. Main-effect GWAS yield a large number of false positives, but genome-wide gene–environment interaction studies that are not properly designed have the potential to produce an even larger number of false positives.

For now, the value of genome-wide association is in relation to the identification of new genetic variants that are unequivocally associated with disease. Subsequently, these variants can be tested for gene–environment interaction as, for example, for the *FTO* variant (see Box 11.3) (Andreasen et al. 2008; Rampersaud et al. 2008).

SUMMARY AND CONCLUSIONS

A key objective of gene–environment interaction studies is to advance our understanding of physiological pathways that underlie disease. Ultimately, understanding gene–environment interactions might allow personalized recommendations about prevention or treatment of disease. Despite a growing evidence base, consistent data are currently still lacking, thus preventing us from making firm recommendations. Large well-designed studies with precisely measured environmental exposures that integrate existing knowledge of molecular biology will be needed to build the foundation on which gene-directed prevention or therapy will be based.

Designing and conducting gene–environment interaction studies hold considerable challenges. To reduce the high probability of false-positive results, studies should be specifically designed to investigate for interaction and tests should be based on a priori hypotheses that are supported by biological evidence. Alternatively, genetic variants detected through GWAS that are robustly associated with disease, yet with unknown biological function, could be tested for gene–environment interaction in an attempt to expand our understanding about their physiological role at the population level.

Designing large cohorts with follow-up data for disease end points holds considerable potential as a framework for establishing future nested case–control studies. Yet, the ability to detect gene–environment interactions will be critically dependent on the measurement precision of all components in the analyses, but, in particular, the environmental exposure. Sample size and measurement precision both contribute to the power in an interdependent way. One can invest time and money in a very large study, but the limitation of imprecise measurement of the environmental factors may prove difficult. Vice versa, one can focus on the precise and expensive measurement of the exposure, but you will have the advantage of sample size that is smaller for a given power. The choice between sample size and measurement precision will be largely driven by the availability and feasibility of objective measures for the given exposure.

Replication of results will be even more important in gene–environment interaction studies than in main-effect studies because the larger number of tests makes these types of studies more prone to false-positive findings. Some degree of coordination will be required to standardize measurements to increase the compatibility between studies allowing the large-scale meta-analyses needed to identify small effect sizes.

False-negative results and publication bias are of particular concern in gene–environment interaction studies. Results of gene–environment interactions are often only reported as add-ons to main-effect genetic papers. However, if not significant, they are likely to be left out, creating the potential for publication bias.

Although the challenges in undertaking gene–environment studies are considerable, new large-scale and well-designed cohorts are under way and promise to address unanswered questions regarding the intricate interactions between genes and environment in the development of disease.

REFERENCES

Andreasen, C.H., Stender-Petersen, K.L., Mogensen, M.S., Torekov, S.S., Wegner, L., Andersen, G., Nielsen, A.L., Albrechtsen, A., Borch-Johnsen, K., Rasmussen, S.S., et al. 2008. Low physical activity accentuates the effect of the FTO rs9939609 polymorphism on body fat accumulation. *Diabetes* **57:** 264–268.

Bouchard, C., Tremblay, A., Despres, J.P., Nadeau, A., Lupien, P.J., Theriault, G., Dussault, J., Moorjani, S., Pinault, S., and Fournier, G. 1990. The response to long-term overfeeding in identical twins. *N. Engl. J. Med.* **322:** 1477–1482.

Bouchard, C., Tremblay, A., Despres, J.P., Theriault, G., Nadeau, A., Lupien, P.J., Moorjani, S., Prud'homme, D., and Fournier, G. 1994. The response to exercise with constant energy intake in identical twins. *Obes. Res.* **2:** 400–410.

Cauchi, S., El Achhab, Y., Choquet, H., Dina, C., Krempler, F., Weitgasser, R., Nejjari, C., Patsch, W., Chikri, M., Meyre, D., et al. 2007. TCF7L2 is reproducibly associated with type 2 diabetes in various ethnic groups: A global meta-analysis. *J. Mol. Med.* **85:** 777–782.

Clayton, D. and McKeigue, P.M. 2001. Epidemiological methods for studying genes and environmental factors in complex diseases. *Lancet* **358:** 1356–1360.

Diabetes Prevention Program Research Group. 2002. Reduction in the incidence of Type 2 diabetes with lifestyle intervention or Metformin. *N. Engl. J. Med.* **346:** 393–403.

Esparza, J., Fox, C., Harper, I.T., Bennett, P.H., Schulz, L.O., Valencia, M.E., and Ravussin, E. 2000. Daily energy expenditure in Mexican and USA Pima indians: Low physical activity as a possible cause of obesity. *Int. J. Obes. Relat. Metab. Disord.* **24:** 55–59.

Florez, J.C., Jablonski, K.A., Bayley, N., Pollin, T.I., de Bakker, P.I.W., Shuldiner, A.R., Knowler, W.C., Nathan, D.M., Altshuler, D., and The Diabetes Prevention Program Research Group. 2006. TCF7L2 polymorphisms and progression to diabetes in the Diabetes Prevention Program. *N. Engl. J. Med.* **355:** 241–250.

Frayling, T.M., Timpson, N.J., Weedon, M.N., Zeggini, E., Freathy, R.M., Lindgren, C.M., Perry, J.R.B., Elliott, K.S., Lango, H., Rayner, N.W., et al. 2007. A common variant in the *FTO* gene is associated with body mass index and predisposes to childhood and adult obesity. *Science* **316:** 889–894.

Friedenreich, C.M., Courneya, K.S., Neilson, H.K., Matthews, C.E., Willis, G., Irwin, M., Troiano, R., and Ballard-Barbash, R. 2006. Reliability and validity of the Past Year Total Physical Activity Questionnaire. *Am. J. Epidemiol.* **163:** 959–970.

Grant, S.F.A., Thorleifsson, G., Reynisdottir, I., Benediktsson, R., Manolescu, A., Sainz, J., Helgason, A., Stefansson, H., Emilsson, V., Helgadottir, A., et al. 2006. Variant of transcription factor 7-like 2 (*TCF7L2*) gene confers risk of type 2 diabetes. *Nat. Genet.* **38:** 320–323.

Hainer, V., Stunkard, A.J., Kunesova, M., Parizkova, J., Stich, V., and Allison, D.B. 2000. Intrapair resemblance in very low calorie diet-induced weight loss in female obese identical twins. *Int. J. Obes. Relat. Metab. Disord.* **24:** 1051–1057.

Hunter, D.J., Kraft, P., Jacobs, K.B., Cox, D.G., Yeager, M., Hankinson, S.E., Wacholder, S., Wang, Z., Welch, R., Hutchinson, A., et al. 2007. A genome-wide association study identifies alleles in FGFR2 associated with risk of sporadic postmenopausal breast cancer. *Nat. Genet.* **39:** 870–874.

Ioannidis, J.P.A., Gwinn, M., Little, J., Higgins, J.P.T., Bernstein, J.L., Boffetta, P., Bondy, M., Bray, M.S., Brenchley, P.E., Buffler, P.A., et al. 2006. A road map for efficient and reliable human genome epidemiology. *Nat. Genet.* **38:** 3–5.

Khoury, M.J. and Flanders, W.D. 1996. Nontraditional epidemiologic approaches in the analysis of gene–environment interaction: Case-control studies with no controls! *Am. J. Epidemiol.* **144:** 207–213.

Kurokawa, N., Young, E.H., Oka, Y., Satoh, H., Wareham, N.J., Sandhu, M.S., and Loos, R.J.F. 2008. The ADRB3 Trp64Arg variant and BMI: A meta-analysis of 44,833 individuals. *Int. J. Obes.* **32:** 1240–1249.

Lissner, L. 2002. Measuring food intake in studies of obesity. *Public Health Nutr.* **5:** 889–892.

Loos, R.J., Lindgren, C.M., Li, S., Wheeler, E., Zhao, J.H., Prokopenko, I., Inouye, M., Freathy, R.M., Abecasis, G.R., Albai, G., et al. 2008. Common variants near MC4R are associated with fat mass, weight and risk of obesity. *Nat. Genet.* **40:** 768–775.

Luan, J.A., Wong, M.Y., Day, N.E., and Wareham, N.J. 2001. Sample size determination for studies of gene–environment interaction. *Int. J. Epidemiol.* **30:** 1035–1040.

Ludovico, O., Pellegrini, F., Di, P.R., Minenna, A., Mastroianno, S., Cardellini, M., Marini, M.A., Andreozzi, F., Vaccaro, O., Sesti, G., et al. 2007. Heterogeneous effect of peroxisome proliferator-activated receptor γ2 Ala12 variant on type 2 diabetes risk. *Obesity* **15:** 1076–1081.

Maes, H.H., Neale, M.C., and Eaves, L.J. 1997. Genetic and environmental factors in relative body weight and human obesity. *Behav. Genet.* **27:** 325–351.

Manolio, T.A., Brooks, L.D., and Collins, F.S. 2008. A HapMap harvest of insights into the genetics of common disease. *J. Clin. Invest.* **118:** 1590–1605.

Mokdad, A.H., Marks, J.S., Stroup, D.F., and Gerberding, J.L. 2004. Actual causes of death in the United States, 2000. *JAMA* **291:** 1238–1245.

Nakano, R., Kurosaki, E., Yoshida, S., Yokono, M., Shimaya, A., Maruyama, T., and Shibasaki, M. 2006. Antagonism of peroxisome proliferator-activated receptor [gamma] prevents high-fat diet-induced obesity in vivo. *Biochem. Pharmacol.* **72:** 42–52.

Piegorsch, W.W., Weinberg, C.R., and Taylor, J.A. 1994. Non-hierarchical logistic models and case-only designs for assessing susceptibility in population-based case-control studies. *Stat. Med.* **13:** 153–162.

Poulsen, P., Kyvik, K.O., and Beck-Nielsen, H. 1999. Heritability of Type II (non-insulin-dependent) diabetes mellitus and abnormal glucose tolerance—A population based twin study. *Diabetologia* **42:** 139–145.

Rampersaud, E., Mitchell, B.D., Pollin, T.I., Fu, M., Shen, H., O'Connell,

J.R., Ducharme, J.L., Hines, S., Sack, P., Naglieri, R., et al. 2008. Physical activity and the association of common FTO gene variants with body mass index and obesity. *Arch. Intern. Med.* **168:** 1791–1797.

Ravussin, E. and Bouchard, C. 2000. Human genomics and obesity: Finding appropriate drug targets. *Eur. J. Pharmacol.* **410:** 131–145.

Ravussin, E., Valencia, M.E., Esparza, J., Bennett, P.H., and Schulz, O. 1994. Effects of a traditional lifestyle on obesity in Pima Indians. *Diabetes Care* **17:** 1067–1074.

Sackett, D.L. 1979. Bias in analytic research. *J. Chronic. Dis.* **32:** 51–63.

Scuteri, A., Sanna, S., Chen, W.-M., Uda, M., Albai, G., Strait, J., Najjar, S., Nagaraja, R., Orru, M., Usala, G., et al. 2007. Genome-wide association scan shows genetic variants in the FTO gene are associated with obesity-related traits. *PLos Genet.* **3:** e115.

Smith, P.G. and Day, N.E. 1984. The design of case-control studies: The influence of confounding and interaction effects. *Int. J. Epidemiol.* **13:** 356–365.

Thomas, D.C. 2004. *Statistical methods in genetic epidemiology.* Oxford University Press, Oxford.

von Elm, E. and Egger, M. 2004. The scandal of poor epidemiological research. *BMJ* **329:** 868–869.

Wang, J., Kuusisto, J., Vanttinen, M., Kuulasmaa, T., Lindstrom, J., Tuomilehto, J., Uusitupa, M., and Laakso, M. 2007. Variants of transcription factor 7-like 2 (*TCF7L2*) gene predict conversion to type 2 diabetes in the Finnish Diabetes Prevention Study and are associated with impaired glucose regulation and impaired insulin secretion. *Diabetologia* **50:** 1192–1200.

Wareham, N.J., Jakes, R.W., Rennie, K.L., Mitchell, J., Hennings, S., and Day, N.E. 2002. Validity and repeatability of the EPIC-Norfolk Physical Activity Questionnaire. *Int. J. Epidemiol.* **31:** 168–174.

Wareham, N.J., van Sluijs, E.M., and Ekelund, U. 2005. Physical activity and obesity prevention: A review of the current evidence. *Proc. Nutr. Soc.* **64:** 229–247.

Wong, M.Y., Day, N.E., Luan, J.A., Chan, K.P., and Wareham, N.J. 2003. The detection of gene-environment interaction for continuous traits: Should we deal with measurement error by bigger studies or better measurement? *Int. J. Epidemiol.* **32:** 51–57.

Young, E.H., Wareham, N.J., Farooqi, S., Hinney, A., Hebebrand, J., Scherag, A., O'Rahilly, S., Barroso, I., and Sandhu, M.S. 2007. The V103I polymorphism of the *MC4R* gene and obesity: Population based studies and meta-analysis of 29,563 individuals. *Int. J. Obes.* **31:** 1437–1441.

Zeggini, E., Scott, L.J., Saxena, R., Voight, B.F., Marchini, J.L., Hu, T., de Bakker, P.I., Abecasis, G.R., Almgren, P., Andersen G., et al. 2008. Meta-analysis of genome-wide association data and large-scale replication identifies additional susceptibility loci for type 2 diabetes. *Nat. Genet.* **40:** 638–645.

WWW RESOURCES

http://www.diogenes-eu.or   The Diogenes Project.
http://www.genevastudy.org   GENEVA (Gene Environment Association) Studies.

http://www.inter-act.eu   InterAct.

12 | Family-Based Genetic Association Tests

Eden R. Martin and Evadnie Rampersaud

Miami Institute for Human Genomics, Leonard M. Miller School of Medicine, University of Miami, Miami, Florida 33136

INTRODUCTION

Genetic studies in families have historically focused on tests of linkage. These tests measure the concordance between the pattern of segregation of a genetic marker allele with the occurrence of a disease phenotype within families and provide information about the amount of genetic recombination between a marker locus and a putative disease locus. Thus, linkage is a statement about transmission of gametes from parents to offspring. In contrast, association is a statement about the population of gametes. Frequencies of marker alleles in disease and nondisease individuals give us information about the amount of association, or correlation, between marker allele and disease status. Family data not only provide us information about linkage by looking at transmissions within families, but also give us information about association by looking at allelic state and the presence of disease phenotype across families. This chapter describes the evolution of family-based tests of association as well as the variety of applications for these tests.

WHAT IS THE RATIONALE FOR CONDUCTING FAMILY-BASED ASSOCIATION STUDIES?

Case–control designs using unrelated individuals are widely used in epidemiologic studies to test for association between a health-related feature, such as exposure to an environmental insult, and disease status. It is well known that cases and controls need to be matched on potential confounders, such as sex or age; otherwise spurious results can emerge. In genetic studies, the features are genetic markers and the primary confounder is the population of origin. The country, region, or ethnic group (e.g., Japanese or Ashkenazi Jewish) that an individual comes from may be associated with both marker genotype and disease status, even when there is no association between marker and disease status within each population, making tests for association in the pooled group invalid unless the confounding factor is taken into account. We use "invalid" in the statistical sense to mean failure to maintain the correct type I error under the null hypothesis (i.e., rejecting the null hypothesis when the null hypothesis is true).

The following examples illustrate how controlling for population of origin is critical for the validity of association tests in case–control samples. Two problems can arise in these studies—population stratification and population admixture. We worry about hidden "population stratification" (i.e., when what we believe to be a single interbreeding population is in fact composed of historically differentiated subpopulations). Population stratification can lead to spurious associations when disease prevalence and allele frequencies differ between subpopulations. An example of the effect of population stratification resulting in spurious associations of the DRD2 A1 allele with alcohol dependence is illustrated in Figure 12.1. The A1 allele is more frequent in Native Americans compared with Caucasians (Fig. 12.1A) (Kidd et al. 1998), and alcohol dependence is more prevalent among certain Native American groups than among Caucasians (Fig. 12.1B). Hence a spurious association between the DRD2 A1 allele and alcohol dependence could be observed in a stratified sample even though there was no association within each stratum (Fig. 12.1C).

A related problem is "population admixture," which can result in associations, even at loci unassociated with disease in the mixing populations, when the degree of admixture differs in cases and controls. An example of population admixture comes from a study of type 2 diabetes mellitus in Native Americans from Pima, Arizona, in which a strong negative association with the GM haplotype Gm3;5,13,14 from the Gm system of human immunoglobin was found to be confounded by the effect of Caucasian admixture (Knowler et al. 1988). The crude odds ratio (OR) for the association between Gm3;5,13,14 and diabetes in a sample of 4920 Native Americans was 0.21 (95% confidence interval [CI] = 0.14–0.32) and was statistically significant, suggesting that the absence of this haplotype (or combination of alleles at multiple loci that are transmitted together on the same chromosome) is a risk factor in the disease. However, Gm3;5,13,14 is also a marker for Caucasian admixture, and therefore the degree of Native American heritage is a potential confounder for this association. Stratum-specific ORs, according to the degree of Native American heritage—either none, half, or full—revealed that there was no significant relationship between Gm3;5,13,14 and diabetes mellitus. The stratum-specific ORs for full, half, and no Native American heritage were 0.63, 0.69, and 0.75, respectively. Further adjustments for confounding by age resulted in an even weaker association between the marker and disease within these strata. Therefore, the relationship between Gm3;5,13,14

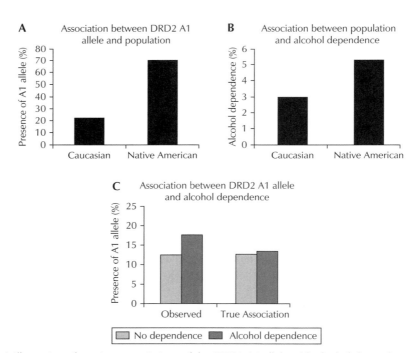

FIGURE 12.1 Illustration of spurious associations of the DRD2 A1 allele with alcohol dependence. An increased frequency of the DRD2 A1 allele in Native Americans compared with Caucasians (*A*), combined with an increased frequency of alcohol dependence among Native Americans compared with Caucasians (*B*), can result in spurious observed associations between the DRD2 A1 allele and alcohol dependence in the stratified sample even though there was no association within each stratum (*C*). (Adapted, with permission, from Hutchison et al. 2004.)

and diabetes mellitus was explained by the inverse relationship between the genetic marker and Caucasian admixture (Knowler et al. 1988).

In contrast, family-based tests of association use family-based controls and avoid the issues of matching inherent in traditional case–control studies. Compared with the case–control method, the family-based study design has several advantages. First, the added family information provides a means for reducing genotyping errors because genotypes of family members can be compared for consistency, although care must be taken not to introduce bias (Gordon et al. 2001). Second, it can provide data for a test for both linkage and association. Third, maternal and paternal genotype effects can also be detected using family-based methods, as well as maternal–fetal genotype interaction effects. Despite the fact that family-based tests generally require more individuals than case–control tests for comparable power (e.g., a family triad provides a matched pair), the protection in the presence of population structure and other advantages was enough to offset concerns of power and lead to extensive developments in this area.

EARLY FAMILY-BASED TESTS FOR ASSOCIATION WERE ALTERNATIVES TO CASE–CONTROL TESTS

The first family-based tests of association were developed as alternatives to traditional case–control tests to alleviate concerns that cases and controls may not come from the same population. The solution, first proposed by Rubinstein and colleagues in 1981, was quite simply to match cases to within-family controls. In its earliest form, the families took the form of a single affected and two parents (family triad, or trio) (Rubinstein et al. 1981). The "case" genotype is then the pair of alleles transmitted from parents to the affected offspring and the "control" genotype is the pair of alleles that was not transmitted from the parents (Fig. 12.2). Rubinstein et al. (1981) and later Falk and Rubinstein (1987) defined the haplotype relative risk (HRR), which compares transmitted marker alleles passed from parents to an affected child with the nontransmitted alleles that were not passed to the child. The HRR is equivalent to the relative risk (RR), a commonly used measure that is often computed in case–control studies and is defined by the probability of an event occurring in one group (e.g., the GG genotype group) versus another (e.g., the AA genotype group). The test based on the HRR was later extended to the more powerful haplotype-based HRR test (HHRR), which considers transmissions from each parent separately (Terwilliger and Ott 1992). The affected family-based controls test (AFBAC), which was developed in 1995 by Thomson (1995), generalized the HHRR to markers with multiple alleles. This was particularly important because microsatellite markers were the predominant genetic marker used in genetic studies at that time.

A $$TDT = \frac{(b-c)^2}{b+c}$$

Where *b* is the number of times a parent with genotype 12 transmits allele 1 to an affected offspring and *c* is the number of times a parent with the 12 genotype transmits allele 2 to an affected offspring

B

	Transmitted	Not Transmitted	
		Allele 1	Allele 2
Allele 1		0	2
Allele 2		0	0

FIGURE 12.2 Illustration of the transmission disequilibrium test (TDT) statistic (*A*) and an example of scoring a TDT family (*B*).

The HRR, HHRR, and AFBAC tests correctly remove bias that may be introduced by inappropriate matching of cases and controls; however, the form of the variance estimator is valid only under random mating (e.g., Hardy–Weinberg equilibrium [HWE]), which means that the statistics can still have an inflated type I error in stratified populations (Spielman and Ewens 1996). Furthermore, each test was designed specifically as a test of association, not of linkage. If either is used as an indirect test of linkage, it will give a false-positive result more frequently than the rate allowed by the nominal type I error (false-positive rate) (Ewens and Spielman 1995; Spielman and Ewens 1996). These properties follow from the mathematical model and insights with respect to the HRR (Ott 1989), which is often cited as the point of departure for the development of the transmission disequilibrium test (TDT) (Spielman and Ewens 1996).

THE TRANSMISSION DISEQUILIBRIUM TEST IS A TEST OF LINKAGE

Unlike the previous family-based tests that focused on testing for association, the TDT was originally designed as a test of linkage between a genetic marker and a disease locus. Specifically, the TDT was designed to verify that associations that had been found were maintained by linkage and not the result of some other force such as population stratification. The development of the TDT by Spielman and colleagues (Spielman et al. 1993) was motivated by the search for insulin-dependent diabetes mellitus (IDDM) genes. From the example given in their paper (Spielman et al. 1993), several studies had found significant evidence of association with the class 1 alleles of a polymorphism in the 5′ region of an insulin gene on 11p in unrelated case–control samples, yet repeated attempts to identify linkage in affected sib-pair (ASP) samples failed. The question then arose as to whether there really was linkage or whether the case–control tests were detecting spurious associations.

As a test of linkage, the TDT can use family triads, ASP families, or even extended families. Like the tests discussed in the previous section, the TDT distinguishes between alleles transmitted and alleles not transmitted from parents to affected offspring, but unlike those tests, the TDT considers only transmissions from heterozygous parents. Conditioning on heterozygous parents is how the test remains valid in stratified populations. Figure 12.2 illustrates the calculations of the TDT. The key observation is that when there is no linkage, a heterozygous parent is equally likely to transmit either allele to an affected offspring. If there is linkage, then an allele in linkage disequilibrium (LD) with the disease allele will tend to be overtransmitted to the affected offspring. Note that the TDT requires association (LD) between the disease and marker allele to have power to detect linkage. For this reason, the TDT is often referred to as a test of linkage in the presence of association.

Spielman and colleagues then applied the TDT to the IDDM data. The data set was a mixture of triad and ASP families. From 124 heterozygous parents, they observed that the class 1 alleles were transmitted 78 times, whereas other alleles were transmitted only 46 times. Thus, there was excess transmission of the class 1 alleles to affected offspring. The p value for the TDT was 0.004, suggesting significant linkage. In contrast, the mean haplotype-sharing test of linkage based on identity by descent (IBD) in ASP families was not significant, with a p value of 0.20. This example illustrates some advantages of the TDT. First, the test can make use of family triad data, which is not possible with traditional linkage methods based on IBD. Second, even if we have all ASP data available, the TDT can have more power to detect linkage, particularly for disease genes with small effects, as we expect in complex diseases (Risch and Merikangas 1996).

THE TRANSMISSION DISEQUILIBRIUM TEST IS A TEST OF ASSOCIATION AND LINKAGE

Although the TDT was originally proposed as a test of linkage in the presence of association, researchers quickly decided that they actually were more interested in detecting association in the presence of linkage. The reason was that until only recently with the emergence of genome-wide association studies (GWAS), a common paradigm was to test for linkage in multiplex family data and then fine-map in regions of linkage with tests of association. Regions of linkage identified through genome screens were typically quite large, often containing hundreds to thousands of genes. There was a need to examine specific candidate genes, but linkage methods often lacked the resolution to discriminate at that level. Association approaches, which require LD between marker and disease loci, have much better resolution and so are more suitable for candidate gene analysis. With the emergence of family-based tests that, like the TDT, depend on LD, investigators recognized the potential to test for association in the same data set used to detect linkage, which could be important in reducing genetic heterogeneity.

In family triads, it is not difficult to show that the TDT follows the appropriate asymptotic distribution when there is either no linkage or no association. Thus, the test really provides a simultaneous test for both linkage and association. This is not generally true, however, when families consist of multiple affected individuals. Specifically, the TDT is not valid as a test of association in the presence of linkage in multiplex families. To see why this is true, consider the following simple example. Suppose that we have a sample of ASP families and that disease is caused by a rare dominant disease allele (D). Because the allele is rare, we can assume that there is exactly one parent heterozygous at the disease locus and this parent must transmit the disease allele to both affected offspring (Fig. 12.3A). Now consider a marker locus (with alleles M_1 and M_2) that is tightly linked to the disease

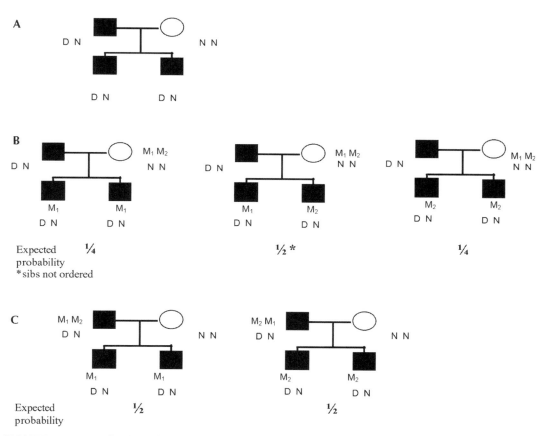

FIGURE 12.3 Example of correlation in transmissions when there is linkage and no association. (*A*) A rare dominant disease allele (D) in ASP families has one heterozygous parent and the disease allele transmitted to both offspring. (*B*) Types of families seen if a marker (with alleles M_1 and M_2) is heterozygous in the parent homozygous for the non-disease allele (N) and associated probabilities. (*C*) Types of families seen if a marker is heterozygous in the parent who is heterozygous at the disease locus and associated probabilities.

locus but not in linkage disequilibrium with the disease allele (the null hypothesis of no association in the presence of linkage). A nontechnical way of visualizing this is to consider that the disease and marker loci are physically very close but not in LD. Only parents who are heterozygous at the marker contribute to the TDT. If hypothetically a parent homozygous at the disease locus is heterozygous at the marker, then clearly either marker allele is equally likely to be transmitted to each affected offspring and the transmissions to the offspring would be independent (Fig. 12.3B). On the other hand, if a parent were heterozygous at the disease locus and heterozygous at the marker, then we expect the two types of families seen in Figure 12.3C. Either marker-disease allele phase is equally likely, but then whichever allele is coupled with the disease allele in the parent is necessarily transmitted to both affected offspring. Thus, in this case, we still expect either marker allele to be transmitted with the same probability on average across the sample, but the transmissions of the marker allele within the family are correlated. It is the correlation between transmissions to affected siblings illustrated in this example that invalidates the TDT statistic as a test of association in regions of linkage, giving rise to excess type I error (i.e., the error of rejecting a null hypothesis when it is actually true).

Because traditional linkage methods require multiplex families, researchers often have available not just family triads but ASP families or even larger pedigrees. Initial suggestions to use a single triad from each family for fine mapping with association tests (Spielman and Ewens 1996) were unsatisfying because this approach omits a fraction of informative offspring. The key was to recognize the need to treat the entire family as the sampling unit. The Tsp statistic (Martin et al. 1997) amended the variance estimator of the TDT statistic to adjust appropriately for the correlation in ASP families. In such families, this statistic reduces to a simple form that depends only on families in which marker allele transmissions to the two affected siblings are concordant. The test maintains the correct type I error when there is no association between marker and disease alleles, regardless of the level of linkage, and can be considerably more powerful than using the TDT with only a single triad chosen per family (Martin et al. 1997). The Tsp test was later generalized to the pedigree disequilibrium test (PDT) (Martin et al. 2000, 2001), which uses an empirical estimator of the variance to provide a valid test in general pedigrees. The PDT is discussed further in the following sections.

FAMILY-BASED ASSOCIATION TESTS HAVE A WIDE RANGE OF APPLICATION

Tests for Late-Onset Diseases

When diseases with onset in adulthood or old age are studied (e.g., Alzheimer's disease, type 2 diabetes, amyotrophic lateral sclerosis), it may be impossible to obtain genotypes for markers in the parents of the affected offspring. This difficulty limited the applicability of the TDT and prompted the development of the "sib TDT" (or S-TDT) method (Spielman and Ewens 1998), which overcomes this problem by using marker data from unaffected siblings instead of parents, thus allowing application of the principle of the TDT to sibships without parental data.

The data appropriate for the S-TDT consist of marker genotypes for sibships that meet two requirements: (1) There must be at least one affected and one unaffected member (a discordant sibling pair [DSP]) in the sibship, and (2) the members of the sibship must not all have the same genotype. The units of observation are the marker genotypes of the offspring, affected and unaffected, for each family. In essence, the S-TDT determines whether the marker allele frequencies among affected offspring differ significantly from the frequencies in their unaffected siblings. Because the comparison between affected and unaffected individuals is performed within families, disease association without linkage will not result in such differences; the marker allele frequencies will, apart from random sampling effects, be the same in the affected and unaffected siblings unless linkage is also present (including the case in which the marker itself is responsible). Thus, the null hypothesis tested by the S-TDT is that disease and marker are unlinked. Like the TDT, the S-TDT can be considered generally as a test of linkage in the presence of association.

Comparing genotype/allele frequencies between affected and unaffected siblings directly is one approach for analyzing families with missing parental data. Alternatively, methods have been developed to estimate the expected parental genotypes given the available genotypes in the family. The approach implemented in the program TRANSMIT (Clayton 1999) is a partial likelihood-based approach that treats parental allele frequencies as nuisance parameters, which are then used to estimate missing parental genotype probabilities and conduct a TDT-like test for excess transmission of a marker allele to affected offspring. In the presence of missing data and when linkage exists between marker and disease loci, TRANSMIT has been shown to have inflated type I error rates (Martin et al. 2003) because the parental genotype inference assumes that transmissions to multiple affected siblings are independent. The association in the presence of linkage (APL) test is designed to correctly infer missing parental data by taking IBD parameters into account (Martin et al. 2003; Chung et al. 2006). The Family-Based Association Tests (FBAT) is another program that handles missing parental genotypes by conditioning on sufficient statistics (e.g., any partially observed parental genotypes and offspring genotype configurations).

Tests for Association and Linkage in General Nuclear Families and Pedigrees

Most genetic studies rarely end up collecting only one particular type of pedigree (e.g., all triads or sibships), but instead gather combinations of nuclear families, sibships, and extended pedigrees containing multiple nuclear families, with different types of missing data. To allow for this variability in sample composition, powerful tests such as the APL, FBAT, PDT, and UNPHASED have been developed to use the information from these general families, and to account properly for the correlations within pedigrees that arise from linkage between disease and marker loci (Martin et al. 2000, 2001, 2003; Rabinowitz and Laird 2000; Dudbridge 2008). Both FBAT (Rabinowitz and Laird 2000) and PDT (Martin et al. 2000, 2001) are similar in structure. They combine families with parental genotypes with DSP families, and account for the correlation among affected siblings within nuclear families using an empirical variance estimator that treats each nuclear family as a cluster of observations. APL can also combine families with and without parental data and can be more powerful in nuclear families when many families are missing parental genotypes; however, APL is not valid as a test of association when there is linkage in extended pedigrees (Martin et al. 2003; Chung et al. 2006).

A general likelihood–based test, UNPHASED, has been developed to handle combinations of nuclear families of any size as well as unrelated singletons or combinations of the two (Dudbridge 2008). Both UNPHASED and APL have been shown to have higher power compared with FBAT; however, both also have higher type I error rates when there are missing genotypes and in the presence of population stratification (Dudbridge 2008).

None of these tests efficiently handles the computational demands of family-based association analysis of binary traits in very large pedigrees, such as founder populations like the Amish or Hutterites. The MQLS (more powerful quasi-likelihood score test) is an association test of binary traits for large complex pedigrees, which requires that familial relationships be specified using kinship coefficients (and inbreeding coefficients for inbred pedigrees) (Thornton and McPeek 2007). A related approach is implemented in a variance components framework in the SOLAR program (Almasy and Blangero 1998). This method uses a liability threshold model, in which the probability that the individual is an affected or unaffected subject is modeled as a function of the individual's genotype, conditional on the correlations in phenotype among relative pairs, specified in a kinship coefficient matrix.

Tests Using Haplotypes

In addition to the traditional single-locus tests that consider each marker independently, several family-based tests for association that consider marker haplotypes have been developed. These tests look for association between disease status and a particular combination, or combinations, of alleles at different loci (i.e., marker haplotypes), which are expected to provide more information than the markers individually. The key assumption is that association arises because of the presence of a disease allele that occurs initially on one, or few, ancestral haplotypes. The amount of association between the disease allele and a particular marker allele or marker haplotype depends strongly on the frequency of the associated allele or haplotype. The strongest associations are found when the positively associated marker allele or haplotype has a frequency similar to that of the risk allele. Thus, haplotypes may serve as good markers for low-frequency risk alleles. Moreover, an analysis based on haplotypes can be advantageous in the presence of multiple risk mutations (Morris and Kaplan 2002).

One of the difficulties with haplotype association tests is that there can be many haplotypes to consider when we use multiple markers or markers with multiple alleles, and often there is no a priori knowledge of which haplotype might be associated. Haplotypes with low frequency will contribute small numbers of observations to the analysis, and caution must thus be used in applying and interpreting the results of haplotype tests. Often, haplotypes with small numbers can be pooled to provide valid asymptotic tests, but this pooling could reduce power if negatively and positively associated haplotypes are pooled. Typically global as well as individual haplotype tests are produced, but individual haplotype tests are generally valid only with likelihood-based approaches that appropriately model the null hypothesis for a specific haplotype. Multiple testing should be considered in interpreting many tests for individual haplotypes. A number of computer programs have been developed for analysis of haplotypes in families including Haploview (Barrett 2005), HBAT (Horvath et al. 2004), TRANSMIT (Clayton 1999), APL (Chung et al. 2006), and UN-PHASED (Schulze and McMahon 2002; Dudbridge 2008).

Tests for Association with Continuous Quantitative Traits

The development of family-based tests for quantitative traits has paralleled the development for binary traits. Several tests have been proposed for family triads, in which the offspring in the triad has a measured quantitative trait, like body mass index or age-at-onset (Allison 1997; Rabinowitz 1997; Lange et al. 2002; Kistner and Weinberg 2004). The purpose of these tests is to determine whether the distribution of a quantitative trait differs between offspring who receive and who do not receive the marker allele of interest from a heterozygous parent. In families in which multiple siblings are available, correlations between quantitative trait values within families must be taken into account (Monks and Kaplan 2000). Association tests for quantitative traits can also be used when there are missing parental data, provided multiple siblings are available (Allison et al. 1999). Recently, association tests for quantitative traits have been extended to general pedigree structure in the QTDT program using a variance components framework (Abecasis et al. 2000a; Rabinowitz and Laird 2000). For very large pedigrees (i.e., in founder populations such as Amish or Hutterites), a variance components approach implemented in the SOLAR program, in which the familial relationships are specified in a kinship coefficient matrix and entered as a random effect in the regression model, can be used (Almasy and Blangero 1998). This component captures the percent of variation in the trait phenotype that can be explained by the pedigree structure.

Tests for X-Linked Association

The development of family-based association tests for the X chromosome has not moved at the same pace as methods for autosomes. The first such methods proposed for the X chromosome were the XS-TDT and the XRC-TDT, proposed by Horvath and colleagues (Horvath et al. 2000). The XS-TDT compares affected and unaffected siblings, whereas the XRC-TDT also includes parental data when available, and in some cases can reconstruct missing parental genotypes. Like the original TDT, XS-TDT, and XRC-TDT are generally tests of linkage in the presence of association but are not valid tests of association in multiplex families when there is linkage. These approaches were not widely used and there were no further developments for X-chromosome association analysis in families until 2006, when Ding and Lin (Ding et al. 2006) introduced the XPDT, which was followed shortly thereafter by the XAPL proposed by Chung et al. (2007). These are extensions of the autosome-based PDT and APL discussed above for X-chromosome markers. Like the PDT, XPDT is valid as a simultaneous test for association and linkage in general pedigrees (nuclear or extended). Like the APL, XAPL uses IBD estimates to correctly estimate parental genotype probabilities and infer missing parental genotypes, leading to more power than XPDT when parental data are missing, for example, in late-onset diseases.

An alternative to XPDT and XAPL recently proposed for triad data is a likelihood-based method XLRT (Zhang et al. 2008). The test can only consider transmissions to a single affected offspring per family, but can use multiple unaffected siblings to infer genotypes if parents are missing. The advantage of the LRT is that it can flexibly test a variety of different hypotheses about genotypic effects and also provide estimates of genotypic relative risks. For quantitative traits, the XQTL test (Zhang et al. 2009) extends the orthogonal model of Abecasis et al. (2000a) to X-chromosome analysis.

Tests of Parent-of-Origin and Maternal Effects

With families, one can test whether transmission of disease-susceptibility alleles is disproportionate and biased toward one parent. For example, distorted maternal transmission to affected offspring suggests differential expression of the maternally derived allele compared with the paternally derived allele (this can occur through genetic imprinting). One can also test whether maternal genotypes contribute to offspring disease risk—for example, through some sort of maternal–fetal genotype interaction in utero. This is not something that can be tested in case–control studies.

An example of imprinting comes from a study of transient neonatal diabetes mellitus (TNDM), in which an association was found between TNDM and paternally inherited, but not maternally inherited, duplications of chromosome 6q24 (Cave et al. 2000). The critical overlapping duplicated region was later found to comprise 440 kb (Gardner et al. 2000) and contained two imprinted genes, *ZAC* and *HYMAI* (Arima et al. 2000). Traditional association methods would not have identified this association if the parent of origin had not been taken into account.

Weinberg and colleagues developed a generalized form of a log-linear model to test appropriately for maternal genotype effects as well as maternal or paternal parent-of-origin effects (POEs) (Weinberg et al. 1998; Weinberg 1999a,b). The expectation–maximization (E-M) algorithm is implemented in this test to handle missing parental genotypes or ambiguity of parental origin of the variant allele inherited by children in families in which all individuals are heterozygous (Weinberg 1999a,b; Rampersaud et al. 2007). Sinsheimer and colleagues (Sinsheimer et al. 2003) further proposed a maternal–fetal incompatibility test that determines whether a combination of maternal–fetal genotypes adversely affects the development of the fetus (Sinsheimer et al. 2003). Conditional logistic regression-based methods including the conditional on parental genotypes (CPG) and extended conditional exchangeable on parental genotypes (CEPG) models were also developed to handle multiallelic markers, multiple linked loci, and multiple linked loci in multiple unlinked regions (Cordell et al. 2004). For quantitative traits, methods for testing POEs and maternal effects have been outlined by Kistner and Weinberg (2004; Kistner et al. 2006) and by Wheeler and Cordell (2007).

Testing for Gene–Gene and Gene–Environment Interactions

Complex diseases are believed to involve multiple genetic and environmental factors that either act independently or interact to influence disease susceptibility. Tests for gene–gene and gene–environment interactions can help shed light on these complex relationships.

One approach to evaluating gene–gene or gene–environment interactions is to use conditional logistic regression (CLR) or generalized estimating equations (GEEs), which allow correlations between related family members (Schaid 1999). Not only can a family-based design protect against false positives caused by population stratification, but it is potentially more efficient than traditional association methods for estimating gene–environment interactions, particularly when the genetic factor is rare (Witte et al. 1999). When multiple affected or unaffected siblings from the same family are used, the variance adjustment proposed by Siegmund can be used to correct for correlations among affected siblings caused by linkage (Siegmund et al. 2000). D. Hancock and colleagues (Hancock et al. 2007) performed simulations to show that although the GEEs provided increased power over the CLR model to detect gene–gene or gene–environment interactions, it had inflated type I error rates when there was association but no linkage (e.g., because of population stratification). A related method using log-linear models has been proposed by Umbach and Weinberg (2000) to test for gene–environment interactions in family triads (Umbach and Weinberg 2000).

To detect higher-order interactions (such as three-locus gene–gene interactions, or interactions between genes and multiple environmental factors), traditional methods are limited because of the large number of interaction terms that need to be estimated. To overcome this limitation, computationally intensive methods have been proposed (e.g., combinatorial partitioning method [CPM], multifactor dimensionality reduction [MDR], and machine-learning methods). One such method, the MDR-PDT (Martin et al. 2006), targets family-based designs by incorporating the PDT statistic into the MDR method (Ritchie et al. 2001). This approach uses data-reduction techniques to improve power to detect high-order interactions. An extension of the MDR, called MDR-Phenomics, was introduced to deal with genetic heterogeneity in pedigree data by integrating discrete phenotypic covariates (Mei et al. 2005, 2007). Current challenges for MDR and similar approaches are (1) the computational burden is expensive when many loci are considered simultaneously; (2) splitting samples into testing and training data sets is likely to introduce many empty cells in contingency tables, especially for high-order interactions; and (3) the interpretation of high-order interaction results may be difficult.

SUMMARY

During the past 20 years, we have seen increases in methods development and applications of family-based tests of association. Families will always form the crux of genetic studies, and although we may see the popularity of family-based association tests ebb and flow, they will always have a place complementary to case–control tests. Both family-based and case–control studies offer advantages. Particularly with the challenges of GWAS analysis, requiring large consortia, family-based and case–control samples may need to be pooled to take full advantage of available data.

ACKNOWLEDGMENTS

We are grateful to Drs. Richard Morris, Todd Ewards, and Adam Naj for their helpful review of this chapter.

REFERENCES

Abecasis, G.R., Cardon, L.R., and Cookson, W.O. 2000a. A general test of association for quantitative traits in nuclear families. *Am. J. Hum. Genet.* **66:** 279–292.

Abecasis, G.R., Cookson, W.O., and Cardon, L.R. 2000b. Pedigree tests of transmission disequilibrium. *Eur. J. Hum. Genet.* **8:** 545–551.

Allison, D.B. 1997. Transmission-disequilibrium tests for quantitative traits. *Am. J. Hum. Genet.* **60:** 676–690.

Allison, D.B., Heo, M., Kaplan, N., and Martin, E.R. 1999. Sibling-based tests of linkage and association for quantitative trials. *Am. J. Hum. Genet.* **64:** 1754–1764.

Almasy, L. and Blangero, J. 1998. Multipoint quantitative-trait linkage analysis in general pedigrees. *Am. J. Hum. Genet.* **62:** 1198–1211.

Arima, T., Drewell, R.A., Oshimura, M., Wake, N., and Surani, M.A. 2000. A novel imprinted gene, *HYMAI*, is located within an imprinted domain on human chromosome 6 containing *ZAC*. *Genomics* **67:** 248–255.

Barrett, J.C., Fry, B., Maller, J., and Daly, M.J. 2005. Haploview: Analysis and visualization of LD and haplotype maps. *Bioinformatics* **21:** 263–265.

Cave, H., Polak, M., Drunat, S., Denamur, E., and Czernichow, P. 2000. Refinement of the 6q chromosomal region implicated in transient neonatal diabetes. *Diabetes* **49:** 108–113.

Chung, R.H., Hauser, E.R., and Martin, E.R. 2006. The APL test: Extension to general nuclear families and haplotypes and the examination of its robustness. *Hum. Hered.* **61:** 189–199.

Chung, R.H., Morris, R.W., Zhang, L., Li, Y.J., and Martin, E.R. 2007. X-APL: An improved family-based test of association in the presence of linkage for the X chromosome. *Am. J. Hum. Genet.* **80:** 59–68.

Clayton, D. 1999. A generalization of the transmission/disequilibrium test for uncertain-haplotype transmission. *Am. J. Hum. Genet.* **65:** 1170–1177.

Cordell, H.J., Barratt, B.J., and Clayton, D.G. 2004. Case/pseudocontrol analysis in genetic association studies: A unified framework for detection of genotype and haplotype associations, gene–gene and gene–environment interactions, and parent-of-origin effects. *Genet. Epidemiol.* **26:** 167–185.

Ding, J., Lin, S., and Liu, Y. 2006. Monte Carlo pedigree disequilibrium test for markers on the X chromosome. *Am. J. Hum. Genet.* **79:** 567–573.

Dudbridge, F. 2003. Pedigree disequilibrium tests for multilocus haplotypes. *Genet. Epidemiol.* **25:** 115–121.

Dudbridge, F. 2008. Likelihood-based association analysis for nuclear families and unrelated subjects with missing genotype data. *Hum. Hered.* **66:** 87–98.

Ewens, W.J. and Spielman, R.S. 1995. The transmission/disequilibrium test: History, subdivision, and admixture. *Am. J. Hum. Genet.* **57:** 455–464.

Falk, C.T. and Rubinstein, P. 1987. Haplotype relative risks: An easy reliable way to construct a proper control sample for risk calculations. *Ann. Hum. Genet.* **51:** 227–233.

Gardner, R.J., Mackay, D.J., Mungall, A.J., Polychronakos, C., Siebert, R., Shield, J.P., Temple, I.K., and Robinson, D.O. 2000. An imprinted locus associated with transient neonatal diabetes mellitus. *Hum. Mol. Genet.* **9:** 589–596.

Gordon, D., Heath, S.C., Liu, X., and Ott, J. 2001. A transmission/disequilibrium test that allows for genotyping errors in the analysis of single-nucleotide polymorphism data. *Am. J. Hum. Genet.* **69:** 371–380.

Hancock, D.B., Martin, E.R., Li, Y.J., and Scott, W.K. 2007. Methods for interaction analyses using family-based case-control data: Conditional logistic regression versus generalized estimating equations. *Genet. Epidemiol.* **31:** 883–893.

Horvath, S., Laird, N.M., and Knapp, M. 2000. The transmission/disequilibrium test and parental-genotype reconstruction for X-chromosomal markers. *Am. J. Hum. Genet.* **66:** 1161–1167.

Horvath, S., Xu, X., Lake, S.L., Silverman, E.K., Weiss, S.T., and Laird, N.M. 2004. Tests for associating haplotypes with general phenotype data: Application to asthma genetics. *Genet. Epidemiol.* **26:** 61–69.

Hutchison, K.E., Stallings, M., McGeary, J., and Bryan, A. 2004. Population stratification in the candidate gene study: Fatal threat or red herring? *Psychol. Bull.* **130:** 66–79.

Kidd, K.K., Morar, B., Castiglione, C.M., Zhao, H., Pakstis, A.J., Speed, W.C., Bonne-Tamir, B., Lu, R.B., Goldman, D., Lee, C., et al. 1998. A global survey of haplotype frequencies and linkage disequilibrium at the DRD2 locus. *Hum. Genet.* **103:** 211–227.

Kistner, E.O. and Weinberg, C.R. 2004. Method for using complete and incomplete trios to identify genes related to a quantitative trait. *Genet. Epidemiol.* **27:** 33–42.

Kistner, E.O., Infante-Rivard, C., and Weinberg, C.R. 2006. A method for using incomplete triads to test maternally mediated genetic effects and parent-of-origin effects in relation to a quantitative trait. *Am. J. Epidemiol.* **163:** 255–261.

Knowler, W.C., William, R.C., Pettitt, D.J., and Steinberg, A. 1988. GM3;5,13,14 and type 2 diabetes mellitus: An association in American Indians by genetic admixture. *Am. J. Hum. Gen.* **43:** 520–526.

Lange, C., Demeo, D.L., and Laird, N.M. 2002. Power and design considerations for a general class of family-based association tests: Quantitative traits. *Am. J. Hum. Genet.* **71:** 1330–1341.

Martin, E.R., Kaplan, N.L., and Weir, B.S. 1997. Tests for linkage and association in nuclear families. *Am. J. Hum. Genet.* **61:** 439–448.

Martin, E.R., Monks, S.A., Warren, L.L., and Kaplan, N.L. 2000. A test for linkage and association in general pedigrees: The pedigree disequilibrium test. *Am. J. Hum. Genet.* **67:** 146–154.

Martin, E.R., Bass, M.P., Hauser, E.R., and Kaplan, N.L. 2003. Accounting for linkage in family-based tests of association with missing parental genotypes. *Am. J. Hum. Genet.* **73:** 1016–1026.

Martin, E.R., Bass, M.P., and Kaplan, N.L. 2001. Correcting for a potential bias in the pedigree disequilibrium test. *Am. J. Hum. Genet.* **68:** 1065–1067.

Martin, E.R., Ritchie, M.D., Hahn, L., Kang, S., and Moore, J.H. 2006. A novel method to identify gene-gene effects in nuclear families: The MDR-PDT. *Genet. Epidemiol.* **30:** 111–123.

Mei, H., Ma, D., Ashley-Koch, A., and Martin, E.E. 2005. Extension of multifactor dimensionality reduction for identifying multilocus effects in the GAW14 simulated data. *BMC Genet. Suppl.* **1:** S145.

Mei, H., Cuccaro, M.L., and Martin, E.R. 2007. Multifactor dimensionality reduction-phenomics: A novel method to capture genetic heterogeneity with use of phenotypic variables. *Am. J. Hum. Genet.* **81:** 1251–1261.

Monks, S.A. and Kaplan, N.L. 2000. Removing the sampling restrictions from family-based tests of association for a quantitative-trait locus. *Am. J. Hum. Genet.* **66:** 576–592.

Morris, R.W. and Kaplan, N.L. 2002. On the advantage of haplotype analysis in the presence of multiple disease susceptibility alleles. *Genet. Epidemiol.* **23:** 221–233.

Ott, J. 1989. Statistical properties of the haplotype relative risk. *Genet. Epidemiol.* **6:** 127–130.

Rabinowitz, D. 1997. A transmission disequilibrium test for quantitative trait loci. *Hum. Hered.* **47:** 342–350.

Rabinowitz, D. and Laird, N. 2000. A unified approach to adjusting association tests for population admixture with arbitrary pedigree structure and arbitrary missing marker information. *Hum. Hered.* **50:** 211–223.

Rampersaud, E., Morris, R.W., Weinberg, C.R., Speer, M.C., and Martin, E.R. 2007. Power calculations for likelihood ratio tests for offspring genotype risks, maternal effects, and parent-of-origin (POO) effects in the presence of missing parental genotypes when unaffected siblings are available. *Genet. Epidemiol.* **31:** 18–30.

Risch, N. and Merikangas, K. 1996. The future of genetic studies of complex human disorders. *Science* **273:** 1516–1517.

Ritchie, M.D., Hahn, L.W., Roodi, N., Bailey, L.R., Dupont, W.D., Parl, F.F., and Moore, J.H. 2001. Multifactor-dimensionality reduction reveals high-order interactions among estrogen-metabolism genes in sporadic breast cancer. *Am. J. Hum. Genet.* **69:** 138–147.

Rubinstein, P., Walker, M., Carpenter, C., Carrier, C., Krassner, J., Falk, C., and Ginsberg, F. 1981. Genetics of HLA disease associations: The use of the haplotype relative risk (HRR) and the "haplo-delta" (dh) estimates in juvenile diabetes from three radical groups. *Hum. Immunol.* **3:** 384.

Schaid, D.J. 1999. Case-parents design for gene–environment interaction. *Genet. Epidemiol.* **16:** 261–273.

Schulze, T.G. and McMahon, F.J. 2002. Genetic association mapping at the crossroads: Which test and why? Overview and practical guidelines. *Am. J. Med. Genet.* **114:** 1–11.

Siegmund, K.D., Langholz, B., Kraft, P., and Thomas, D.C. 2000. Testing linkage disequilibrium in sibships. *Am. J. Hum. Genet.* **67:** 244–248.

Sinsheimer, J.S., Palmer, C.G., and Woodward, J.A. 2003. Detecting genotype combinations that increase risk for disease: Maternal-fetal genotype incompatibility test. *Genet. Epidemiol.* **24:** 1–13.

Spielman, R.S. and Ewens, W.J. 1996. The TDT and other family-based tests for linkage disequilibrium and association. *Am. J. Hum. Genet.* **59:** 983–989.

Spielman, R.S. and Ewens, W.J. 1998. A sibship test for linkage in the presence of association: The sib transmission/disequilibrium test. *Am. J. Hum. Genet.* **62:** 450–458.

Spielman, R.S., McGinnis, R.E., and Ewens, W.J. 1993. Transmission test for linkage disequilibrium: The insulin gene region and insulin-dependent diabetes mellitus (IDDM). *Am. J. Hum. Genet.* **52:** 506–516.

Terwilliger, J.D. and Ott, J. 1992. A haplotype-based "haplotype relative risk" approach to detecting allelic associations. *Hum. Hered.* **42:** 337–346.

Thomson, G. 1995. Mapping disease genes: Family-based association studies. *Am. J. Hum. Genet.* **57:** 487–498.

Thornton, T. and McPeek, M.S. 2007. Case–control association testing with related individuals: A more powerful quasi-likelihood score test. *Am. J. Hum. Genet.* **81:** 321–337.

Umbach, D.M. and Weinberg, C. 2000. The use of case-parent triads to study joint effects of genotype and exposure. *Am. J. Hum. Genet.* **66:** 251–261.

Weinberg, C.R. 1999a. Allowing for missing parents in genetic studies of case-parent triads. *Am. J. Hum. Genet.* **64:** 1186–1193.

Weinberg, C.R. 1999b. Methods for detection of parent-of-origin effects in genetic studies of case-parents triads. *Am. J. Hum. Genet.* **65:** 229–235.

Weinberg, C.R., Wilcox, A.J., and Lie, R.T. 1998. A log-linear approach to case-parent-triad data: Assessing effects of disease genes that act either directly or through maternal effects and that may be subject to parental imprinting. *Am. J. Hum. Genet.* **62:** 969–978.

Wheeler, E. and Cordell, H.J. 2007. Quantitative trait association in parent offspring trios: Extension of case/pseudocontrol method and comparison of prospective and retrospective approaches. *Genet. Epidemiol.* **31:** 813–833.

Witte, J.S., Gauderman, W.J., and Thomas, D.C. 1999. Asymptotic bias and efficiency in case–control studies of candidate genes and gene–environment interactions: Basic family designs. *Am. J. Epidemiol.* **149:** 693–705.

Zhang, L., Martin, E.R., Chung, R.H., Li, Y.J., and Morris, R.W. 2008. X-LRT: A likelihood approach to estimate genetic risks and test association with X-linked markers using a case-parents design. *Genet. Epidemiol.* **32:** 370–380.

Zhang, L., Martin, E.R., Morris, R.W., and Li, Y.J. 2009. Association test for X-linked QTL in family-based designs. *Am. J. Hum. Genet.* **84:** 431–444.

WEB RESOURCES

http://www.biostat.harvard.edu/~fbat/fbat.htm   Rabinowitz and Laird 2000. FBAT (family-based association test).

http://www.broad.mit.edu/personal/jcbarret/haploview/   Barrett et al. 2005. HAPLOVIEW.

http://www.chg.duke.edu/research/apl.html   Chung et al. 2006, APL; Chung et al. 2007, XAPL; Zhang et al. 2008, XLRT; Zhang et al. 2009, XQTL.

http://chgr.mc.vanderbilt.edu/ritchielab/   Ritchie et al. 2001, MDR; Martin et al. 2006, MDR-PDT.

http://www-gene.cimr.cam.ac.uk/clayton/software/   Clayton 1999. TRANSMIT.

http://www.mihg.org and http://www.chg.duke.edu/software/pdt.html   Martin et al. 2000. PDT (pedigree disequilibrium test).

http://www.mrc-bsu.cam.ac.uk/personal/frank/software/unphased ftp://ftp.hgmp.mrc.ac.uk/pub/linkage/   Dudbridge 2003. UNPHASED.

http://www.sph.umich.edu/csg/abecasis/QTDT/   Abecasis et al. 2000a,b. QTDT (quantitative [trait] transmission/disequilibrium test).

http://www.vipbg.vcu.edu/software_docs/solar/doc/00.contents.html   Almasy and Blangero et al. 1998. SOLAR (sequential oligogenic linkage analysis routines).

13 Copy-Number Variation and Common Human Diseases

Dheeraj Malhotra and Jonathan Sebat

Cold Spring Harbor Laboratory, Cold Spring Harbor, New York 11724

INTRODUCTION

An understanding of the genetic basis of complex disease requires a fundamental appreciation of the underlying genetic variation. Recent technological innovations that have enabled a more complete characterization of variation in the human genome have led to new discoveries. The development of microarray-based methods and, more recently, next-generation sequencing platforms has enabled comprehensive surveys of different forms of genetic variation prevalent in the human genome. These advances have allowed genome-wide association studies (GWAS) on thousands of samples to be performed routinely, and considerable success has been achieved in identifying many loci that are associated with increased risk of complex diseases (Chanock and Hunter 2008; Manolio et al. 2008). These new technological capabilities have revealed surprising things about human genetic variation and have allowed us to approach disease in new ways. For example, two initial genome-wide studies (Iafrate et al. 2004; Sebat et al. 2004) showed that structural variation, mainly copy-number variation (CNV), is abundant in the human genome. Copy-number variants are defined as stretches of DNA larger than 1 kb that have a variable number of copies in the genomes of different individuals (Feuk et al. 2006).

Following these initial studies, a series of large-scale scans for CNVs in normal populations has revealed that CNV in the human genome is extensive and complex. Thus, CNVs, in principle, could account for a significant fraction of variation in disease risk. The importance of CNV in human disease is well supported by the growing number of disease associations detected through targeted assessment of the copy-number status of individual genes (Gonzalez et al. 2005; Hollox et al. 2008; McCarroll et al. 2008a; de Cid et al. 2009) and whole-genome association studies (International Schizophrenia Consortium 2008; Marshall et al. 2008; Stefansson et al. 2008; Walsh et al. 2008; Glessner et al. 2009).

This chapter discusses our current understanding of CNV in the human genome, how recent studies of CNV in common human diseases have influenced our understanding of the genetic architecture of these diseases, and finally, how advances in new microarray and sequencing technologies can be used to study the genetics of complex human diseases.

COPY-NUMBER VARIATION IN THE HUMAN GENOME

Recently, extensive studies have been performed to establish whole-genome CNV catalogs in humans using two the following primary approaches: (1) microarray-based comparative genomic hybridization (aCGH), either with bacterial artificial chromosomes (BACs) or oligonucleotide probes (de Vries et al. 2005; Sharp et al. 2005; Conrad et al. 2006; Hinds et al. 2006; McCarroll et al. 2006; Redon et al. 2006; de Smith et al. 2007; Wong et al. 2007), and (2) bioinformatics mining of whole-genome shotgun (WGS) sequence data from individual genomes (Tuzun et al. 2005; Khaja et al. 2006; Korbel et al. 2007; Kidd et al. 2008). Although both of these approaches have their inherent limitations and no single detection method currently is capable of capturing the entire spectrum of CNV, these methods have been complementary in contributing to the ever-growing catalog of CNV regions in the human genome. Currently, the Database of Genomic Variants (http://projects.tcag.ca/variation) provides the most comprehensive curated catalog of structural variation in the human genome reported in normal healthy controls. According to its current release (March 11, 2009), 6558 copy-number variable loci cover a total of 29% (850 Mb) of the human genome. Nonetheless, our current view of structural variation lacks precision because the vast majority of these CNVs remain ill defined at the sequence level owing to the limitations of both sequence- and microarray-based methods.

In the case of end sequence pair (ESP) approaches, which rely on detecting length and orientation discrepancies after alignment of end sequences against the reference genome, such studies have less power to identify CNVs in the duplicated regions of the genome. This is because end sequence reads cannot be aligned unambiguously with a reference genome, and the insert size of the clone library used for sequencing introduces theoretical size limitations for detection of insertions (Tuzun et al. 2005; Korbel et al. 2007; Kidd et al. 2008). Likewise, aCGH-based approaches are dependent on the maximum resolution that can be achieved by the spacing of probes on the microarray chip.

In an effort to overcome these limitations and to map the full spectrum of CNVs prevalent in the human genome comprehensively, several collaborative initiatives are under way, mainly the Genome Structural Variation Consortium, the National Human Genome Research Institute (NHGRI) structural variation initiative (Eichler et al. 2007), and the 1000 Genomes Project (http://www.1000genomes.org/page.php). Each of these efforts has made significant progress in this direction.

Furthermore, the application of next-generation low-cost sequencing techniques already has provided full sequences of four personal genomes from different ethnicities and of both sexes (Benteley et al. 2008; Wang et al. 2008; Wheeler et al. 2008), as well as sequence resolution of structural variants greater than 8 kb in eight human genomes (Kidd et al. 2008). These studies suggest that structural variation accounts for ≥20% of all genetic variation in humans and involves ≥70% of the bases that are variable between any two individuals. They also highlight the fact that the sequencing of large numbers of individuals is necessary to capture fully the diversity and complexity of the genome. On the basis of these studies, it is estimated that any two individuals differ by an average of 9–24 Mb (0.5%–1% of the genome) of DNA sequence between them, underscoring the important role of this class of variation in genome evolution and in human health and disease.

It is becoming increasingly clear that a structurally dynamic genome is present not only in humans but is also commonly found across other species. Genome-wide CNV catalogs have been established for the mouse (Li et al. 2004; Adams et al. 2005; Snijders et al. 2005; Cutler et al. 2007; Egan et al. 2007; Graubert et al. 2007; She et al. 2008; Henrichsen et al. 2009), rat (Guryev et al. 2008), chimpanzee (Perry et al. 2006; Kehrer-Sawatzki and Cooper 2007), rhesus macaque (Lee et al. 2008), and *Drosophila melanogaster* (Dopman and Hartl 2007; Emerson et al. 2008). CNVs identified in all of these species are associated consistently with segmental duplications (SDs), or they cluster in other repeat regions of the genome. What is notably interesting in inbred strains of mice is the occurrence of spontaneous de novo rearrangements, which highlight the dynamic nature of CNV regions and challenge the concept of isogenicity of inbred strains (Egan et al. 2007).

MECHANISMS OF STRUCTURAL MUTATION

It has become evident from various studies of different genomic disorders (Lupski and Stankiewicz 2005) and from two genome-wide, fine-scale structural variation studies (Korbel et al. 2007; Kidd et al. 2008) that three major rearrangement mechanisms can explain the majority of structural variants in the human genome: nonallelic homologous rearrangement (NAHR), nonhomologous end joining (NHEJ), and fork stalling and template switching (FoSTeS) (Lee et al. 2007). NAHR-mediated genomic rearrangements are associated directly with specific DNA elements called low-copy repeats (LCRs) or SDs (segments of DNA >1 kb with >90% sequence identity that have become fixed in human populations). SDs constitute ~5% of the human genome (Bailey et al. 2002), and there is about fourfold to tenfold enrichment of CNVs and a significant enrichment of genes within the CNV regions flanked by SDs (Sharp et al. 2005). NAHR can occur during both meiosis and mitosis (Gu et al. 2008). This rearrangement requires two SDs arranged in tandem and of sufficient length and high homology to act as recombination substrates. Using the requirements highlighted by examination of this mechanism, several new genomic disorders have been predicted and uncovered, as described in the following studies: Sharp et al. 2006; Shaw-Smith et al. 2006; Koolen et al. 2006; Mefford et al. 2007; Sharp et al. 2007; Sharp et al. 2008.

Recent evidence from fine-scale structural variation discovery studies suggests that most of the nonpathological CNVs present in any individual arise by nonrecurrent genomic rearrangement events (Korbel et al. 2007; Kidds et al. 2008). A microhomology-mediated break-induced replication model (MMBIR) has been postulated to explain the mechanism underlying the origin of these nonrecurrent CNVs (Hastings et al. 2009); however, not enough experimental data is available at present to predict the specific DNA elements that are associated with NHEJ-, MMBIR-, or FoSTeS-based DNA rearrangements. The precise characterization of sequence features of CNVs in the near future by more sophisticated and sensitive genome-wide assays will provide additional clues about the known and as yet unknown mutational mechanisms that generate the CNVs and their distribution in the genome.

The evolutionary drive responsible for the extent of CNV prevalent in the human genome is not yet very clear. The pattern of distribution of copy-number diversity in the human genome is indicative of adaptive selection because of the functional bias in genes associated with these CNVs (Johnson et al. 2001; Ciccarelli et al. 2005; Stefansson et al. 2005; Nguyen et al. 2006; Perry et al. 2006; Popesco et al. 2006). However, recent studies (de Vries et al. 2005; Koolen et al. 2006; Sharp et al. 2006; Shaw-Smith et al. 2006) also support the notion that it is the interaction of biased mutational effects and selective pressures (both positive and negative) that are the likely dominant forces shaping this variation.

COPY-NUMBER VARIATION AND ITS IMPACT ON GENE EXPRESSION

Given the extent of CNV in the human genome, it is certain that some CNVs will have phenotypic consequences. Structural variants can alter major functional elements (genes and their regulatory regions) in a number of different ways. This may include changes in gene dosage, disruption of gene structure, creation of fusion genes, unmasking of a deleterious recessive allele, and/or modifying gene regulatory elements (Feuk et al. 2006). Effects of CNVs on the expression levels of genes have been studied in human lymphoblastoid cell lines (Stranger et al. 2007) and more extensively in different tissues in the mouse (Heinrichsen et al. 2009) and rat (Guryev et al. 2008). Although CNV regions account for >15% of total detected genetic variation in human gene expression phenotypes (Stranger et al. 2007), it is likely that the contribution of CNVs to interindividual changes in gene expression levels will be shown to be substantial as the full spectrum of CNVs in the genome is elucidated.

POPULATION CHARACTERISTICS OF CNVs

A requirement for understanding the phenotypic consequences of copy-number polymorphism is an appreciation of the basic population genetics properties of CNVs. In one sense, CNVs—like single-nucleotide polymorphisms (SNPs)—are simply another flavor of genetic variation. However, some fundamental differences between the two forms of variation have important implications for their relative contribution to disease. The most important difference lies in the mutation rate of SNPs and CNVs. Direct and indirect estimates of the human per nucleotide substitution rate show that spontaneous new point mutations occur at rate of 10^{-8} per nucleotide per generation (Kondrashov 2003). In contrast, the de novo locus-specific mutation rates for genomic rearrangements generating large CNVs is between 10^{-4} and 10^{-6}, which is at least two to four orders of magnitude greater than those for point mutations (Lupski 2007). Although we still do not have clear estimates of the mutation rates of intermediate-sized CNVs, sequencing of parent–child trios with dense coverage using next-generation sequencing will give us direct estimates of their mutation rates in the near future.

The comparatively high structural mutation rate in the genome and the propensity for recurrent mutations at some loci raises the possibility that some common structural variants could arise independently on different haplotype backgrounds. This has been offered as one explanation for the apparently reduced linkage disequilibrium (LD) between CNVs and SNP haplotypes (Conrad et al. 2006; Redon et al. 2006). However, a recent large-scale, high-resolution CNV study found that most of the common CNVs are, in fact, stable, inherited, ancestral polymorphisms that occurred only once in an ancient haplotype during human evolution, and these show LD with flanking SNPs (McCarroll et al. 2008b). Using new array platforms and computational methods with improved accuracy for determining the allelic state of a CNV, common biallelic CNVs now can be fairly accurately genotyped from experimental data (Barnes et al. 2008; Cooper et al. 2008; McCarroll et al. 2008b), and these studies have shown that biallelic variants significantly outnumber multiallelic CNVs in the genome (McCarroll et al. 2008b).

GENETIC ARCHITECTURE OF COMMON HUMAN DISEASES: LESSONS FROM SNP- AND CNV-BASED GWAS

An appreciation of the extent to which different genetic variants contribute to the risk of common diseases requires a thorough understanding of the allelic architecture of these variants in the disease. Allelic architecture refers to the number of existing disease variants, their frequencies, the risks that they confer, and their modes of interaction. Allelic spectra (i.e., the number of disease loci and the frequency of each disease allele) can vary enormously from locus to locus and disease to disease. The identification of genetic risk factors in common complex diseases has been challenging because of the underlying genetic heterogeneity (allelic or locus) in most diseases, complexity in disease phenotypes, and the intriguing interaction of multiple genetic factors coupled with the etiological heterogeneity of environmental risk factors that add another layer of complexity to dissect disease susceptibility genes.

The most widely used recent approach to delineate the allelic spectra of most common diseases is GWAS using common tag SNPs (a representative SNP in a region of the genome with high LD) as genetic markers for screening in large numbers of cases and controls. This methodology is driven by the common disease/common variant (CDCV) hypothesis (Reich and Lander 2001; Wang et al. 2005), which posits that susceptibility to common disease is conferred by multiple common alleles with small effects (e.g., genotype relative risks of 1.1–1.5). The past 2 to 3 years have seen tremendous success with GWAS in identifying reproducible statistical associations of hundreds of loci across the genome with common complex traits (Chanock and Hunter 2008; Manolio et al. 2008).

The results of these studies no doubt have provided useful insights into new molecular targets and pathways of many diseases (Xavier and Rioux 2008).

However, GWAS have shown certain limitations that are important in designing future studies aimed at identifying causal disease variants, the most important of which is that these studies have not explained any more than 2%–10% of the heritable genetic variation involved in the diseases. For most of the implicated variants, the magnitude of effect in disease susceptibility/protection is small. Furthermore, there is great difficulty in identifying the true causal variants for most of the common diseases from the identified statistical associations between indirect proxies (tag SNPs) and a disease trait. Although it may be premature to conclude that common variants contribute little to overall risk, there is growing recognition of the limitations of the GWAS approach (Frazer et al. 2009; Goldstein 2009), and there is growing enthusiasm for the alternative multiple rare variant model of common disease (MRV/CD).

The MRV/CD hypothesis holds that the significant proportion of susceptibility to most common diseases could be attributable to the accumulation of the effects of a series of low-frequency, dominant, and independently acting variants in a variety of different genes, each conferring a detectable, large increase in relative risk. Several recent reports support this hypothesis, highlighting the value of lower-frequency common variants (between 0.5% and 1%) and of extremely rare SNPs (<0.5% frequency) in influencing different common disease traits (Hirschhorn and Altshuler 2002; Cohen et al. 2004, 2005, 2006; Fearnhead et al. 2005; Kotowski et al. 2006; Ji et al. 2008).

HOW HAVE CNV STUDIES CHANGED OUR PERSPECTIVE ON THE GENETIC ARCHITECTURE OF COMMON DISEASES?

Our knowledge of CNVs in influencing specific human traits initially came from genomic disorders, a heterogeneous group of clinically distinct human genetic diseases, caused by genomic rearrangements that result in loss or gain of dosage-sensitive gene(s) specific to each disease (Lupski and Stankiewicz 2005). Almost all genomic disorders are linked with a single locus that is almost 100% penetrant. In addition, they share a common underlying mechanism for genomic rearrangement, which is NAHR between LCRs that flank the rearranged unique genomic segment. Genomic disorders are rare in the population, but they do provide critical insights into directly identifying the risk factors for common complex diseases. Apart from their role in sporadic genomic disorders, CNVs have been implicated in several Mendelian diseases (Lupski et al. 1991; Lakich et al. 1993; Konrad et al. 1996), and they suggest new hypotheses to predict the overall penetrance and/or variations in the severity of phenotype in several dominant Mendelian disorders that are caused by genes present within CNV regions (Beckmann et al. 2007).

Copy-number analysis in disease association studies is currently approached by analyzing common CNVs and rare CNVs separately for association testing. Progress in the development of genotyping algorithms for common biallelic CNVs now allows testing of the association of common CNVs with disease, similar to SNP association tests. However, to date no GWAS have tested and identified common biallelic CNV association with any complex disease trait directly. Using information on the LD pattern of common CNVs with SNPs, a 20-kb deletion polymorphism in the regulatory region of the *IRGM* gene has been strongly associated with Crohn's disease (McCarroll et al. 2008a). Common CNVs at other loci have been implicated in complex diseases using candidate gene approaches (Gonzalez et al. 2005; Aitman et al. 2006; Fellermann et al. 2006; Fanciulli et al. 2007; Yang et al. 2007; Hollox et al. 2008; Mamtani et al. 2008; McCarroll et al. 2008a; Willcocks et al. 2008; de Cid et al. 2009). Intriguingly, some of the reported associations involve multiallelic CNVs. These are regions that must be genotyped directly because disease risk is associated primarily with dosage of the gene rather than with any single allele.

GWAS of CNV have focused on large, rare CNVs and have tested the MRV/CD hypothesis, which proposes that spontaneously recurring CNV regions, each individually rare and highly penetrant and occurring at multiple loci across the genome, can contribute to disease risk significantly, and that the prevalence of the disease is a reflection of this underlying genetic heterogeneity. This hypothesis posits that rare alleles of large effect may have an important role in common diseases that have high heritability and a strong effect on the fecundity of affected individuals (Sebat 2007). Multiple, individual, recent, genome-wide studies strongly support this hypothesis and suggest a causal role of individually rare CNVs in schizophrenia (International Schizophrenia Consortium 2008; Stefansson et al. 2008; Walsh et al. 2008; Xu et al. 2008) and in autism (Sebat et al. 2007; Glessner et al. 2009), both diseases with highly heritability and strong fitness effects. These studies also highlight several important points. First, the genome-wide burden of rare CNVs is enriched significantly in patients compared with healthy controls (Sebat et al. 2007; International Schizophrenia Consortium 2008; Walsh et al. 2008; Xu et al. 2008). Second, at some genomic hotspots, recurrent deletions or duplications occur with significantly increased frequency in patients (International Schizophrenia Consortium 2008; Stefansson et al. 2008), and third, most of the CNVs are distributed throughout the genome and involve many different genes.

The success of the CNV-based approach has been possible because of high-resolution microarray techniques that have allowed direct detection of rare CNVs in the human genome. Rare CNVs can be tested for disease association in multiple ways, including comparing the frequencies of individual variants (single-marker association), comparing the frequencies of multiple individually rare mutations in the same genes (gene-based mutational burden analysis), or comparing the collective frequencies of all rare CNVs in cases and controls (genome-wide mutational burden analysis) (Fig. 13.1). In studies of rare CNVs, it is important to avoid the confounding effects of

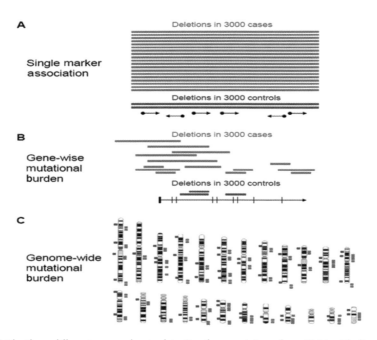

FIGURE 13.1 The three different approaches to detecting the association of rare CNVs with disease are illustrated. The association of CNVs with disease can be determined by comparing cases and controls with respect to the frequency of an individual mutation (single marker association) (A), the aggregate frequency of multiple mutations within a single gene or region (gene-wise mutational burden) (B), or the aggregate frequency of all rare structural variants (genome-wide mutational burden) (C). (A and B) A typical "genome browser" view displaying tracks for CNVs and annotated genes; (C) ideogram displaying the distribution of CNVs genome wide.

nonuniform sensitivity of different microarray platforms and the accompanying differential bias in CNV detection in cases and controls.

Genome-wide CNV studies of neurodevelopmental and psychiatric disorders have provided new insights into their genetic architecture. It is becoming clear that mutations of large effect do exist for these disorders. For instance, the relative risks for microdeletions of 1q21.1, 15q13.3, and 22q11.2 in schizophrenia are 7–15, 11–18, and 30, respectively (Karayiorgou et al. 1995; International Schizophrenia Consortium 2008; Stefansson et al. 2008). These alleles presumably act in a dominant fashion; however, they are rarely observed segregating in large multigeneration pedigrees. On the contrary, all have been observed as rare de novo events (Mefford et al. 2008; Sharp et al. 2008). The observations of similar CNV hotspots in genomic loci in multiple different disorders add another dimension in understanding the genetic etiology of underlying diseases (O'Donovan et al. 2008).

A critical factor in the search for large, highly penetrant CNVs in complex diseases will be the frequency of new mutation events and to test if these CNVs simply are the result of mutation processes or if they occur on any specific genetic background. Moreover, it will be important to assess the penetrance and expressivity of de novo rare variants, which can be variable depending on the severity of disease, age of onset, etc. (Pulver et al. 1994; Murphy et al. 1999), and therefore necessary to include other independent risk factors and models that could explain the pathogenicity of these variants.

If the current paradigm for the role of rare, highly penetrant CNVs in disease risk holds true for other diseases, then finding statistically significant associations of the rare variants that are recurrently associated with disease risk will require comprehensive screening of several thousands of samples to establish a causal relationship with a specific disease. This will be more relevant to rare recurrent variants, as exemplified by two recent studies regarding schizophrenia risk (International Schizophrenia Consortium 2008; Stefansson et al. 2008). The sample sizes of many of the cohorts now collected for GWAS (Database of Genotypes and Phenotypes, dbGAP, http://www.ncbi.nlm.nih.gov/sites/entrez?Db=gap) provide a well-powered resource to test disease associations of rare recurrent CNVs. Alternatively, family-based methods allow greater power to detect nonrecurrent, highly penetrant, rare variants. The overall picture emerging from these studies indicates that the genetic risk conferred by some rare genetic variants is high but may or may not be sufficient to manifest disease in all carriers. Study designs and CNV discovery platforms that allow for comprehensive testing of these hypotheses will give a more complete picture of the role of CNVs in complex disease.

TECHNOLOGICAL ADVANCES IN DETECTING CNVs AND THEIR IMPLICATIONS FOR DISEASE STUDIES

Our current knowledge of the role of CNVs in common diseases has come largely from low-resolution aCGH using BAC probes or oligonucleotide probes in representational oligonucleotide microarray analysis (ROMA) and commercially available low-resolution SNP platforms (Affy 500K, Illumina 550K). In addition to microarrays, CNV identification and CNV association with disease have been studied with various methods, such as quantitative polymerase chain reaction (PCR) (Gonzalez et al. 2005), multiplex ligation-dependent probe amplification (MLPA) (Breunis et al. 2009), and multiplex amplifiable probe hybridization (MAPH) (Kousoulidou et al. 2007), but with restricted loci at a high-throughput scale.

Sequencing approaches, such as paired-end mapping (Korbel et al. 2007; Kidd et al. 2008) and complete sequence assembly comparisons (Bentley et al. 2008; Wang et al 2008; Wheeler et al. 2008), are now being used to identify most forms of sequence variation (in addition to CNVs, translocations, and inversions). The integration of sequence variation studies in large-scale sequencing of multiple genomes using next-generation sequencing platforms (Mardis 2008) undoubtedly will provide a clear picture about the sequence variation landscape in the near future. However, these technologies are currently cost prohibitive for laboratories involved in studying hundreds or thousands of samples for disease association or population genetics studies.

Microarrays are still, and will remain at least for the next few years, the preferred choice for robust and economical analysis of CNVs in multiple genomes. The next generation of high-density aCGH and SNP-based platforms, including high-density aCGH and SNP genotyping platforms, currently consist of >1 million probes, and products are currently in development that will provide feature densities of 5–10 million probes. These high-resolution platforms provide improved sensitivity to detect CNVs and allow testing for association of common and rare CNVs with disease risk.

Finally, the validation and fine mapping of candidate genes identified from genome-wide CNV scans can be achieved by comprehensive analysis of these regions using tiling resolution arrays and high-throughput sequencing. The entire spectrum of mutations identified from the validated candidate gene(s) will provide better estimates of the genetic architecture of disease, and the spectrum of mutations identified may provide new insights into the underlying molecular basis of diseases.

CONCLUSIONS

Our understanding of the genetic etiology of complex diseases is maturing constantly, thanks mainly to advances in genetic variation discovery platforms. The ability to map major risk alleles for most highly heritable common disorders has been confounded by genetic heterogeneity, de novo rare mutations, and/or epigenetic changes. Success in mapping these risk factors lies in study designs and methods that directly address the problem of heterogeneity. The degree of genetic heterogeneity can vary from disease to disease and in some diseases may be such that multiple individually rare mutations involving many genes can produce the disease phenotype. In the latter case, current methods relying on GWAS will fail to identify the relevant genes.

The success of the CNV-based approach suggests that a similar methodology could be applied to identify rare point mutations by deep resequencing of a large number of candidate genes using next-generation sequencing technologies. CNV studies are only a first step toward such comprehensive analysis, and the genetic findings emerging from these studies have shown credible evidence that the relationship to disease of mutations in candidate genes could result also from their involvement in different key pathways that are known to be important in disease pathophysiology. These findings will help in the development of new therapies. Resequencing of complete genomes is a realistic possibility in the near future, and these efforts will provide a more complete catalog of mutations. During the next 5 to 6 years, systematic investigation of different classes of genetic variants and their interactions undoubtedly will provide useful insights into the genetic architecture of common complex human diseases.

REFERENCES

Adams, D.J., Dermitzakis, E.T., Cox, T., Smith, J., Davies, R., Banerjee, R., Bonfield, J., Mullikin, J.C., Chung, Y.J., Rogers, J., et al. 2005. Complex haplotypes, copy number polymorphisms and coding variation in two recently divergent mouse strains. *Nat. Genet.* **37:** 532–536.

Aitman, T.J., Dong, R., Vyse, T.J., Norsworthy, P.J., Johnson, M.D., Smith, J., Mangion, J., Roberton-Lowe, C., Marshall, A.J., Petretto, E., et al. 2006. Copy number polymorphism in Fcgr3 predisposes to glomerulonephritis in rats and humans. *Nature* **439:** 851–855.

Bailey, J.A., Gu, Z., Clark, R.A., Reinert, K., Samonte, R.V., Schwartz, S., Adams, M.D., Myers, E.W., Li, P.W., and Eichler, E.E. 2002. Recent segmental duplications in the human genome. *Science* **297:** 1003–1007.

Barnes, C., Plagnol, V., Fitzgerald, T., Redon, R., Marchini, J., Clayton, D., and Hurles, M.E. 2008. A robust statistical method for case-control association testing with copy number variation. *Nat. Genet.* **40:** 1245–1252.

Beckmann, J.S., Estivill, X., and Antonarakis, S.E. 2007. Copy number variants and genetic traits: Closer to the resolution of phenotypic to genotypic variability. *Nat. Rev. Genet.* **8:** 639–646.

Bentley, D.R., Balasubramanian, S., Swerdlow, H.P., Smith, G.P., Milton, J., Brown, C.G., Hall, K.P., Evers, D.J., Barnes, C.L., Bignell, H.R., et al. 2008. Accurate whole human genome sequencing using reversible terminator chemistry. *Nature* **456:** 53–59.

Breunis, W.B., van Mirre, E., Geissler, J., Laddach, N., Wolbink, G., van der Schoot, E., de Haas, M., de Boer, M., Roos, D., and Kuijpers, T.W. 2009. Copy number variation at the FCGR locus includes FCGR3A, FCGR2C and FCGR3B but not FCGR2A and FCGR2B. *Hum. Mutat.* **30:** E640–E650.

Chanock, S.J. and Hunter, D.J. 2008. Genomics: When the smoke clears. *Nature* **452:** 537–538.

Ciccarelli, F.D., von Mering, C., Suyama, M., Harrington, E.D., Izaurralde, E., and Bork, P. 2005. Complex genomic rearrangements lead to novel primate gene function. *Genome Res.* **15:** 343–351.

Cohen, J., Pertsemlidis, A., Kotowski, I.K., Graham, R., Garcia, C.K., and Hobbs, H.H. 2005. Low LDL cholesterol in individuals of African descent resulting from frequent nonsense mutations in PCSK9. *Nat. Genet.* **37:** 161–165.

Cohen, J.C., Kiss, R.S., Pertsemlidis, A., Marcel, Y.L., McPherson, R., and Hobbs, H.H. 2004. Multiple rare alleles contribute to low plasma levels of HDL cholesterol. *Science* **305:** 869–872.

Cohen, J.C., Boerwinkle, E., Mosley, Jr., T.H., and Hobbs, H.H. 2006. Sequence variations in PCSK9, low LDL, and protection against coronary heart disease. *N. Engl. J. Med.* **354:** 1264–1272.

Conrad, D.F., Andrews, T.D., Carter, N.P., Hurles, M.E., and Pritchard, J.K. 2006. A high-resolution survey of deletion polymorphism in the human genome. *Nat. Genet.* **38:** 75–81.

Cooper, G.M., Zerr, T., Kidd, J.M., Eichler, E.E., and Nickerson, D.A. 2008. Systematic assessment of copy number variant detection via genome-wide SNP genotyping. *Nat. Genet.* **40:** 1199–1203.

Cutler, G., Marshall, L.A., Chin, N., Baribault, H., and Kassner, P.D. 2007. Significant gene content variation characterizes the genomes of inbred mouse strains. *Genome Res.* **17:** 1743–1754.

de Cid, R., Riveira-Munoz, E., Zeeuwen, P.L., Robarge, J., Liao, W., Dannhauser, E.N., Giardina, E., Stuart, P.E., Nair, R., Helms, C., et al. 2009. Deletion of the late cornified envelope *LCE3B* and *LCE3C* genes as a susceptibility factor for psoriasis. *Nat. Genet.* **41:** 211–215.

de Smith, A.J., Tsalenko, A., Sampas, N., Scheffer, A., Yamada, N.A., Tsang, P., Ben-Dor, A., Yakhini, Z., Ellis, R.J., Bruhn, L., et al. 2007. Array CGH analysis of copy number variation identifies 1284 new genes variant in healthy white males: Implications for association studies of complex diseases. *Hum. Mol. Genet.* **16:** 2783–2794.

de Vries, B.B., Pfundt, R., Leisink, M., Koolen, D.A., Vissers, L.E., Janssen, I.M., Reijmersdal, S., Nillesen, W.M., Huys, E.H., Leeuw, N., et al. 2005. Diagnostic genome profiling in mental retardation. *Am. J. Hum. Genet.* **77:** 606–616.

Dopman, E.B. and Hartl, D.L. 2007. A portrait of copy-number polymorphism in *Drosophila melanogaster*. *Proc. Natl. Acad. Sci.* **104:** 19920–19925.

Egan, C.M., Sridhar, S., Wigler, M., and Hall, I.M. 2007. Recurrent DNA copy number variation in the laboratory mouse. *Nat. Genet.* **39:** 1384–1389.

Eichler, E.E., Nickerson, D.A., Altshuler, D., Bowcock, A.M., Brooks, L.D., Carter, N.P., Church, D.M., Felsenfeld, A., Guyer, M., Lee, C., et al. 2007. Completing the map of human genetic variation. *Nature* **447:** 161–165.

Emerson, J.J., Cardoso-Moreira, M., Borevitz, J.O., and Long, M. 2008. Natural selection shapes genome-wide patterns of copy-number polymorphism in *Drosophila melanogaster*. *Science* **320:** 1629–1631.

Fanciulli, M., Norsworthy, P.J., Petretto, E., Dong, R., Harper, L., Kamesh, L., Heward, J.M., Gough, S.C., de Smith, A., Blakemore, A.I., et al. 2007. FCGR3B copy number variation is associated with susceptibility to systemic, but not organ-specific, autoimmunity. *Nat. Genet.* **39:** 721–723.

Fearnhead, N.S., Winney, B., and Bodmer, W.F. 2005. Rare variant hypothesis for multifactorial inheritance: Susceptibility to colorectal adenomas as a model. *Cell Cycle* **4:** 521–525.

Fellermann, K., Stange, D.E., Schaeffeler, E., Schmalzl, H., Wehkamp, J., Bevins, C.L., Reinisch, W., Teml, A., Schwab, M., Lichter, P., et al. 2006. A chromosome 8 gene-cluster polymorphism with low human β-defensin 2 gene copy number predisposes to Crohn disease of the colon. *Am. J. Hum. Genet.* **79:** 439–448.

Feuk, L., Carson, A.R., and Scherer, S.W. 2006. Structural variation in the human genome. *Nat. Rev. Genet.* **7:** 85–97.

Frazer, K.A., Murray, S.S., Schork, N.J., and Topol, E.J. 2009. Human genetic variation and its contribution to complex traits. *Nat. Rev. Genet.* **10:** 241–251.

Glessner, J.T., Wang, K., Cai, G., Korvatska, O., Kim, C.E., Wood, S., Zhang, H., Estes, A., Brune, C.W., Bradfield, J.P., et al. 2009. Autism genome-wide copy number variation reveals ubiquitin and neuronal genes. *Nature* **459:** 569–573.

Goldstein, D.B. 2009. Common genetic variation and human traits. *N. Engl. J. Med.* **360:** 1696–1698.

Gonzalez, E., Kulkarni, H., Bolivar, H., Mangano, A., Sanchez, R., Catano, G., Nibbs, R.J., Freedman, B.I., Quinones, M.P., Bamshad, M.J., et al. 2005. The influence of CCL3L1 gene-containing segmental duplications on HIV-1/AIDS susceptibility. *Science* **307:** 1434–1440.

Graubert, T.A., Cahan, P., Edwin, D., Selzer, R.R., Richmond, T.A., Eis, P.S., Shannon, W.D., Li, X., McLeod, H.L., Cheverud, J.M., et al. 2007. A high-resolution map of segmental DNA copy number variation in the mouse genome. *PLoS Genet.* **3:** e3.

Gu, W., Zhang, F., and Lupski, J.R. 2008. Mechanisms for human genomic rearrangements. *Pathogenetics* **1:** 4.

Guryev, V., Saar, K., Adamovic, T., Verheul, M., van Heesch, S.A., Cook, S., Pravenec, M., Aitman, T., Jacob, H., Shull, J.D., et al. 2008. Distribution and functional impact of DNA copy number variation in the rat. *Nat. Genet.* **40:** 538–545.

Hastings, P.J., Ira, G., and Lupski, J.R. 2009. A microhomology-mediated break-induced replication model for the origin of human copy number variation. *PLoS Genet.* **5:** e1000327.

Henrichsen, C.N., Vinckenbosch, N., Zollner, S., Chaignat, E., Pradervand, S., Schutz, F., Ruedi, M., Kaessmann, H., and Reymond, A. 2009. Segmental copy number variation shapes tissue transcriptomes. *Nat. Genet.* **41:** 424–429.

Hinds, D.A., Kloek, A.P., Jen, M., Chen, X., and Frazer, K.A. 2006. Common deletions and SNPs are in linkage disequilibrium in the human genome. *Nat. Genet.* **38:** 82–85.

Hirschhorn, J.N. and Altshuler, D. 2002. Once and again issues surrounding replication in genetic association studies. *J. Clin. Endocrinol. Metab.* **87:** 4438–4441.

Hollox, E.J., Huffmeier, U., Zeeuwen, P.L., Palla, R., Lascorz, J., Rodijk-Olthuis, D., van de Kerkhof, P.C., Traupe, H., de Jongh, G., den Heijer, M., et al. 2008. Psoriasis is associated with increased β-defensin genomic copy number. *Nat. Genet.* **40:** 23–25.

Iafrate, A.J., Feuk, L., Rivera, M.N., Listewnik, M.L., Donahoe, P.K., Qi, Y., Scherer, S.W., and Lee, C. 2004. Detection of large-scale variation in the human genome. *Nat. Genet.* **36:** 949–951.

International Schizophrenia Consortium. 2008. Rare chromosomal deletions and duplications increase risk of schizophrenia. *Nature* **455:** 237–241.

Ji, W., Foo, J.N., O'Roak, B.J., Zhao, H., Larson, M.G., Simon, D.B., Newton-Cheh, C., State, M.W., Levy, D., and Lifton, R.P. 2008. Rare independent mutations in renal salt handling genes contribute to blood pressure variation. *Nat. Genet.* **40:** 592–599.

Johnson, M.E., Viggiano, L., Bailey, J.A., Abdul-Rauf, M., Goodwin, G., Rocchi, M., and Eichler, E.E. 2001. Positive selection of a gene family during the emergence of humans and African apes. *Nature* **413:** 514–519.

Karayiorgou, M., Morris, M.A., Morrow, B., Shprintzen, R.J., Goldberg,

R., Borrow, J., Gos, A., Nestadt, G., Wolyniec, P.S., Lasseter, V.K., et al. 1995. Schizophrenia susceptibility associated with interstitial deletions of chromosome 22q11. *Proc. Natl. Acad. Sci.* **92:** 7612–7616.

Kehrer-Sawatzki, H. and Cooper, D.N. 2007. Structural divergence between the human and chimpanzee genomes. *Hum. Genet.* **120:** 759–778.

Khaja, R., Zhang, J., MacDonald, J.R., He, Y., Joseph-George, A.M., Wei, J., Rafiq, M.A., Qian, C., Shago, M., Pantano, L., et al. 2006. Genome assembly comparison identifies structural variants in the human genome. *Nat. Genet.* **38:** 1413–1418.

Kidd, J.M., Cooper, G.M., Donahue, W.F., Hayden, H.S., Sampas, N., Graves, T., Hansen, N., Teague, B., Alkan, C., Antonacci, F., et al. 2008. Mapping and sequencing of structural variation from eight human genomes. *Nature* **453:** 56–64.

Kondrashov, A.S. 2003. Direct estimates of human per nucleotide mutation rates at 20 loci causing Mendelian diseases. *Hum. Mutat.* **21:** 12–27.

Konrad, M., Saunier, S., Heidet, L., Silbermann, F., Benessy, F., Calado, J., Le Paslier, D., Broyer, M., Gubler, M.C., and Antignac, C. 1996. Large homozygous deletions of the 2q13 region are a major cause of juvenile nephronophthisis. *Hum. Mol. Genet.* **5:** 367–371.

Koolen, D.A., Vissers, L.E., Pfundt, R., de Leeuw, N., Knight, S.J., Regan, R., Kooy, R.F., Reyniers, E., Romano, C., Fichera, M., et al. 2006. A new chromosome 17q21.31 microdeletion syndrome associated with a common inversion polymorphism. *Nat. Genet.* **38:** 999–1001.

Korbel, J.O., Urban, A.E., Affourtit, J.P., Godwin, B., Grubert, F., Simons, J.F., Kim, P.M., Palejev, D., Carriero, N.J., Du, L., et al. 2007. Paired-end mapping reveals extensive structural variation in the human genome. *Science* **318:** 420–426.

Kotowski, I.K., Pertsemlidis, A., Luke, A., Cooper, R.S., Vega, G.L., Cohen, J.C., and Hobbs, H.H. 2006. A spectrum of PCSK9 alleles contributes to plasma levels of low-density lipoprotein cholesterol. *Am. J. Hum. Genet.* **78:** 410–422.

Kousoulidou, L., Parkel, S., Zilina, O., Palta, P., Puusepp, H., Remm, M., Turner, G., Boyle, J., van Bokhoven, H., de Brouwer, A., et al. 2007. Screening of 20 patients with X-linked mental retardation using chromosome X-specific array-MAPH. *Eur. J. Med. Genet.* **50:** 399–410.

Lakich, D., Kazazian, Jr., H.H., Antonarakis, S.E., and Gitschier, J. 1993. Inversions disrupting the factor VIII gene are a common cause of severe haemophilia A. *Nat. Genet.* **5:** 236–241.

Lee, J.A., Carvalho, C.M., and Lupski, J.R. 2007. A DNA replication mechanism for generating nonrecurrent rearrangements associated with genomic disorders. *Cell* **131:** 1235–1247.

Lee, A.S., Gutierrez-Arcelus, M., Perry, G.H., Vallender, E.J., Johnson, W.E., Miller, G.M., Korbel, J.O., and Lee, C. 2008. Analysis of copy number variation in the rhesus macaque genome identifies candidate loci for evolutionary and human disease studies. *Hum. Mol. Genet.* **17:** 1127–1136.

Li, J., Jiang, T., Mao, J.H., Balmain, A., Peterson, L., Harris, C., Rao, P.H., Havlak, P., Gibbs, R., and Cai, W.W. 2004. Genomic segmental polymorphisms in inbred mouse strains. *Nat. Genet.* **36:** 952–954.

Lupski, J.R. 2007. Genomic rearrangements and sporadic disease. *Nat. Genet.* (suppl. 7) **39:** S43–S47.

Lupski, J.R., de Oca-Luna, R.M., Slaugenhaupt, S., Pentao, L., Guzzetta, V., Trask, B.J., Saucedo-Cardenas, O., Barker, D.F., Killian, J.M., Garcia, C.A., et al. 1991. DNA duplication associated with Charcot-Marie-Tooth disease type 1A. *Cell* **66:** 219–232.

Lupski, J.R. and Stankiewicz, P. 2005. Genomic disorders: Molecular mechanisms for rearrangements and conveyed phenotypes. *PLoS Genet.* **1:** e49.

Mamtani, M., Rovin, B., Brey, R., Camargo, J.F., Kulkarni, H., Herrera, M., Correa, P., Holliday, S., Anaya, J.M., and Ahuja, S.K. 2008. CCL3L1 gene-containing segmental duplications and polymor-

phisms in CCR5 affect risk of systemic lupus erythaematosus. *Ann. Rheum. Dis.* **67:** 1076–1083.

Manolio, T.A., Brooks, L.D., and Collins, F.S. 2008. A HapMap harvest of insights into the genetics of common disease. *J. Clin. Invest.* **118:** 1590–1605.

Mardis, E.R. 2008. The impact of next-generation sequencing technology on genetics. *Trends Genet.* **24:** 133–141.

Marshall, C.R., Noor, A., Vincent, J.B., Lionel, A.C., Feuk, L., Skaug, J., Shago, M., Moessner, R., Pinto, D., Ren, Y., et al. 2008. Structural variation of chromosomes in autism spectrum disorder. *Am. J. Hum. Genet.* **82:** 477–488.

McCarroll, S.A., Hadnott, T.N., Perry, G.H., Sabeti, P.C., Zody, M.C., Barrett, J.C., Dallaire, S., Gabriel, S.B., Lee, C., Daly, M.J., et al. 2006. Common deletion polymorphisms in the human genome. *Nat. Genet.* **38:** 86–92.

McCarroll, S.A., Huett, A., Kuballa, P., Chilewski, S.D., Landry, A., Goyette, P., Zody, M.C., Hall, J.L., Brant, S.R., Cho, J.H., et al. 2008a. Deletion polymorphism upstream of IRGM associated with altered IRGM expression and Crohn's disease. *Nat. Genet.* **40:** 1107–1112.

McCarroll, S.A., Kuruvilla, F.G., Korn, J.M., Cawley, S., Nemesh, J., Wysoker, A., Shapero, M.H., de Bakker, P.I., Maller, J.B., Kirby, A., et al. 2008b. Integrated detection and population-genetic analysis of SNPs and copy number variation. *Nat. Genet.* **40:** 1166–1174.

Mefford, H.C., Clauin, S., Sharp, A.J., Moller, R.S., Ullmann, R., Kapur, R., Pinkel, D., Cooper, G.M., Ventura, M., Ropers, H.H., et al. 2007. Recurrent reciprocal genomic rearrangements of 17q12 are associated with renal disease, diabetes, and epilepsy. *Am. J. Hum. Genet.* **81:** 1057–1069.

Mefford, H.C., Sharp, A.J., Baker, C., Itsara, A., Jiang, Z., Buysse, K., Huang, S., Maloney, V.K., Crolla, J.A., Baralle, D., et al. 2008. Recurrent rearrangements of chromosome 1q21.1 and variable pediatric phenotypes. *N. Engl. J. Med.* **359:** 1685–1699.

Murphy, K.C., Jones, L.A., and Owen, M.J. 1999. High rates of schizophrenia in adults with velo-cardio-facial syndrome. *Arch. Gen. Psychiatry* **56:** 940–945.

Nguyen, D.Q., Webber, C., and Ponting, C.P. 2006. Bias of selection on human copy-number variants. *PLoS Genet.* **2:** e20.

O'Donovan, M.C., Kirov, G., and Owen, M.J. 2008. Phenotypic variations on the theme of CNVs. *Nat. Genet.* **40:** 1392–1393.

Perry, G.H., Tchinda, J., McGrath, S.D., Zhang, J., Picker, S.R., Caceres, A.M., Iafrate, A.J., Tyler-Smith, C., Scherer, S.W., Eichler, E.E., et al. 2006. Hotspots for copy number variation in chimpanzees and humans. *Proc. Natl. Acad. Sci.* **103:** 8006–8011.

Popesco, M.C., Maclaren, E.J., Hopkins, J., Dumas, L., Cox, M., Meltesen, L., McGavran, L., Wyckoff, G.J., and Sikela, J.M. 2006. Human lineage-specific amplification, selection, and neuronal expression of DUF1220 domains. *Science* **313:** 1304–1307.

Pulver, A.E., Nestadt, G., Goldberg, R., Shprintzen, R.J., Lamacz, M., Wolyniec, P.S., Morrow, B., Karayiorgou, M., Antonarakis, S.E., Housman, D., et al. 1994. Psychotic illness in patients diagnosed with velo-cardio-facial syndrome and their relatives. *J. Nerv. Ment. Dis.* **182:** 476–478.

Redon, R., Ishikawa, S., Fitch, K.R., Feuk, L., Perry, G.H., Andrews, T.D., Fiegler, H., Shapero, M.H., Carson, A.R., Chen, W., et al. 2006. Global variation in copy number in the human genome. *Nature* **444:** 444–454.

Reich, D.E. and Lander, E.S. 2001. On the allelic spectrum of human disease. *Trends Genet.* **17:** 502–510.

Sebat, J. 2007. Major changes in our DNA lead to major changes in our thinking. *Nat. Genet.* (suppl. 7) **39:** S3–S5.

Sebat, J., Lakshmi, B., Troge, J., Alexander, J., Young, J., Lundin, P., Maner, S., Massa, H., Walker, M., Chi, M., et al. 2004. Large-scale copy number polymorphism in the human genome. *Science* **305:** 525–528.

Sebat, J., Lakshmi, B., Malhotra, D., Troge, J., Lese-Martin, C., Walsh,

T., Yamrom, B., Yoon, S., Krasnitz, A., Kendall, J., et al. 2007. Strong association of de novo copy number mutations with autism. *Science* **316:** 445–449.

Sharp, A.J., Locke, D.P., McGrath, S.D., Cheng, Z., Bailey, J.A., Vallente, R.U., Pertz, L.M., Clark, R.A., Schwartz, S., Segraves, R., et al. 2005. Segmental duplications and copy-number variation in the human genome. *Am. J. Hum. Genet.* **77:** 78–88.

Sharp, A.J., Hansen, S., Selzer, R.R., Cheng, Z., Regan, R., Hurst, J.A., Stewart, H., Price, S.M., Blair, E., Hennekam, R.C., et al. 2006. Discovery of previously unidentified genomic disorders from the duplication architecture of the human genome. *Nat. Genet.* **38:** 1038–1042.

Sharp, A.J., Selzer, R.R., Veltman, J.A., Gimelli, S., Gimelli, G., Striano, P., Coppola, A., Regan, R., Price, S.M., Knoers, N.V., et al. 2007. Characterization of a recurrent 15q24 microdeletion syndrome. *Hum. Mol. Genet.* **16:** 567–572.

Sharp, A.J., Mefford, H.C., Li, K., Baker, C., Skinner, C., Stevenson, R.E., Schroer, R.J., Novara, F., De Gregori, M., Ciccone, R., et al. 2008. A recurrent 15q13.3 microdeletion syndrome associated with mental retardation and seizures. *Nat. Genet.* **40:** 322–328.

Shaw-Smith, C., Pittman, A.M., Willatt, L., Martin, H., Rickman, L., Gribble, S., Curley, R., Cumming, S., Dunn, C., Kalaitzopoulos, D., et al. 2006. Microdeletion encompassing MAPT at chromosome 17q21.3 is associated with developmental delay and learning disability. *Nat. Genet.* **38:** 1032–1037.

She, X., Cheng, Z., Zollner, S., Church, D.M., and Eichler, E.E. 2008. Mouse segmental duplication and copy number variation. *Nat. Genet.* **40:** 909–914.

Snijders, A.M., Nowak, N.J., Huey, B., Fridlyand, J., Law, S., Conroy, J., Tokuyasu, T., Demir, K., Chiu, R., Mao, J.H., et al. 2005. Mapping segmental and sequence variations among laboratory mice using BAC array CGH. *Genome Res.* **15:** 302–311.

Stefansson, H., Helgason, A., Thorleifsson, G., Steinthorsdottir, V., Masson, G., Barnard, J., Baker, A., Jonasdottir, A., Ingason, A., Gudnadottir, V.G., et al. 2005. A common inversion under selection in Europeans. *Nat. Genet.* **37:** 129–137.

Stefansson, H., Rujescu, D., Cichon, S., Pietilainen, O.P., Ingason, A., Steinberg, S., Fossdal, R., Sigurdsson, E., Sigmundsson, T., Buizer-Voskamp, J.E., et al. 2008. Large recurrent microdeletions associated with schizophrenia. *Nature* **455:** 232–236.

Stranger, B.E., Forrest, M.S., Dunning, M., Ingle, C.E., Beazley, C., Thorne, N., Redon, R., Bird, C.P., de Grassi, A., Lee, C., et al. 2007.

Relative impact of nucleotide and copy number variation on gene expression phenotypes. *Science* **315:** 848–853.

Tuzun, E., Sharp, A.J., Bailey, J.A., Kaul, R., Morrison, V.A., Pertz, L.M., Haugen, E., Hayden, H., Albertson, D., Pinkel, D., et al. 2005. Fine-scale structural variation of the human genome. *Nat. Genet.* **37:** 727–732.

Walsh, T., McClellan, J.M., McCarthy, S.E., Addington, A.M., Pierce, S.B., Cooper, G.M., Nord, A.S., Kusenda, M., Malhotra, D., Bhandari, A., et al. 2008. Rare structural variants disrupt multiple genes in neurodevelopmental pathways in schizophrenia. *Science* **320:** 539–543.

Wang, W.Y., Barratt, B.J., Clayton, D.G., and Todd, J.A. 2005. Genome-wide association studies: Theoretical and practical concerns. *Nat. Rev. Genet.* **6:** 109–118.

Wang, J., Wang, W., Li, R., Li, Y., Tian, G., Goodman, L., Fan, W., Zhang, J., Li, J., Guo, Y., et al. 2008. The diploid genome sequence of an Asian individual. *Nature* **456:** 60–65.

Wheeler, D.A., Srinivasan, M., Egholm, M., Shen, Y., Chen, L., McGuire, A., He, W., Chen, Y.J., Makhijani, V., Roth, G.T., et al. 2008. The complete genome of an individual by massively parallel DNA sequencing. *Nature* **452:** 872–876.

Willcocks, L.C., Lyons, P.A., Clatworthy, M.R., Robinson, J.I., Yang, W., Newland, S.A., Plagnol, V., McGovern, N.N., Condliffe, A.M., Chilvers, E.R., et al. 2008. Copy number of FCGR3B, which is associated with systemic lupus erythematosus, correlates with protein expression and immune complex uptake. *J. Exp. Med.* **205:** 1573–1582.

Wong, K.K., deLeeuw, R.J., Dosanjh, N.S., Kimm, L.R., Cheng, Z., Horsman, D.E., MacAulay, C., Ng, R.T., Brown, C.J., Eichler, E.E., and Lam, W.L. 2007. A comprehensive analysis of common copy-number variations in the human genome. *Am. J. Hum. Genet.* **80:** 91-104.

Xavier, R.J. and Rioux, J.D. 2008. Genome-wide association studies: A new window into immune-mediated diseases. *Nat. Rev. Immunol.* **8:** 631–643.

Xu, B., Roos, J.L., Levy, S., van Rensburg, E.J., Gogos, J.A., and Karayiorgou, M. 2008. Strong association of de novo copy number mutations with sporadic schizophrenia. *Nat. Genet.* **40:** 880–885.

Yang, Y., Chung, E.K., Wu, Y.L., Savelli, S.L., Nagaraja, H.N., Zhou, B., Hebert, M., Jones, K.N., Shu, Y., Kitzmiller, K., et al. 2007. Gene copy-number variation and associated polymorphisms of complement component C4 in human systemic lupus erythematosus (SLE): Low copy number is a risk factor for and high copy number is a protective factor against SLE susceptibility in European Americans. *Am. J. Hum. Genet.* **80:** 1037–1054.

WWW RESOURCES

http://www.1000genomes.org/page.php   1000 Genomes Project.
http://www.ncbi.nlm.nih.gov/sites/entrez?Db=gap   Database of Genotypes and Phenotypes, dbGAP.

http://projects.tcag.ca/variation   The Database of Genomic Variants.

14 | Oncogenomics

Simon J. Furney,[1-3] Gunes Gundem,[1,3] and Nuria Lopez-Bigas[1,3]

[1]*Biomedical Genomics Group, Research Unit on Biomedical Informatics—GRIB, University Pompeu Fabra, Barcelona Biomedical Research Park, 08003 Barcelona, Spain*

INTRODUCTION

The role of genetic alterations in tumor cells was first introduced in the early 20th century (Ponder 2001). By the early 1970s, it had been demonstrated that viruses could promote cellular transformation in vitro, and Alfred Knudson (1971) had described his hypothesis of two genetic events in the rare cancer retinoblastoma, eventually shown to be due to loss of both alleles of the tumor suppressor gene (Cavenee et al. 1983). Further experimental studies showed point mutations to be the mechanism of activation in oncogenes (Reddy et al. 1982; Tabin et al. 1982). However, at this stage, environmental influences were still viewed as the cause of common cancers (Doll and Peto 1981; Peto 2001).

Today, cancer is viewed as a genetic disease and many genetic mechanisms of oncogenesis have been described (Vogelstein and Kinzler 2004). The progression from normal tissue to invasive cancer is thought to occur over a timescale of 5–20 years. This transformation is driven by both inherited genetic factors and somatic genetic alterations and mutations, and it results in uncontrolled cell growth and, in many cases, death.

In this chapter, we review the main types of genomic and genetic alterations involved in cancer, namely, copy-number changes, genomic rearrangements, somatic mutations, polymorphisms, and epigenomic alterations in cancer. We then discuss the transcriptomic consequences of these alterations in tumor cells. The use of "next-generation" sequencing methods in cancer research is described in the relevant sections. Finally, we discuss different approaches for candidate prioritization and integration and analysis of these complex data.

THE GENETIC BASIS OF CANCER

The transformation of a normal cell into a cancer cell is a multistep process, with each intermediate stage conferring a selective advantage on the cell (Vogelstein and Kinzler 1993). These changes result primarily from irreversible aberrations in the DNA sequence or structure (e.g., translocations, mutations, and copy-number alterations) (Fig. 14.1). However, cancer alterations also include potentially reversible changes, known as epigenetic modifications, to the DNA and/or histone proteins, which are closely associated to the DNA in chromatin (Esteller 2008). Normal cellular homeostasis and division are tightly controlled processes that incorporate signals from many pathways to reg-

[2]*Present address*: NIHR Biomedical Research Centre for Mental Health, Institute of Psychiatry, King's College London, United Kingdom.
[3]All authors contributed equally to this chapter.

143

ulate the expression of the appropriate genes. Mutations or alterations to genes involved in these processes can contribute to cellular transformation by unbalancing the natural physiological equilibrium of a cell. Indeed, cancer progression is the accumulation of a series of genetic alterations in a somatic cell (Vogelstein and Kinzler 2004).

Although there are many different types of cancer (and even subtypes within the same tissue) that result from the action of different sets of genes (Dyrskjot et al. 2003), it has been suggested that the combinations of genes required for oncogenesis can be reduced to six essential changes in cellular physiology (Hanahan and Weinberg 2000): self-sufficiency in growth signals, insensitivity to growth inhibitory signals, evasion of apoptosis, limitless replicative potential, sustained angiogenesis, and tissue invasion and metastasis. The requirement for uncoupling of the normal processes that results in these six alterations demonstrates the complex genetic nature of cancer. It has further been proposed that the chronological order of these alterations is not fixed and can vary among different cancer types (Hanahan and Weinberg 2000).

The genetic alterations that lead to cancer occur only in certain genes. Cancer-causing genes have been traditionally classified as either proto-oncogenes (e.g., the genes for MYC, ERBB2 [Her-2/neu], and EGFR) or tumor suppressor genes such as the genes that encode TP53, CDKN2A, and RB. Proto-oncogenes normally function as proliferative agents, and when mutated or misregulated in cancer, they promote uncontrolled cell growth. Usually they are phenotypically dominant, requiring a gain-of-function mutation or chromosomal gain to become oncogenic. Conversely, tumor suppressor genes are endowed with antiproliferative properties and generally require inactivation of both alleles to induce cancer. This can occur, for example, by point mutation, deletion, or epigenetic silencing. In addition to proto-oncogenes and tumor suppressor genes, stability genes (e.g., base excision repair and mismatch repair genes), which keep genetic alteration to a minimum, have been proposed more recently as an additional type of cancer gene (Vogelstein and Kinzler 2004).

In the last decade, the study of the genetic basis of cancer has undergone a profound transformation. Until recently, most cancer genes had been identified by positional cloning (Futreal et al. 2004), and scientists were focused on studying particular candidate genes involved in oncogenesis. Today, high-throughput techniques allow scientists to simultaneously analyze a large number of genes and their alterations. Cytogenetic methods such as comparative genome hybridization (CGH) have been used to analyze structural changes and genome-wide gains and losses. The use of cDNA microarrays to simultaneously analyze the expression of thousands of genes in tumor samples has become preva-

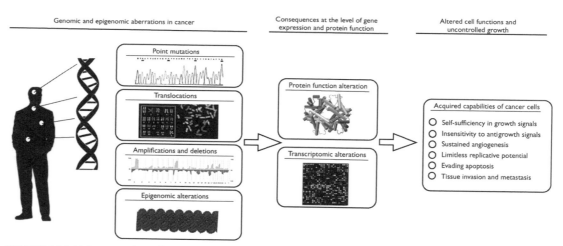

FIGURE 14.1 Main genomic and epigenomic alterations identified in tumor samples. These alterations have consequences at the levels of gene expression and alteration of protein functions. These aberrations and their consequences allow cancer cells to acquired key capabilities for their status (Hanahan and Weinberg 2000).

lent in cancer research. Studies have shown that gene-expression data from tumors are clinically relevant in breast cancer and lymphoma prognosis (van't Veer et al. 2002; Dave et al. 2004) and are able to define cancer subtypes and response to therapies (Ramaswamy and Golub 2002). The use of mutational profiling of tumor genomes has yielded important results during the past few years (Benvenuti et al. 2005). Large-scale exon resequencing of human tumors has been used to identify point mutations in candidate cancer genes in a variety of different tumors (Davies et al. 2002, 2005; Bardelli et al. 2003; Stephens et al. 2004; Sjoblom et al. 2006; Greenman et al. 2007; Wood et al. 2007; Jones et al. 2008; McLendon et al. 2008; Parsons et al. 2008), and new high-throughput methods for DNA methylation and histone modification profiling are being used to identify epigenomic alterations in cancer (American Association for Cancer Research and the European Union Network of Excellence Scientific Advisory Board 2008; Esteller 2008). In addition, several major projects that aim to identify all genetic alterations in common tumor types using genome-wide, high-throughput techniques are in progress, for example, The Cancer Genome Atlas (http://cancergenome.nih.gov) of the National Institutes of Health (NIH), the Cancer Genome Project (http://www.sanger.ac.uk/genetics/CGP/) at the Sanger Institute, and the International Cancer Genome Consortium (http://www.icgc.org/).

These genome-wide, high-throughput technologies have transformed the field of cancer research and have provided powerful ways to understand the mechanism of disease pathogenesis. They also have the potential to identify possible targets for therapy, discover molecular biomarkers that allow early detection of cancer, improve the diagnosis and prognosis or certain tumors, and predict the response to therapies (Baak et al. 2005; Chin and Gray 2008). However, these technologies also yield large volumes of data of multiple types. One of the main challenges is to distinguish between alterations that are causative (driver alterations) from those that are the consequences of the large number of cell divisions coupled with genome instability and checkpoint errors characteristic of cancer cells (passenger alterations) and are not directly involved in tumor development. New methods are needed to be able to prioritize the more promising candidates from genes that are unlikely to be contributing to tumorigenesis (Haber and Settleman 2007; Higgins et al. 2007; Furney et al. 2008a).

Analysis at the level of individual genes is informative, but it does not capture the full complexity of biological systems. Thus, it is also important to study the alterations identified in cancer cells at a more general level. One way to approach this is by embedding genes into functional or regulatory modules and focusing on the study of altered modules instead of single genes. Some of these approaches have been used in the analysis of microarray data, for example, the "module maps" (Ihmels et al. 2002; Segal et al. 2004; Tanay et al. 2004) and "molecular concept maps" (Tomlins et al. 2007).

Recently, more sophisticated studies exploiting data from different techniques and different types of alterations are becoming common in cancer research. A number of studies have revealed the effectiveness of integrative functional genomics in cancer research, in which information from complementary experimental data sources is combined to provide greater insight into the process of tumorigenesis (Rhodes et al. 2004; Lu et al. 2005b; Bild et al. 2006; Carter et al. 2006; Liu et al. 2006; Stransky et al. 2006; Tomlins et al. 2007; Jones et al. 2008; McLendon et al. 2008; Parsons et al. 2008).

A SURVEY OF GENOMIC AND GENETIC ALTERATIONS IN CANCER

Analyzing Copy-Number Changes in Cancer

Aneuploidy in tumor cells, particularly in human cancers, is observed frequently. Cytogenetic methods such as karyotyping, fluorescence in situ hybridization (FISH), and CGH (Kallioniemi et al. 1992) have been used with great effect to analyze large structural chromosomal changes, gains and losses of specific genes, and genome-wide gains and losses in cancer. In the last decade, array-based CGH (aCGH) (Pinkel et al. 1998) has become the technique of choice for investigating copy-number changes in cancer research and has been used to classify tumors, identify markers, and delineate the structure of chromosomal aneuploidies (Kallioniemi 2008). Recently, meta-analyses of CGH data from tumors have shown that tumors can be classified using these data (Baudis 2007; Jong et al. 2007). Access to the results of many CGH studies is provided in curated online databases such as the NCBI/NCI's Cancer Chromosomes (Knutsen et al. 2005) and Progenetix (Baudis and Cleary 2001).

The development of high-resolution single-nucleotide polymorphism (SNP) arrays has facilitated surveying of copy-number changes at a higher resolution and the detection of loss of heterozygosity (Mulligan et al. 2007; Weir et al. 2007). For instance, Mulligan et al. (2007) have applied this technology in more than 200 cases of pediatric acute lymphoblastic leukemia to identify a range of somatic deletions and amplifications.

Finding Genomic Rearrangements

Chromosomal translocations and subsequent gene fusion events have an important role in the initial steps of tumorigenesis. About 360 different gene fusion events have been identified (Mitelman et al. 2007). Translocations are recognized as a common mechanism of oncogenesis in leukemias and lymphomas (Mitelman et al. 1997; Rowley 1998), whereas relatively few translocations have been detected in solid tumors (Mitelman 2000). This is probably not because they are uncommon in solid tumors but due to technical and analytical limitations reflecting the complex genomic profiles and heterogeneous nature of these malignancies (Mitelman et al. 2007). Perhaps the most well-known chromosomal translocation in cancer is the Philadelphia chromosome discovered by Peter Nowell and David Hungerford in 1960 (Nowell 2007). Prior cytogenetic and molecular studies showed that it consisted of a translocation between chromosomes 9 and 22, resulting in a chimeric, constitutively active tyrosine kinase BCR-ABL fusion protein that is responsible for chronic myeloid leukemia (Groffen et al. 1984; Shtivelman et al. 1985).

Effects of Translocations in Cancer

At the molecular level, the effect of most of the translocations involved in cancer can be attributed to one of the following mechanisms: (1) Translocations can create chimeric proteins due to the fusion of parts of two genes, one in each breakpoint, as in case of the BCR-ABL fusion protein (Groffen et al. 1984; Shtivelman et al. 1985). As a result of this fusion, the activity of the nonreceptor tyrosine kinase ABL is misregulated. This case is particularly relevant due to the effectiveness of the drug Gleevec (imatinib mesylate), which inhibits tyrosine kinase activity, in combating this type of cancer (Druker 2002). Numerous other translocations resulting in fusion proteins have also been described (Rabbitts 1994; Mitelman 2000; Rowley 2001). (2) Translocations can result in the misregulation of one of the genes involved in the fusion event by placing it close to the regulatory elements of another gene. This usually results in the ectopic expression of an apparently normal gene. Examples of these cases are common translocations (between chromosomes 8 and 2, 14, or 22) present in Burkitt's lymphoma that place the *MYC* gene close to an immunoglobulin gene, encoding either the heavy chain (IGH) or the kappa (IGK) or lambda (IGL) light chains. As a consequence of the translocation, the *MYC* gene becomes constitutively expressed due to the influence of regulatory elements of the immunoglobulins (Kuppers 2005).

The Mitelman Database of Chromosome Aberrations in Cancer (now part of the NCI/NCBI Cancer Chromosomes Database) catalogs chromosomal aberrations and relates them to tumor characteristics (Mitelman 2008). This database is manually curated from published literature by its authors.

Methods for Detecting Chromosomal Rearrangements

Numerous methods exist for the detection of chromosomal rearrangements (for review, see Morozova and Marra 2008). The earliest methods applied involved examination of chromosomes and chromosome banding patterns by microscopy. An important advance in molecular cytogenetics was the development of in situ hybridization techniques (Buongiorno-Nardelli and Amaldu 1970). This procedure is based on the hybridization of a labeled probe to a complementary target where probe copy number is assessed by microscopy. Some developments of the classical FISH methods are multiplex FISH (M-FISH) (Speicher et al. 1996; Speicher and Ward 1996), spectral karyotyping (SKY) (Schrock et al. 1996), and combined binary ratio labeling (COBRA) (Tanke et al. 1999), which allow the simultaneous display of all chromosomes in 24 colors. FISH techniques are adequate to detect gross chromosomal aberrations; however, they are limited for smaller-scale chromosomal aberrations.

Recently, Arul Chinnaiyan and colleagues have applied a new integrative analytical methodology called cancer outlier profile analysis (MacDonald and Ghosh 2006). This method, which identifies associations between genomic and transcriptional abnormalities, allowed them to identify a family of common translocations in prostate cancer that brings ETS family genes under the control of TMPRSS2, in effect placing the expression of these genes under androgen-mediated regulation (Tomlins et al. 2005, 2006).

Sequencing approaches have also been developed for the detection of chromosomal aberrations. In this case, DNA from a tumor is cloned into a large insert, and the ends of the resultant clones are sequenced and then mapped onto the reference human DNA sequence. Paired ends that map farther apart than the maximum size tolerated by the clone indicate the presence of a structural aberration (Volik et al. 2003, 2006; Krzywinski et al. 2007). More recently, the combination of ultrafast DNA sequencing and bioinformatics allows high-resolution and massive paired-end mapping (PEM) (Korbel et al. 2007). This technique consists of the isolation of 3-kb sequence fragments and then end sequencing with 454/Roche technology, followed by mapping of paired-end reads back to the reference sequence using a computational algorithm developed by the authors. Campbell and colleagues have used this approach to identify structural variants in the genome of germ-line and lung cancer cells of two individuals. This analysis allowed the identification of 306 germ-line structural variants and 103 somatic rearrangements to the base-pair level of resolution (Campbell et al. 2008). In addition, Maher et al. (2009) have used a combination of high-throughput long- and short-read transcriptome sequencing to identify known and novel fusion transcripts in cancer cell lines and tumors.

Somatic Mutations in Cancer

Somatic mutations are alterations in the nucleotide sequence of a gene, such as single base-pair changes as well as those creating small insertions or deletions. Mutations can be classified in a variety of ways: (1) silent (no net effect on the amino acid code), missense (change of the original amino acid codon to another), or nonsense (change of the original amino acid codon to a stop codon); (2) loss of function (the function is lost or weakened) or gain of function (the protein becomes more active or gains a new or abnormal function); or (3) transition and transversion.

Mutational Patterns

Different types of mutations affect genes altered in cancers. However, one can draw some generalizations from the mutational patterns observed. For example, oncogenes usually undergo gain-of-function mutations. A typical example is BRAF. One of the most common changes observed in this kinase is the conversion of a valine to a glutamate at codon 599 within the activation loop of the kinase domain. This substitution leads to the constitutive activation of the protein product even in the absence of an activating signal. The "turned-on" BRAF kinase phosphorylates downstream targets leading to abnormal growth (Wan et al. 2004). On the other hand, tumor suppressor genes are usually rendered nonfunctional by loss-of-function mutations. A point mutation in TP53 inactivates its capacity to bind to the sequences it regulates transcriptionally (Vogelstein et al. 2000). "Disabled" TP53 cannot do its normal job of inhibiting cell growth and stimulating cell death in times of stress.

Databases

As data have accumulated, the results from mutational analysis studies have been stored in online databases. Some of these focus on a specific gene (p53 database, http://www-p53.iarc.fr/; Olivier et al. 2002); EGFR (http://www.somaticmutations-egfr.org/), whereas others are tissue-specific (Breast Cancer Mutations Database, http://research.nhgri.nih.gov/bic/). COSMIC (Catalogue of Somatic Mutations in Cancer, http://www.sanger.ac.uk/genetics/CGP/cosmic), on the other hand, stores somatic mutations that have been reported in the literature regarding many cancer types (Forbes et al. 2006).

Sequencing and Mutational Screens

Initial large-scale sequencing efforts focused on signaling pathways previously known to be mutated in at least one gene (Davies et al. 2002; Rajagopalan et al. 2002). In addition to well-known pathways, specific gene families have been scrutinized: the tyrosine kinases (Bardelli et al. 2003), lipid kinases (Samuels et al. 2004), tyrosine phosphatases (Wang et al. 2004), and tyrosine kinase receptors (Paez et al. 2004). These and similar studies pointed to the importance of kinase and phosphatase mutations and led to the identification of some important genes such as *PI3KCA*, *BRAF*, *EGFR*, and *JAK2* in many tumors.

The first report on the genomic landscape of somatic mutations focused on human breast and colorectal cancers (Sjoblom et al. 2006). A two-stage strategy was followed in this study. In the discovery screen, the authors performed mutational screens for the consensus coding sequences (CCDS) in 11 breast and 11 colorectal tumors. The putative mutations were filtered to exclude silent changes, changes present in normal samples, known polymorphisms from dbSNP (Single Nucleotide Polymorphism Database, http://www.ncbi.nlm.nih.giv/projects/SNP), false-positive calls upon visual inspection of sequence chromatograms, and confirmation by resequencing. The mutations passing all these criteria were sequenced again in a validation screen in 24 additional breast and colorectal tumors. After filtering as before, 921 and 751 mutations were identified in breast and colorectal cancers, respectively. In all, 92% of the mutations were single-base substitutions, the majority of which were missense. There were significant differences in the mutational spectra of the two tumor types at CG base pairs: Colorectal cancer samples were biased in TA transitions, whereas breast cancer samples were prone to GC transversions. A total of 44% and 11% of the colorectal mutations occurred in 5′-CG-3′ and 5′-TpC-3′ sites, respectively; these numbers were 17% and 31% for breast mutations. This result implies that there might be differences in the mechanisms of mutagenesis in the two tumor types.

To discriminate the "driver" mutations from the "passenger" mutations, a cancer mutation prevalence (CaMP) score was calculated as follows: Mutations were divided into different categories taking into account the type of the base mutated, the resulting base change, the 5′ and 3′ neighbors, and codon usage. This resulted in the identification of 122 and 69 candidate genes for the breast and colorectal tumors, respectively. In these genes, some biological functions were overrepresented in the candidate genes, such as transcription factors and cell-adhesion- and signal-transduction-related genes.

Overall, this first large-scale sequencing effort revealed that the majority of the genes identified had not been previously known to have been mutated. In addition, different genes were mutated in breast and colorectal cancers. These genes also showed different biases in the type of nucleotide substitutions. Moreover, even the samples of the same cancer type were very heterogeneous, which might be the reason why gene sets related in a biologically meaningful way can explain prognosis, response to therapy, etc., better than individual genes.

Difficulties in Predicting Candidate Cancer Genes

The Sjoblom et al. study, however, raised a lot of questions. Some were skeptical about the usefulness of the brute-force sequencing projects, emphasizing the importance of focusing on reverse engineering approaches (Loeb and Bielas 2007; Strauss 2007). Critics compared the high costs of such large-scale projects with the limited results obtained. It is true that high-throughput approaches cannot replace functional studies, but such bioinformatic screenings can guide experimental studies in more efficient directions, especially with the advent of more cost-efficient technologies. Other discussions centered around the robustness of the statistical methods, the background mutations rates, and the small sample sizes used (Forrest and Cavet 2007; Getz et al. 2007; Rubin and Green 2007). All of these are important factors that can affect the resulting genes found to be significantly mutated.

In another study, the Sjoblom et al. (2006) analysis was extended to include all RefSeq genes (Wood et al. 2007). Using the same methods, 1718 genes with at least one nonsynonymous mutation in either breast or colorectal cancer were identified. The mutation spectra of the two tumor types were similar to those of the previous analysis. Comparison of these with the sequencing of pancreas and brain tumors (Jones et al. 2008; Parsons et al. 2008) indicated that breast tumors have a somatic mutation spectrum different from that of the other three, with a relatively high number of mutations at 5′-TpG sites and a small number at 5′CpG sites.

Of the 1718 genes with nonsynonymous mutations, 280 were predicted to be candidate cancer genes. One of the conclusions the authors reached was that very few genes are mutated at high frequencies in human cancers ("mountains" in the mutational landscape). These genes (e.g., *TP53*, *PTEN*, and *PIK3CA*) might have critical roles in tumorigenesis. On the other hand, a much larger number of genes are mutated at low frequencies. This indicates that a large number of the mutations confer only small advantages to the tumorigenic phenotype. However, this view also points to the difficulty of discrimination of driver mutations from passengers. Recently, Ding et al. (2008) sequenced 623 genes in 188 human lung adenocarcinomas, identifying 26 genes that were mutated at significantly high frequencies.

Common Variants in Cancer

The International HapMap project (International HapMap Consortium 2005) has facilitated the recent explosion of genome-wide association studies (GWAS) attempting to determine common variants (in general, SNPs) that contribute to common diseases. Many of these GWAS have identified SNPs associated with different tumor types. We mention only some of these studies below because it is not feasible to provide a comprehensive review of the field within the scope of this chapter.

In breast cancer, Cox et al. (2007) found a common coding variant in caspase 8 to be associated with an increased risk of the disease. Easton et al. (2007) identified five novel loci, including FGFR2 and TOX3, showing genome-wide significant association with breast cancer. The CASP8 and TOX3 associations were independently confirmed by Tapper et al. (2008), who also identified SNPs in six genes associated with disease prognosis. Further recent studies have found associations at a number of genomic loci (Ahmed et al. 2009; Thomas et al. 2009; Zheng et al. 2009).

Studies of prostate cancer have also identified a number of loci associated with the disease, including independent replications of a risk locus at chromosome 8q24 (Gudmundsson et al. 2007, 2008; Haiman et al. 2007a,b; Yeager et al. 2007; Thomas et al. 2008). In addition, genome-wide associations have been identified in a number of other tumor types such as lung cancer (Amos et al. 2008; Y. Wang et al. 2008), chronic lymphocytic leukemia (Di Bernardo et al. 2008), colorectal cancer (Houlston et al. 2008; Tenesa et al. 2008), urinary bladder cancer (Kiemeney et al. 2008), diffuse cancer-type gastric cancer (Sakamoto et al. 2008), and basal cell carcinoma (Stacey et al. 2008), and also in multiple tumor types (Rafnar et al. 2009).

Epigenomic Alterations in Cancer

Epigenetic alterations are increasingly being recognized as central mechanisms of tumor development. Modifications of the DNA methylation landscape as well as of histone modifications seem to be a common feature of many tumor samples (Esteller 2007, 2008).

Types of Epigenetic Changes

The low level of DNA methylation in tumors compared to that in normal tissue counterparts was one of the first epigenetic alterations to be found in human cancer (Feinberg and Vogelstein 1983). This hypomethylation occurs mainly in gene-poor areas (Weber et al. 2005). The proposed mechanisms by which genome hypomethylation can contribute to the development of a cancer cell are generation of chromosomal instability (Eden et al. 2003), reactivation of transposable elements (Bestor 2005), and loss of imprinting (Feinberg 1999; Cui et al. 2003; Kaneda and Feinberg 2005).

In contrast, hypermethylation of CpG islands in promoter regions of certain genes (tumor suppressor genes) is an important event in many cancers. This is the case regarding the retinoblastoma tumor suppressor gene (*Rb*) (Greger et al. 1989; Sakai et al. 1991), *P16*[INK4a] (Herman et al. 1994, 1995; Gonzalez-Zulueta et al. 1995; Merlo et al. 1995), *hMLH1* (Herman and Baylin 2003), and *BRCA1* (breast cancer susceptibility gene 1) inactivation (Herman and Baylin 2003).

Histone modifications (such as acetylations or methylations) have direct effects on the regulation of gene transcription. Generally, histone acetylation is associated with transcriptional activation (Bernstein et al. 2007; Mikkelsen et al. 2007); however, the histone methylation effect depends on the residue modified (Bernstein et al. 2007; Mikkelsen et al. 2007). It is becoming clear that combinations of histone modifications have an effect on transcriptional regulation (Z. Wang et al. 2008).

Several lines of evidence point to the importance of alterations in histone modification as relevant steps in the transformation process. Examples include the association between CpG island hypermethylation in cancer and a particular combination of histones markers, namely, deacetylation of histones H3 and H4, loss of histone H3 lysine K4 (H3K4) trimethylation, and gain of H3K9 methylation and H3K27 trimethylation (Fahrner et al. 2002; Ballestar et al. 2003; Vire et al. 2006). In addition, it has been observed that cancer cells undergo a general loss of monoacetylated and trimethylated forms of histone H4 (Fraga et al. 2005). However, it is thought that the main findings on the extent and implications of epigenomics in cancer are still to come in the future with the development of the international Human Epigenome Project (American Association for Cancer Research Human Epigenetic Task Force 2008) (http://www.epigenome.org/).

Methods for Detecting Epigenetic Modifications

Several approaches are available to study epigenetic modifications in normal and cancer cells. Some of these profile epigenetic alteration in a genome-wide manner, whereas others are centered in gene-specific alterations.

High-performance liquid chromatography (HPLC) and high-performance capillary electrophoresis (HPCE) allow the quantification of the total amount of 5-methylcytosine (Fraga and Esteller 2002; Esteller 2007). The study of DNA methylation at particular sequences has classically been based on the action of restriction enzymes that can distinguish between methylated and unmethylated recognition sites (Esteller 2007). Later, methods based on the use of bisulfite treatment of DNA, which changes unmethylated cytosines to uracil and leaves methylated cytosines unchanged, were developed (Clark et al. 1994; Herman et al. 1996). These methods can be coupled with polymerase chain reaction (PCR) and sequencing of candidate genes. They can also be combined with genomic approaches to detect genome-wide DNA methylation patterns, for example, by using promoter microarrays or arbitrary primed PCR, in which no prior sequence information is required for amplification.

In addition, techniques can be used that are based on chromatin immunoprecipitation (ChIP), with the ChIP-on-chip approach using antibodies against methyl-CpG-binding domain proteins (MBDs) (Lopez-Serra et al. 2006), which have a great affinity for binding to methylated cytosines. An antibody directly against 5-methylcytosine (methyl-DIP) can also be used (Weber et al. 2005; Keshet et al. 2006).

Another way of assessing genome-wide DNA methylation patterns is by using gene-expression profiling microarrays comparing mRNA levels from cancer cell lines before and after treatment with a demethylating drug (Suzuki et al. 2002; Yamashita et al. 2002). However, this method yields a significant amount of false positives, requiring confirmation by bisulfate genomic sequencing.

The profiling of histone modification marks is typically studied by ChIP using antibodies against specific histone modifications. The immunoprecipitated DNA is then analyzed by PCR with specific primers to investigate the presence of a candidate DNA sequence or on a microarray chip (ChIP-on-chip) to profile an extensive map of histone modifications (Azuara et al. 2006; Bernstein et al. 2007). More recently, ChIP has been combined with ultrasequencing techniques (ChIP-seq) to obtain higher-resolution chromatin modification maps (Z. Wang et al. 2008).

Databases

Several databases have been created to collect and annotate alterations in DNA methylation (Table 14.1). DNA Methylation Database (MethDB, http://www.methdb.de) is a well-maintained resource that stores DNA methylation data in a standard format (Grunau et al. 2001). In addition, specialized databases focus on methylation aberrations detected in cancer samples: PubMeth (http://www.pubmeth.org; Ongenaert et al. 2008), MeInfoText (http://mit.lifescience.ntu.edu.tw/index.html; Fang et al. 2008), and MethyCancer (http://methycancer.psych.ac.cn; He et al. 2008). MethyCancer collects data from other public databases and resources, including MethDB, and integrates this information with CpG island prediction and expression data. PubMeth and MeInfoText extract information from MedLine publications using text mining and manual curation.

TRANSCRIPTOMIC CHANGES IN TUMORS

The result of the cumulative effect of the different alteration types we have described is observed at the level of expression of the gene product. For example, genomic copy-number loss and epigenetic silencing may account for the down-regulation of the micro RNA (miRNA) gene expression, which further contributes to a genome-wide transcriptional deregulation at the level of mRNAs (Zhang et al. 2008). Therefore, to paint a complete picture of tumorigenesis, it is crucial to include changes at the expression level of both miRNAs and mRNAs. Actually, the use of high-throughput gene-expression profiling studies of tumorigenic cells has been used extensively and has changed cancer research substantially.

Methods for Detecting Transcriptomic Changes

Although it has long been known that tumor cells express some genes at abnormal levels, these large-scale expression studies showed that large numbers of genes are differentially expressed in cancer cells. Given that changes in expression are a reflection of the underlying complexity of different alterations, it is no surprise that high-throughput expression analysis is extremely difficult. How should long lists of deregulated genes be interpreted? How should one decide which of the transcriptionally deregulated genes are causally implicated in cancer?

Expression Analysis

One suggestion has come from "gene signature" studies. Instead of a single gene, tumorigenic phenotypes can be explained by the signature defined by the expression level of a list of genes. To identify groups of genes that change in expression, "unsupervised methods" have proved to be very useful. Without any a priori information, these methods can help to discover patterns in the data. These methods led to the characterization of previously unknown, but clinically significant, subtypes of cancer in breast cancer (Perou et al. 2000; Sorlie et al. 2003), B-cell lymphoma (Alizadeh et al. 2000), Burkitt's lymphoma (Dave et al. 2006), prostate cancer (Lapointe et al. 2004), and lung cancer (Hayes

TABLE 14.1 Resources and databases for oncogenomics

Name	Description	Web address
Mutations in cancer		
IARC TP53 Mutation Database	Compiles all TP53 gene variations identified in human populations and tumor samples. Data compiled from peer-reviewed literature and generalist databases.	http://www-p53.iarc.fr
EGFR Mutations Database	Comprehensive compendium of all somatic mutations in EGFR that have been identified in human cancers and reported in peer-reviewed literature.	http://www.somaticmutations-egfr.org/
Breast Cancer Mutations Database	Central repository for information regarding mutations and polymorphisms in breast cancer susceptibility genes.	http://research.nhgri.nih.gov/bic/
COSMIC	Catalogue of Somatic Mutations in Cancer.	http://www.sanger.ac.uk/genetics/CGP/cosmic
Structural alterations in cancer		
Progenetix	Overview of copy-number abnormalities in human cancer from comparative genomic hybridization (CGH) experiments. A curated database, it collects genomic gain/loss information of individual cancer and leukemia cases published in peer-reviewed journals.	http://www.progenetix.net/progenetix
Mitelman Database of Chromosome Aberrations in Cancer	Relates chromosomal aberrations to tumor characteristics, based on either individual cases or associations.	http://cgap.nci.nih.gov/Chromosomes/Mitelman
NCBI/NCI's Cancer Chromosomes	Integrates NCI/NCBI SKY/M-FISH and CGH Database, NCI Mitelman Database of Chromosome Aberrations in Cancer, and NCI Recurrent Aberrations in Cancer Database.	http://www.ncbi.nlm.nih.gov/sites/entrez?db=cancerchromosomes
DNA methylation alterations in cancer		
MethyCancer	Database of human DNA methylation and cancer; it collects data from other public databases and resources and integrates this information with CpG island prediction and expression data.	http://methycancer.psych.ac.cn
Transcriptomic alterations in cancer		
GEO	Gene-expression/molecular abundance repository supporting MIAME-compliant data submissions and a curated, online resource for gene-expression data browsing, query, and retrieval. Not specific for cancer, but contains many cancer data sets.	http://www.ncbi.nlm.nih.gov/geo/
ArrayExpress	Public archive for transcriptomics data aimed at storing MIAME- and MINSEQE-compliant data in accordance with Microarray Gene Expression Database (MGED) recommendations. ArrayExpress Warehouse stores gene-indexed expression profiles from a curated subset of experiments in the archive. Like GEO, not specific for cancer but contains many cancer data sets.	http://www.ebi.ac.uk/microarray-as/ae/

Oncomine	Cancer microarray database and web-based data-mining platform aimed at facilitating discovery from genome-wide expression analyses.	http://www.oncomine.org/
Integrative projects		
The Cancer Genome Atlas	Comprehensive and coordinated effort to accelerate our understanding of the molecular basis of cancer through application of genome analysis technologies, including large-scale genome sequencing.	http://tcga.cancer.gov/
Cancer Genome Project	Sanger Center Project aimed at identifying somatically acquired sequence variants/ mutations and, hence, genes critical in the development of human cancers.	http://www.sanger.ac.uk/genetics/CGP/
International Cancer Genome Consortium	International consortium with the goal of obtaining comprehensive description of genomic, transcriptomic, and epigenomic changes in 50 different tumor types and/or subtypes of clinical and societal importance across the globe.	http://www.icgc.org/
NCI-CGAP	Interdisciplinary program established and administered by the National Cancer Institute (NCI) to generate information and technological tools needed to decipher molecular anatomy of the cancer cell.	http://www.ncbi.nlm.nih.gov/ncicgap/
Integrative resources		
IntOGen	Discovery tool for cancer researchers; integrates multidimensional OncoGenomics Data for the identification of genes and groups of genes (biological modules) involved in cancer development.	http://www.intogen.org
UCSC Cancer Genomics Browser	Suite of web-based tools to integrate, visualize, and analyze cancer genomics and clinical data. Displays a whole-genome and pathway-oriented view of genome-wide experimental measurements for individual and sets of samples alongside their associated clinical information.	http://genome-cancer.ucsc.edu
Other resources		
Cancer Gene Census	Ongoing effort to catalog those genes for which mutations have been causally implicated in cancer.	http://www.sanger.ac.uk/genetics/CGP/Census/
CancerGenes	Resource to simplify the process of gene selection and prioritization in large collaborative projects. Combines gene lists annotated by experts with information from key public databases.	http://cbio.mskcc.org/CancerGenes
CGPrio	Resource for the prioritization of candidate cancer genes after genomic experiments. Prioritization of oncogenes and tumor suppressor genes is based on computational classifiers that use different combinations of sequence and functional data, including sequence conservation, protein domains and interactions, and regulatory data.	http://bg.upf.edu/cgprio

et al. 2006). In addition to mRNA expression, even miRNA expression information has been proven to be helpful in dissecting cancer (He et al. 2005; Lu et al. 2005b; Volinia et al. 2006).

There are also other methods used in expression analysis that make use of "supervised methods." Using existing biological information as a guide, this approach has been successfully used to predict recurrence, metastasis, outcome, response to drugs, etc. (Beer et al. 2002; Pomeroy et al. 2002; Shipp et al. 2002; van de Vijver et al. 2002; van't Veer et al. 2002; Ramaswamy et al. 2003; Paik et al. 2004; Potti et al. 2006).

Databases

Accumulation of large amounts of expression data prompted the generation of public databases such as NCBI's Gene Expression Omnibus (http://www.ncbi.nlm.nih.gov/geo; Barrett et al. 2005), the European Bioinformatics Institute's ArrayExpress (http://www.ebi.ac.uk/microarray-as/ae/; Parkinson et al. 2007), the Stanford Microarray Database (http://smd.stanford.edu; Marinelli et al. 2008), and Oncomine (http://www.oncomine.org; Rhodes et al. 2007) (see Table 14.1). The first three mainly serve as data storage platforms and also provide data analysis options. Oncomine, on the other hand, is designed as a data-mining tool specific to cancer-related expression analysis. Such repositories make it possible to compare microarray results among one another. A higher level of information can be extracted by the meta-analysis of expression data from different studies.

Module Maps and Molecular Concept Maps

Given the heterogeneity of cancer and the noisy nature of expression data, however, being able to make discoveries at such a level necessitates the adoption of "gene-set-centered" approaches. An example of this is "module maps" (Ihmels et al. 2002; Tanay et al. 2004). This method was used in the meta-analysis of 2000 microarray experiments using 300 gene sets (Segal et al. 2004). A total of 456 gene modules were identified and were later used to compare different types of cancers. The authors found that previously unrelated tumor types could have similar expression patterns when analyzed at the level of modules. For example, a bone osteoblastic module (consisting of genes associated with proliferation and differentiation in bones) was found to be up-regulated in some breast cancers and down-regulated in lung cancer, hepatocellular carcinoma, and acute lymphoblastic leukemia.

Another example of integrative approaches to cancer expression data is the "molecular concept map" (Tomlins et al. 2007). Molecular concepts are sets of biologically related genes coming from gene annotations from external databases, computationally derived regulatory networks, and microarray gene-expression profiles coming from the Oncomine database. The gene signatures were obtained using a method called COPA (cancer outlier profile analysis). This method was developed to identify "outlier" gene sets, even if their expression level is low or a small number of samples show overexpression (Tomlins et al. 2005).

PRIORITIZATION OF CANDIDATE CANCER GENES

Cancer Gene Census

In 2004, Futreal et al. (2004) published a census of human cancer genes gleaned from published literature. Subsequent additions to the initial census of 291 genes have increased the total to more than 370 genes in 2008. A number of criteria were used for inclusion in the census. Only genes in which cancer-causing mutations have been reported were included, and a requirement for two independent reports of mutations in primary clinical samples was used. Genes involved in translocation or copy-number change were included. However, genes for which there was only evidence of differential expression level or aberrant promoter DNA methylation in tumors were excluded.

The survey also included various data about each gene. For instance, the mutation type evident in the cancer gene (somatic, germ line, or both), neoplasm types associated with the gene (leukemias/lymphomas, mesenchymal, epithelial, etc.), the phenotypic nature of the mutated gene (dominant or recessive), and the mechanism of mutation affecting each gene (e.g., translocation, deletion, and frameshift) were recorded.

A number of general trends were highlighted in the analysis of the compiled list of genes. Approximately 90% of the genes had somatic mutations, 20% had germ-line mutations, and 10% were susceptible to both types of mutation. The most common somatic genetic changes seen were chromosomal translocations, with recurrent events frequently taking place in leukemias and lymphomas. A total of 90% of somatic mutations were phenotypically dominant in tumors, whereas 90% of germ-line mutations were found to be recessive.

In addition, the study examined the distribution of Pfam protein domains (Finn et al. 2006) in the proteins encoded by the cancer genes compared to the entire human proteome. Protein kinase domains, domains involved in transcriptional regulation, and DNA maintenance and repair-associated domains were overrepresented in the group of cancer genes.

Computational Prioritization of Cancer Genes

Many issues remain to be determined in understanding oncogenesis in different tumor types, for example, elucidation of candidate causative agents, distinguishing between driver and passenger alterations (Haber and Settleman 2007; Higgins et al. 2007), and characterization of the function of cancer genes in the oncogenic process (Hu et al. 2007). Oncogenomic experiments are now providing the cancer research community with numerous candidate causative genes. However, it is imperative to prioritize the more promising candidates from genes that are unlikely to be contributing to tumorigenesis. A number of previous computational studies have aimed at predicting cancer-associated missense mutations (Kaminker et al. 2007a,b).

Recently, we have described a number of different approaches for candidate cancer prioritization, irrespective of the oncogenic alteration. We have shown before that it is possible to develop an accurate classifier for distinguishing between Cancer Gene Census genes and other human genes (Furney et al. 2006). However, it is evident from cancer biology that altered proto-oncogenes and tumor suppressor genes promote oncogenesis in different ways. Furthermore, we have also shown that differences in sequence and regulatory properties exist between these two types of cancer genes (Furney et al. 2008b). These issues prompted us to devise separate classifiers for proto-oncogenes and tumor suppressor genes (Furney et al. 2008a). We constructed computational classifiers using different combinations of sequence and functional data including sequence conservation, protein domains and interactions, and regulatory data. We found that these classifiers are able to distinguish between known cancer genes and other human genes. Furthermore, the classifiers also discriminate candidate cancer genes from a recent mutational screen from other human genes. We have provided a web-based facility (CGPrio) through which cancer biologists may access our results (http://bg.upf.edu/cgprio).

INTEGRATION OF ONCOGENOMIC DATA TYPES

An integrative approach is necessary to obtain a more complete view of the deregulation of normal cellular processes that occurs during oncogenesis. During the past few years, a number of studies have revealed the effectiveness of integrative functional genomics in cancer research, whereby information from complementary experimental data sources is combined to provide greater insight into the process of tumorigenesis (Rhodes et al. 2004; Lu et al. 2005a; Bild et al. 2006; Carter et al. 2006; Stransky et al. 2006; Tomlins et al. 2007).

Study Approaches

Integrative studies have combined data from different microarray experiments (Rhodes et al. 2004; Tomlins et al. 2007), expression and copy-number change data (Carter et al. 2006; Stransky et al. 2006), and expression of mRNAs and miRNAs (Lu et al. 2005a). Other recent studies have used a comparative oncogenomic approach to identify genes contributing to oncogenesis and metastasis (Kim et al. 2006; Zender et al. 2006; Maser et al. 2007). For example, Kim et al. (2006) found a 850-kb amplicon from an array CGH analysis of a melanoma mouse model equivalent to a section of a much larger amplification observed in human melanoma. Using expression analysis, they were able to identify *NEDD9* as the gene most likely to be responsible for driving metastasis.

Zender and colleagues (2006) identified syntenic amplifications in human liver carcinomas and a mouse model of hepatocellular carcinoma by array CGH of tumors from both species. A subset of candidate oncogenes was identified by excluding those genes absent in the amplified regions in either mouse or human tumors. RNA and protein expression analyses in both species of the remaining genes pinpointed *cIAP1* and *Yap* as oncogenes.

Maser et al. (2007) engineered murine lymphomas with destabilized genomes to mimic the far more prevalent chromosomal instability associated with human tumors. They generated a mouse lymphoma that was deficient for Atm, Terc, and p53 and assessed these tumors and human T-cell acute lymphoblastic leukemias/lymphomas using array CGH. The authors found recurrent syntenic amplifications and deletions in the human and mouse lymphomas and, upon targeted resequencing of candidate genes within syntenic regions, discovered frequent somatic mutations in PTEN and FBXW7.

Recently, three large-scale collaborative projects have resulted in the integrative analysis of human glioblastomas and pancreatic cancers (Jones et al. 2008; McLendon et al. 2008; Parsons et al. 2008). The Cancer Genome Atlas Research Network (http://cancergenome.nih.gov) presented an integrative analysis of DNA copy number, DNA methylation, and mRNA expression in more than 200 human glioblastomas (McLendon et al. 2008). In addition, they determined the nucleotide sequence in 91 of the tumors. This wealth of data allowed the authors to identify core signaling pathways that are affected in glioblastoma, including receptor tyrosine kinase (RTK) signaling and the p53 and retinoblastoma tumor suppressor pathways.

Parsons et al. (2008) interrogated the same tumor type in 22 samples by sequencing >20,000 protein-coding genes, analyzing copy-number changes, and performing serial analysis of gene expression (SAGE) on 16 samples. This study found that the majority of tumors showed alterations in genes belonging to each of the p53, retinoblastoma, and PI3K pathways. In addition, the candidate cancer genes identified by the authors included several genes previously associated with glioblastoma (e.g., *p53*, *EGFR*, and *NF1*).

Jones et al. (2008) surveyed pancreatic tumors using a similar strategy of transcript nucleotide sequence determination, copy-number change evaluation, and gene-expression analysis. On average, they detected 63 alterations per tumor, most of which were point mutations. Through pathway analysis, they found a core set of 12 signaling pathways/processes in which at least one gene had a genetic alteration in 67%–100% of the tumors.

In addition, in a study by Parsons et al. (2008), IDH1 (isocytrate dehydrogenase 1) was detected to be mutated in all secondary glioblastomas and was linked to a better prognosis. In a McLendon et al. (2008) study, on the other hand, integration of methylation profiling led to the identification of how MGMT (O^6-methylguanine–DNA methyltransferase) promoter methylation status has substantial influence on the overall frequency and pattern of mutations in glioblastoma. This has clinical implications for alkylating agents, and one such agent is temozolomide, which is used in the clinical treatment of this cancer. The current standard practice for patients is surgical intervention followed by adjuvant radiation therapy or chemotherapy with temozolomide. However, this treatment only produces a median survival of 15 months.

These studies underline the need to investigate different types of aberrations in cancer and highlight how crucial the integration of these different methods is in understanding oncogenesis.

Integrative Oncogenomic Projects and Resources

The International Cancer Genome Consortium

The International Cancer Genome Consortium (ICGC; http://www.icgc.org/), launched in 2008, is a collaboration designed to produce high-quality genomic data in multiple cancer types. This international consortium has three primary goals: (1) to coordinate projects to generate comprehensive catalogs of somatic mutations in tumors in 50 different cancer types and/or subtypes that are of global, clinical, and societal significance; (2) to generate transcriptomic and epigenomic data sets from the same tumors; and (3) to ensure that these data are available to the research community at large as quickly as possible and with minimal restrictions.

The Cancer Genome Atlas

The Cancer Genome Atlas (TCGA; http://tcga.cancer.gov/) is a U.S. National Institutes of Health initiative involving the National Cancer Institute and the National Human Genome Research Institute. The goal of the project is to increase the understanding of cancer through the systematic use of various genome-wide technologies. The Cancer Genome Atlas Pilot Project (http://cancergenome.nih.gov) was undertaken as a feasability study. Three tumor types—brain (glioblastoma multiforme), lung (squamous carcinoma), and ovarian (serous cystadenocarcinoma)—were selected for analysis in this pilot phase. The project entails collaboration among a central Biospecimen Core Resource, Cancer Genome Characterization Centers, Genome Sequencing Centers, and a Data Coordinating Center. Data produced by TCGA are available through the TCGA Data Portal. Initial fruit of TCGA's labor is the publication of their analysis to date of glioblastomas (see above for details; McLendon et al. 2008).

The Cancer Genome Project

The Cancer Genome Project (http://www.sanger.ac.uk/genetics/CGP/) comprises various endeavors at the Sanger Institute, including the Cancer Gene Census (http://www.sanger.ac.uk/genetics/CGP/Census/), COSMIC (http://www.sanger.ac.uk/genetics/CGP/cosmic), and a number of other cancer-related projects.

IntOGen

IntOGen (Integrative OncoGenomics, http://www.intogen.org) is a resource that integrates different types of oncogenomics data. At the moment, this resource includes genomic alterations (amplifications and deletions), microarray expression profiles, and mutation screenings. The experiments are collected from different public databases or directly from the authors, and the type of cancer from the samples is annotated using the controlled vocabulary of the International Classification of Diseases (ICD-10 and ICD-O). All of the experiments are processed in a standard way and then analyzed statistically to identify genes that are significantly altered. Groups of experiments annotated with the same ICD term are combined to identify genes significantly altered in this cancer type.

IntOGen is designed to be a discovery tool for cancer researchers. Users interested in a particular gene can easily see whether their gene of interest has been found to be altered (e.g., overexpressed, mutated, or deleted) in different cancer types and subtypes. On the other hand, researchers interested in a particular tumor type are able to search for the genes that are more significantly altered (with mutations or genomic or transcriptomic alterations) in this type of cancer. Additionally, this resource is highly useful for prioritization of candidate cancer genes. The probabilities given by our prioritization method (CGPrio, described above) are integrated in IntOGen (Furney et al. 2008a). Users can upload a list of candidate cancer genes and prioritize them, taking into account evidence of oncogenomic alterations detected in other experiments and the probabilities of being a cancer gene given by CGPrio.

IntOGen not only is focused on individual gene analysis, but also studies the implication of functional and regulatory modules in different cancer types. For example, users can search IntOGen for the biological pathways with a higher proportion of genes altered in a particular cancer type or have a wide view of the alterations of genes in a particular pathway in different tumor types.

Overall, the integration of a large compendium of oncogenomic experiments together with genomic data and statistical integrative analysis provides a powerful tool for online discovery of genes involved in different types of cancer.

SUMMARY

The last decade has witnessed profound changes in how cancer research is conducted. First, emerging technologies have allowed surveys of alterations in tumor cells on a genome-wide scale, giving rise to the field of oncogenomics. In tandem with this is the realization that, due to the complex nature of oncogenesis, methods that integrate multiple types of data are required to understand and elucidate the tumorigenic process. Recognition of this has led to the formation of the International Cancer Genome Consortium (ICGC), which will endeavor to use existing and improving technologies to perform integrative oncogenomic research in multiple cancer types. Projects under the aegis of the ICGC will occupy much of the cancer research field for years to come.

ACKNOWLEDGMENTS

We acknowledge funding from the International Human Frontier Science Program Organization (HFSPO) and from the Spanish Ministerio de Educacion y Ciencia (MEC) grant number SAF2006-0459. N.L.-B. is the recipient of a Ramon y Cajal contract of the MEC and acknowledges support from Instituto Nacional de Bioinformatica.

REFERENCES

Ahmed, S., Thomas, G., Ghoussaini, M., Healey, C.S., Humphreys, M.K., Platte, R., Morrison, J., Maranian, M., Pooley, K.A., Luben, R., et al. 2009. Newly discovered breast cancer susceptibility loci on 3p24 and 17q23.2. *Nat. Genet.* **41:** 585–590.

Alizadeh, A.A., Eisen, M.B., Davis, R.E., Ma, C., Lossos, I.S., Rosenwald, A., Boldrick, J.C., Sabet, H., Tran, T., Yu, X., et al. 2000. Distinct types of diffuse large B-cell lymphoma identified by gene expression profiling. *Nature* **403:** 503–511.

American Association for Cancer Research Human Epigenome Task Force; European Union, Network of Excellence, Scientific Advisory Board. 2008. Moving AHEAD with an international human epigenome project. *Nature* **454:** 711–715.

Amos, C.I., Wu, X., Broderick, P., Gorlov, I.P., Gu, J., Eisen, T., Dong, Q., Zhang, Q., Gu, X., Vijayakrishnan, J., et al. 2008. Genome-wide association scan of tag SNPs identifies a susceptibility locus for lung cancer at 15q25.1. *Nat. Genet.* **40:** 616–622.

Azuara, V., Perry, P., Sauer, S., Spivakov, M., Jorgensen, H.F., John, R.M., Gouti, M., Casanova, M., Warnes, G., Merkenschlager, M., et al. 2006. Chromatin signatures of pluripotent cell lines. *Nat. Cell Biol.* **8:** 532–538.

Baak, J.P., Janssen, E.A., Soreide, K., and Heikkilae, R. 2005. Genomics and proteomics—The way forward. *Ann. Oncol.* (suppl. 2) **16:** ii30–ii44.

Ballestar, E., Paz, M.F., Valle, L., Wei, S., Fraga, M.F., Espada, J., Cigudosa, J.C., Huang, T.H., and Esteller, M. 2003. Methyl-CpG binding proteins identify novel sites of epigenetic inactivation in human cancer. *EMBO J.* **22:** 6335–6345.

Bardelli, A., Parsons, D.W., Silliman, N., Ptak, J., Szabo, S., Saha, S., Markowitz, S., Willson, J.K., Parmigiani, G., Kinzler, K.W., et al. 2003. Mutational analysis of the tyrosine kinome in colorectal cancers. *Science* **300:** 949.

Barrett, T., Suzek, T.O., Troup, D.B., Wilhite, S.E., Ngau, W.C., Ledoux, P., Rudnev, D., Lash, A.E., Fujibuchi, W., and Edgar, R. 2005. NCBI GEO: Mining millions of expression profiles—Database and tools. *Nucleic Acids Res.* **33:** D562–D566.

Baudis, M. 2007. Genomic imbalances in 5918 malignant epithelial tumors: An explorative meta-analysis of chromosomal CGH data. *BMC Cancer* **7:** 226.

Baudis, M. and Cleary, M.L. 2001. Progenetix.net: An online repository for molecular cytogenetic aberration data. *Bioinformatics* **17:** 1228–1229.

Beer, D.G., Kardia, S.L., Huang, C.C., Giordano, T.J., Levin, A.M., Misek, D.E., Lin, L., Chen, G., Gharib, T.G., Thomas, D.G., et al. 2002. Gene-expression profiles predict survival of patients with lung adenocarcinoma. *Nat. Med.* **8:** 816–824.

Benvenuti, S., Arena, S., and Bardelli, A. 2005. Identification of cancer genes by mutational profiling of tumor genomes. *FEBS Lett.* **579:** 1884–1890.

Bernstein, B.E., Meissner, A., and Lander, E.S. 2007. The mammalian epigenome. *Cell* **128:** 669–681.

Bestor, T.H. 2005. Transposons reanimated in mice. *Cell* **122:** 322–325.

Bild, A.H., Yao, G., Chang, J.T., Wang, Q., Potti, A., Chasse, D., Joshi, M.B., Harpole, D., Lancaster, J.M., Berchuck, A., et al. 2006. Oncogenic pathway signatures in human cancers as a guide to targeted therapies. *Nature* **439:** 353–357.

Buongiorno-Nardelli, M. and Amaldi, F. 1970. Autoradiographic detection of molecular hybrids between RNA and DNA in tissue sections. *Nature* **225:** 946–948.

Campbell, P.J., Stephens, P.J., Pleasance, E.D., O'Meara, S., Li, H., Santarius, T., Stebbings, L.A., Leroy, C., Edkins, S., Hardy, C., et al. 2008. Identification of somatically acquired rearrangements in cancer using genome-wide massively parallel paired-end sequencing. *Nat. Genet.* **40:** 722–729.

Carter, S.L., Eklund, A.C., Kohane, I.S., Harris, L.N., and Szallasi, Z. 2006. A signature of chromosomal instability inferred from gene expression profiles predicts clinical outcome in multiple human cancers. *Nat. Genet.* **38:** 1043–1048.

Cavenee, W.K., Dryja, T.P., Phillips, R.A., Benedict, W.F., Godbout, R., Gallie, B.L., Murphree, A.L., Strong, L.C., and White, R.L. 1983. Expression of recessive alleles by chromosomal mechanisms in retinoblastoma. *Nature* **305:** 779–784.

Chin, L. and Gray, J.W. 2008. Translating insights from the cancer genome into clinical practice. *Nature* **452:** 553–563.

Clark, S.J., Harrison, J., Paul, C.L., and Frommer, M. 1994. High sensitivity mapping of methylated cytosines. *Nucleic Acids Res.* **22:** 2990–2997.

Cox, A., Dunning, A.M., Garcia-Closas, M., Balasubramanian, S., Reed, M.W., Pooley, K.A., Scollen, S., Baynes, C., Ponder, B.A., Chanock, S., et al. 2007. A common coding variant in CASP8 is associated with breast cancer risk. *Nat. Genet.* **39:** 352–358.

Cui, H., Cruz-Correa, M., Giardiello, F.M., Hutcheon, D.F., Kafonek, D.R., Brandenburg, S., Wu, Y., He, X., Powe, N.R., and Feinberg, A.P. 2003. Loss of IGF2 imprinting: A potential marker of colorectal cancer risk. *Science* **299:** 1753–1755.

Dave, S.S., Wright, G., Tan, B., Rosenwald, A., Gascoyne, R.D., Chan, W.C., Fisher, R.I., Braziel, R.M., Rimsza, L.M., Grogan, T.M., et al. 2004. Prediction of survival in follicular lymphoma based on molecular features of tumor-infiltrating immune cells. *N. Engl. J. Med.* **351:** 2159–2169.

Dave, S.S., Fu, K., Wright, G.W., Lam, L.T., Kluin, P., Boerma, E.J., Greiner, T.C., Weisenburger, D.D., Rosenwald, A., Ott, G., et al. 2006. Molecular diagnosis of Burkitt's lymphoma. *N. Engl. J. Med.* **354:** 2431–2442.

Davies, H., Bignell, G.R., Cox, C., Stephens, P., Edkins, S., Clegg, S., Teague, J., Woffendin, H., Garnett, M.J., Bottomley, W., et al. 2002. Mutations of the BRAF gene in human cancer. *Nature* **417:** 949–954.

Davies, H., Hunter, C., Smith, R., Stephens, P., Greenman, C., Bignell, G., Teague, J., Butler, A., Edkins, S., Stevens, C., et al. 2005. Somatic mutations of the protein kinase gene family in human lung cancer. *Cancer Res.* **65:** 7591–7595.

Di Bernardo, M.C., Crowther-Swanepoel, D., Broderick, P., Webb, E., Sellick, G., Wild, R., Sullivan, K., Vijayakrishnan, J., Wang, Y., Pittman, A.M., et al. 2008. A genome-wide association study identifies six susceptibility loci for chronic lymphocytic leukemia. *Nat. Genet.* **40:** 1204–1210.

Ding, L., Getz, G., Wheeler, D.A., Mardis, E.R., McLellan, M.D., Cibulskis, K., Sougnez, C., Greulich, H., Muzny, D.M., Morgan, M.B., et al. 2008. Somatic mutations affect key pathways in lung adenocarcinoma. *Nature* **455:** 1069–1075.

Doll, R. and Peto, R. 1981. The causes of cancer: Quantitative estimates of avoidable risks of cancer in the United States today. *J. Natl. Cancer Inst.* **66:** 1191–1308.

Druker, B.J. 2002. STI571 (Gleevec) as a paradigm for cancer therapy. *Trends Mol. Med.* (suppl. 4) **8:** S14–S18.

Dyrskjot, L., Thykjaer, T., Kruhoffer, M., Jensen, J.L., Marcussen, N., Hamilton-Dutoit, S., Wolf, H., and Orntoft, T.F. 2003. Identifying distinct classes of bladder carcinoma using microarrays. *Nat. Genet.* **33:** 90–96.

Easton, D.F., Pooley, K.A., Dunning, A.M., Pharoah, P.D., Thompson, D., Ballinger, D.G., Struewing, J.P., Morrison, J., Field, H., Luben, R., et al. 2007. Genome-wide association study identifies novel breast cancer susceptibility loci. *Nature* **447:** 1087–1093.

Eden, A., Gaudet, F., Waghmare, A., and Jaenisch, R. 2003. Chromosomal instability and tumors promoted by DNA hypomethylation. *Science* **300:** 455.

Esteller, M. 2007. Cancer epigenomics: DNA methylomes and histone-modification maps. *Nat. Rev. Genet.* **8:** 286–298.

Esteller, M. 2008. Epigenetics in cancer. *N. Engl. J. Med.* **358:** 1148–1159.

Fahrner, J.A., Eguchi, S., Herman, J.G., and Baylin, S.B. 2002. Dependence of histone modifications and gene expression on DNA hypermethylation in cancer. *Cancer Res.* **62:** 7213–7218.

Fang, Y.C., Huang, H.C., and Juan, H.F. 2008. MeInfoText: Associated gene methylation and cancer information from text mining. *BMC Bioinformatics* **9:** 22.

Feinberg, A.P. 1999. Imprinting of a genomic domain of 11p15 and loss of imprinting in cancer: An introduction. *Cancer Res.* (suppl. 7) **59:** 1743s–1746s.

Feinberg, A.P. and Vogelstein, B. 1983. Hypomethylation distinguishes genes of some human cancers from their normal counterparts. *Nature* **301:** 89–92.

Finn, R.D., Mistry, J., Schuster-Bockler, B., Griffiths-Jones, S., Hollich, V., Lassmann, T., Moxon, S., Marshall, M., Khanna, A., Durbin, R., et al. 2006. Pfam: Clans, web tools and services. *Nucleic Acids Res.* **34:** D247–D251.

Forbes, S., Clements, J., Dawson, E., Bamford, S., Webb, T., Dogan, A., Flanagan, A., Teague, J., Wooster, R., Futreal, P.A., et al. 2006. Cosmic 2005. *Br. J. Cancer* **94:** 318–322.

Forrest, W.F. and Cavet, G. 2007. Comment on "The consensus coding sequences of human breast and colorectal cancers." *Science* **317:** 1500; author reply 1500.

Fraga, M.F. and Esteller, M. 2002. DNA methylation: A profile of methods and applications. *Biotechniques* **33:** 632, 634, 636–649.

Fraga, M.F., Ballestar, E., Villar-Garea, A., Boix-Chornet, M., Espada, J., Schotta, G., Bonaldi, T., Haydon, C., Ropero, S., Petrie, K., et al. 2005. Loss of acetylation at Lys16 and trimethylation at Lys20 of histone H4 is a common hallmark of human cancer. *Nat. Genet.* **37:** 391–400.

Furney, S.J., Higgins, D.G., Ouzounis, C.A., and Lopez-Bigas, N. 2006. Structural and functional properties of genes involved in human cancer. *BMC Genomics* **7:** 3.

Furney, S.J., Madden, S.F., Higgins, D.G., and Lopez-Bigas, N. 2008b. Distinct patterns in the regulation and evolution of human cancer genes. *In Silico Biol.* **8:** 33–46.

Furney, S.J., Calvo, B., Larrañaga, P., Lozano, J.A., and Lopez-Bigas, N. 2008a. Prioritization of candidate cancer genes—An aid to oncogenomic studies. *Nucleic Acids Res.* **36:** e115.

Futreal, P.A., Coin, L., Marshall, M., Down, T., Hubbard, T., Wooster, R., Rahman, N., and Stratton, M.R. 2004. A census of human cancer genes. *Nat. Rev. Cancer* **4:** 177–183.

Getz, G., Hofling, H., Mesirov, J.P., Golub, T.R., Meyerson, M., Tibshirani, R., and Lander, E.S. 2007. Comment on "The consensus coding sequences of human breast and colorectal cancers." *Science* **317:** 1500.

Gonzalez-Zulueta, M., Bender, C.M., Yang, A.S., Nguyen, T., Beart, R.W., Van Tornout, J.M., and Jones, P.A. 1995. Methylation of the 5′ CpG island of the p16/CDKN2 tumor suppressor gene in normal and transformed human tissues correlates with gene silencing. *Cancer Res.* **55:** 4531–4535.

Greenman, C., Stephens, P., Smith, R., Dalgliesh, G.L., Hunter, C., Bignell, G., Davies, H., Teague, J., Butler, A., Stevens, C., et al. 2007. Patterns of somatic mutation in human cancer genomes. *Nature* **446:** 153–158.

Greger, V., Passarge, E., Hopping, W., Messmer, E., and Horsthemke, B. 1989. Epigenetic changes may contribute to the formation and spontaneous regression of retinoblastoma. *Hum. Genet.* **83:** 155–158.

Groffen, J., Stephenson, J.R., Heisterkamp, N., Bartram, C., de Klein, A., and Grosveld, G. 1984. The human *c-abl* oncogene in the Philadelphia translocation. *J. Cell. Physiol.* **3:** 179–191.

Grunau, C., Renault, E., Rosenthal, A., and Roizes, G. 2001. MethDB—A public database for DNA methylation data. *Nucleic Acids Res.* **29:** 270–274.

Gudmundsson, J., Sulem, P., Manolescu, A., Amundadottir, L.T., Gudbjartsson, D., Helgason, A., Rafnar, T., Bergthorsson, J.T., Agnarsson, B.A., Baker, A., et al. 2007. Genome-wide association study identifies a second prostate cancer susceptibility variant at 8q24. *Nat. Genet.* **39:** 631–637.

Gudmundsson, J., Sulem, P., Rafnar, T., Bergthorsson, J.T., Manolescu, A., Gudbjartsson, D., Agnarsson, B.A., Sigurdsson, A., Benediktsdottir, K.R., Blondal, T., et al. 2008. Common sequence variants on 2p15 and Xp11.22 confer susceptibility to prostate cancer. *Nat. Genet.* **40:** 281–283.

Haber, D.A. and Settleman, J. 2007. Cancer: Drivers and passengers. *Nature* **446:** 145–146.

Haiman, C.A., Le Marchand, L., Yamamato, J., Stram, D.O., Sheng, X., Kolonel, L.N., Wu, A.H., Reich, D., and Henderson, B.E. 2007a. A common genetic risk factor for colorectal and prostate cancer. *Nat. Genet.* **39:** 954–956.

Haiman, C.A., Patterson, N., Freedman, M.L., Myers, S.R., Pike, M.C., Waliszewska, A., Neubauer, J., Tandon, A., Schirmer, C., McDonald, G.J., et al. 2007b. Multiple regions within 8q24 independently affect risk for prostate cancer. *Nat. Genet.* **39:** 638–644.

Hanahan, D. and Weinberg, R.A. 2000. The hallmarks of cancer. *Cell* **100:** 57–70.

Hayes, D.N., Monti, S., Parmigiani, G., Gilks, C.B., Naoki, K., Bhattacharjee, A., Socinski, M.A., Perou, C., and Meyerson, M. 2006. Gene expression profiling reveals reproducible human lung adenocarcinoma subtypes in multiple independent patient cohorts. *J. Clin. Oncol.* **24:** 5079–5090.

He, L., Thomson, J.M., Hemann, M.T., Hernando-Monge, E., Mu, D., Goodson, S., Powers, S., Cordon-Cardo, C., Lowe, S.W., Hannon, G.J., et al. 2005. A microRNA polycistron as a potential human oncogene. *Nature* **435:** 828–833.

He, X., Chang, S., Zhang, J., Zhao, Q., Xiang, H., Kusonmano, K., Yang, L., Sun, Z.S., Yang, H., and Wang, J. 2008. MethyCancer: The database of human DNA methylation and cancer. *Nucleic Acids Res.* **36:** D836–D841.

Herman, J.G. and Baylin, S.B. 2003. Gene silencing in cancer in association with promoter hypermethylation. *N. Engl. J. Med.* **349:** 2042–2054.

Herman, J.G., Graff, J.R., Myohanen, S., Nelkin, B.D., and Baylin, S.B. 1996. Methylation-specific PCR: A novel PCR assay for methylation status of CpG islands. *Proc. Natl. Acad. Sci.* **93:** 9821–9826.

Herman, J.G., Merlo, A., Mao, L., Lapidus, R.G., Issa, J.P., Davidson, N.E., Sidransky, D., and Baylin, S.B. 1995. Inactivation of the *CDKN2/p16/MTS1* gene is frequently associated with aberrant DNA methylation in all common human cancers. *Cancer Res.* **55:** 4525–4530.

Herman, J.G., Latif, F., Weng, Y., Lerman, M.I., Zbar, B., Liu, S., Samid, D., Duan, D.S., Gnarra, J.R., Linehan, W.M., et al. 1994. Silencing of the VHL tumor-suppressor gene by DNA methylation in renal carcinoma. *Proc. Natl. Acad. Sci.* **91:** 9700–9704.

Higgins, M.E., Claremont, M., Major, J.E., Sander, C., and Lash, A.E. 2007. CancerGenes: A gene selection resource for cancer genome projects. *Nucleic Acids Res.* **35:** D721–D726.

Houlston, R.S., Webb, E., Broderick, P., Pittman, A.M., Di Bernardo, M.C., Lubbe, S., Chandler, I., Vijayakrishnan, J., Sullivan, K., Penegar, S., et al. 2008. Meta-analysis of genome-wide association data identifies four new susceptibility loci for colorectal cancer. *Nat. Genet.* **40:** 1426–1435.

Hu, P., Bader, G., Wigle, D.A., and Emili, A. 2007. Computational prediction of cancer-gene function. *Nat. Rev. Cancer* **7:** 23–34.

Ihmels, J., Friedlander, G., Bergmann, S., Sarig, O., Ziv, Y., and Barkai, N. 2002. Revealing modular organization in the yeast transcriptional network. *Nat. Genet.* **31:** 370–377.

International HapMap Consortium. 2005. A haplotype map of the human genome. *Nature* **437:** 1299–1320.

Jones, S., Zhang, X., Parsons, D.W., Lin, J.C., Leary, R.J., Angenendt, P., Mankoo, P., Carter, H., Kamiyama, H., Jimeno, A., et al. 2008. Core signaling pathways in human pancreatic cancers revealed by global genomic analyses. *Science* **321:** 1801–1806.

Jong, K., Marchiori, E., van der Vaart, A., Chin, S.F., Carvalho, B., Tijssen, M., Eijk, P.P., van den Ijssel, P., Grabsch, H., Quirke, P., et al. 2007. Cross-platform array comparative genomic hybridization meta-analysis separates hematopoietic and mesenchymal from epithelial tumors. *Oncogene* **26:** 1499–1506.

Kallioniemi, A. 2008. CGH microarrays and cancer. *Curr. Opin. Biotechnol.* **19:** 36–40.

Kallioniemi, A., Kallioniemi, O.P., Sudar, D., Rutovitz, D., Gray, J.W., Waldman, F., and Pinkel, D. 1992. Comparative genomic hybridization for molecular cytogenetic analysis of solid tumors. *Science* **258:** 818–821.

Kaminker, J.S., Zhang, Y., Watanabe, C., and Zhang, Z. 2007a. CanPredict: A computational tool for predicting cancer-associated missense mutations. *Nucleic Acids Res.* **35:** W595–W598.

Kaminker, J.S., Zhang, Y., Waugh, A., Haverty, P.M., Peters, B., Sebisanovic, D., Stinson, J., Forrest, W.F., Bazan, J.F., Seshagiri, S., et al. 2007b. Distinguishing cancer-associated missense mutations from common polymorphisms. *Cancer Res.* **67:** 465–473.

Kaneda, A. and Feinberg, A.P. 2005. Loss of imprinting of IGF2: A common epigenetic modifier of intestinal tumor risk. *Cancer Res.* **65:** 11236–11240.

Keshet, I., Schlesinger, Y., Farkash, S., Rand, E., Hecht, M., Segal, E., Pikarski, E., Young, R.A., Niveleau, A., Cedar, H., et al. 2006. Evidence for an instructive mechanism of de novo methylation in cancer cells. *Nat. Genet.* **38:** 149–153.

Kiemeney, L.A., Thorlacius, S., Sulem, P., Geller, F., Aben, K.K., Stacey, S.N., Gudmundsson, J., Jakobsdottir, M., Bergthorsson, J.T., Sigurdsson, A., et al. 2008. Sequence variant on 8q24 confers susceptibility to urinary bladder cancer. *Nat. Genet.* **40:** 1307–1312.

Kim, M., Gans, J.D., Nogueira, C., Wang, A., Paik, J.H., Feng, B., Brennan, C., Hahn, W.C., Cordon-Cardo, C., Wagner, S.N., et al. 2006. Comparative oncogenomics identifies *NEDD9* as a melanoma metastasis gene. *Cell* **125:** 1269–1281.

Knudson, Jr., A.G. 1971. Mutation and cancer: Statistical study of retinoblastoma. *Proc. Natl. Acad. Sci.* **68:** 820–823.

Knutsen, T., Gobu, V., Knaus, R., Padilla-Nash, H., Augustus, M., Strausberg, R.L., Kirsch, I.R., Sirotkin, K., and Ried, T. 2005. The interactive online SKY/M-FISH & CGH database and the Entrez cancer chromosomes search database: Linkage of chromosomal aberrations with the genome sequence. *Genes Chromosomes Cancer* **44:** 52–64.

Korbel, J.O., Urban, A.E., Affourtit, J.P., Godwin, B., Grubert, F., Simons, J.F., Kim, P.M., Palejev, D., Carriero, N.J., Du, L., et al. 2007. Paired-end mapping reveals extensive structural variation in the human genome. *Science* **318:** 420–426.

Krzywinski, M., Bosdet, I., Mathewson, C., Wye, N., Brebner, J., Chiu, R., Corbett, R., Field, M., Lee, D., Pugh, T., et al. 2007. A BAC clone fingerprinting approach to the detection of human genome rearrangements. *Genome Biol.* **8:** R224.

Kuppers, R. 2005. Mechanisms of B-cell lymphoma pathogenesis. *Nat. Rev.* **5:** 251–262.

Lapointe, J., Li, C., Higgins, J.P., van de Rijn, M., Bair, E., Montgomery, K., Ferrari, M., Egevad, L., Rayford, W., Bergerheim, U., et al. 2004. Gene expression profiling identifies clinically relevant subtypes of prostate cancer. *Proc. Natl. Acad. Sci.* **101:** 811–816.

Liu, E.T., Kuznetsov, V.A., and Miller, L.D. 2006. In the pursuit of complexity: Systems medicine in cancer biology. *Cancer Cell* **9:** 245–247.

Loeb, L.A. and Bielas, J.H. 2007. Limits to the Human Cancer Genome Project? *Science* **315:** 762; author reply 764–765.

Lopez-Serra, L., Ballestar, E., Fraga, M.F., Alaminos, M., Setien, F., and Esteller, M. 2006. A profile of methyl-CpG binding domain protein occupancy of hypermethylated promoter CpG islands of tumor suppressor genes in human cancer. *Cancer Res.* **66:** 8342–8346.

Lu, J., Getz, G., Miska, E.A., Alvarez-Saavedra, E., Lamb, J., Peck, D., Sweet-Cordero, A., Ebert, B.L., Mak, R.H., Ferrando, A.A., et al. 2005a. MicroRNA expression profiles classify human cancers. **435:** 834–838.

MacDonald, J.W. and Ghosh, D. 2006. COPA—Cancer outlier profile analysis. *Bioinformatics* **22:** 2950–2951.

Maher, C.A., Kumar-Sinha, C., Cao, X., Kalyana-Sundaram, S., Han, B., Jing, X., Sam, L., Barrette, T., Palanisamy, N., and Chinnaiyan, A.M. 2009. Transcriptome sequencing to detect gene fusions in cancer. *Nature* **458:** 97–101.

Marinelli, R.J., Montgomery, K., Liu, C.L., Shah, N.H., Prapong, W., Nitzberg, M., Zachariah, Z.K., Sherlock, G.J., Natkunam, Y., West, R.B., et al. 2008. The Stanford Tissue Microarray Database. *Nucleic Acids Res.* **36:** D871–D877.

Maser, R.S., Choudhury, B., Campbell, P.J., Feng, B., Wong, K.K., Protopopov, A., O'Neil, J., Gutierrez, A., Ivanova, E., Perna, I., et al. 2007. Chromosomally unstable mouse tumors have genomic alterations similar to diverse human cancers. *Nature* **447:** 966–971.

McLendon, R., Friedman, A., Bigner, D., Van Meir, E.G., Brat, D.J., Mastrogianakis, M., Olson, J.J., Mikkelsen, T., Lehman, N., Aldape, K., et al. 2008. Comprehensive genomic characterization defines human glioblastoma genes and core pathways. *Nature* **455:** 1061–1068.

Merlo, A., Herman, J.G., Mao, L., Lee, D.J., Gabrielson, E., Burger, P.C., Baylin, S.B., and Sidransky, D. 1995. 5′ CpG island methylation is associated with transcriptional silencing of the tumor suppressor p16/CDKN2/MTS1 in human cancers. *Nat. Med.* **1:** 686–692.

Mikkelsen, T.S., Ku, M., Jaffe, D.B., Issac, B., Lieberman, E., Giannoukos, G., Alvarez, P., Brockman, W., Kim, T.K., Koche, R.P., et al. 2007. Genome-wide maps of chromatin state in pluripotent and lineage-committed cells. *Nature* **448:** 553–560.

Mitelman, F. 2000. Recurrent chromosome aberrations in cancer. *Mutation Res.* **462:** 247–253.

Mitelman, F., Mertens, F., and Johansson, B. 1997. A breakpoint map of recurrent chromosomal rearrangements in human neoplasia. *Nat. Genet.* **15:** 417–474.

Mitelman, F., Johansson, B., and Mertens, F. 2007. The impact of translocations and gene fusions on cancer causation. *Nat. Rev. Cancer* **7:** 233–245.

Mitelman, F., Johansson, B., and Mertens, F., eds. 2009. *Mitelman database of chromosome aberrations in cancer.* The Cancer Genome Anatomy Project, Washington, D.C.

Morozova, O. and Marra, M.A. 2008. From cytogenetics to next-generation sequencing technologies: Advances in the detection of genome rearrangements in tumors. *Biochem. Cell Biol.* **86:** 81–91.

Mulligan, C.G., Goorha, S., Radtke, I., Miller, C.B., Coustan-Smith, E., Dalton, J.D., Girtman, K., Mathew, S., Ma, J., Pounds, S.B., et al. 2007. Genome-wide analysis of genetic alterations in acute lymphoblastic leukaemia. *Nature* **446:** 758–764.

Nowell, P.C. 2007. Discovery of the Philadelphia chromosome: A personal perspective. *J. Clin. Invest.* **117:** 2033–2035.

Olivier, M., Eeles, R., Hollstein, M., Khan, M.A., Harris, C.C., and Hainaut, P. 2002. The IARC TP53 Database: New online mutation analysis and recommendations to users. *Human Mutation* **19:** 607–614.

Ongenaert, M., Van Neste, L., De Meyer, T., Menschaert, G., Bekaert, S., and Van Criekinge, W. 2008. PubMeth: A cancer methylation database combining text-mining and expert annotation. *Nucleic Acids Res.* **36:** D842–D846.

Paez, J.G., Janne, P.A., Lee, J.C., Tracy, S., Greulich, H., Gabriel, S., Herman, P., Kaye, F.J., Lindeman, N., Boggon, T.J., et al. 2004. EGFR mutations in lung cancer: Correlation with clinical response to gefitinib therapy. *Science* **304:** 1497–1500.

Paik, S., Shak, S., Tang, G., Kim, C., Baker, J., Cronin, M., Baehner, F.L., Walker, M.G., Watson, D., Park, T., et al. 2004. A multigene assay to predict recurrence of tamoxifen-treated, node-negative breast cancer. *N. Engl. J. Med.* **351:** 2817–2826.

Parkinson, H., Kapushesky, M., Shojatalab, M., Abeygunawardena, N., Coulson, R., Farne, A., Holloway, E., Kolesnykov, N., Lilja, P., Lukk, M., et al. 2007. ArrayExpress—A public database of microarray experiments and gene expression profiles. *Nucleic Acids Res.* **35:** D747–D750.

Parsons, D.W., Jones, S., Zhang, X., Lin, J.C., Leary, R.J., Angenendt, P., Mankoo, P., Carter, H., Siu, I.M., Gallia, G.L., et al. 2008. An integrated genomic analysis of human glioblastoma multiforme. *Science* **321:** 1807–1812.

Perou, C.M., Sorlie, T., Eisen, M.B., van de Rijn, M., Jeffrey, S.S., Rees, C.A., Pollack, J.R., Ross, D.T., Johnsen, H., Akslen, L.A., et al. 2000. Molecular portraits of human breast tumors. *Nature* **406:** 747–752.

Peto, J. 2001. Cancer epidemiology in the last century and the next decade. *Nature* **411:** 390–395.

Pinkel, D., Segraves, R., Sudar, D., Clark, S., Poole, I., Kowbel, D., Collins, C., Kuo, W.L., Chen, C., Zhai, Y., et al. 1998. High resolution analysis of DNA copy number variation using comparative genomic hybridization to microarrays. *Nat. Genet.* **20:** 207–211.

Pomeroy, S.L., Tamayo, P., Gaasenbeek, M., Sturla, L.M., Angelo, M., McLaughlin, M.E., Kim, J.Y., Goumnerova, L.C., Black, P.M., Lau, C., et al. 2002. Prediction of central nervous system embryonal tumor outcome based on gene expression. *Nature* **415:** 436–442.

Ponder, B.A. 2001. Cancer genetics. *Nature* **411:** 336–341.

Potti, A., Mukherjee, S., Petersen, R., Dressman, H.K., Bild, A., Koontz, J., Kratzke, R., Watson, M.A., Kelley, M., Ginsburg, G.S., et al. 2006. A genomic strategy to refine prognosis in early-stage non-small-cell lung cancer. *N. Engl. J. Med.* **355:** 570–580.

Rabbitts, T.H. 1994. Chromosomal translocations in human cancer. *Nature* **372:** 143–149.

Rafnar, T., Sulem, P., Stacey, S.N., Geller, F., Gudmundsson, J., Sigurdsson, A., Jakobsdottir, M., Helgadottir, H., Thorlacius, S., Aben, K.K., et al. 2009. Sequence variants at the *TERT-CLPTM1L* locus associate with many cancer types. *Nat. Genet.* **41:** 221–227.

Rajagopalan, H., Bardelli, A., Lengauer, C., Kinzler, K.W., Vogelstein, B., and Velculescu, V.E. 2002. Tumorigenesis: *RAF/RAS* oncogenes and mismatch-repair status. *Nature* **418:** 934.

Ramaswamy, S. and Golub, T.R. 2002. DNA microarrays in clinical oncology. *J. Clin. Oncol.* **20:** 1932–1941.

Ramaswamy, S., Ross, K.N., Lander, E.S., and Golub, T.R. 2003. A molecular signature of metastasis in primary solid tumors. *Nat. Genet.* **33:** 49–54.

Reddy, E.P., Reynolds, R.K., Santos, E., and Barbacid, M. 1982. A point mutation is responsible for the acquisition of transforming properties by the T24 human bladder carcinoma oncogene. *Nature* **300:** 149–152.

Rhodes, D.R., Yu, J., Shanker, K., Deshpande, N., Varambally, R., Ghosh,

D., Barrette, T., Pandey, A., and Chinnaiyan, A.M. 2004. Large-scale meta-analysis of cancer microarray data identifies common transcriptional profiles of neoplastic transformation and progression. *Proc. Natl. Acad. Sci.* **101:** 9309–9314.

Rhodes, D.R., Kalyana-Sundaram, S., Mahavisno, V., Varambally, R., Yu, J., Briggs, B.B., Barrette, T.R., Anstet, M.J., Kincead-Beal, C., Kulkarni, P., et al. 2007. Oncomine 3.0: Genes, pathways, and networks in a collection of 18,000 cancer gene expression profiles. *Neoplasia* **9:** 166–180.

Rowley, J.D. 1998. The critical role of chromosome translocations in human leukemias. *Annu. Rev. Genet.* **32:** 495–519.

Rowley, J.D. 2001. Chromosome translocations: Dangerous liaisons revisited. *Nat. Rev.* **1:** 245–250.

Rubin, A.F. and Green, P. 2007. Comment on "The consensus coding sequences of human breast and colorectal cancers." *Science* **317:** 1500.

Sakai, T., Toguchida, J., Ohtani, N., Yandell, D.W., Rapaport, J.M., and Dryja, T.P. 1991. Allele-specific hypermethylation of the retinoblastoma tumor-suppressor gene. *Am. J. Hum. Genet.* **48:** 880–888.

Sakamoto, H., Yoshimura, K., Saeki, N., Katai, H., Shimoda, T., Matsuno, Y., Saito, D., Sugimura, H., Tanioka, F., Kato, S., et al. 2008. Genetic variation in PSCA is associated with susceptibility to diffuse-type gastric cancer. *Nat. Genet.* **40:** 730–740.

Samuels, Y., Wang, Z., Bardelli, A., Silliman, N., Ptak, J., Szabo, S., Yan, H., Gazdar, A., Powell, S.M., Riggins, G.J., et al. 2004. High frequency of mutations of the *PIK3CA* gene in human cancers. *Science* **304:** 554.

Schrock, E., du Manoir, S., Veldman, T., Schoell, B., Wienberg, J., Ferguson-Smith, M.A., Ning, Y., Ledbetter, D.H., Bar-Am, I., Soenksen, D., et al. 1996. Multicolor spectral karyotyping of human chromosomes. *Science* **273:** 494–497.

Segal, E., Friedman, N., Koller, D., and Regev, A. 2004. A module map showing conditional activity of expression modules in cancer. *Nat. Genet.* **36:** 1090–1098.

Shipp, M.A., Ross, K.N., Tamayo, P., Weng, A.P., Kutok, J.L., Aguiar, R.C., Gaasenbeek, M., Angelo, M., Reich, M., Pinkus, G.S., et al. 2002. Diffuse large B-cell lymphoma outcome prediction by gene-expression profiling and supervised machine learning. *Nat. Med.* **8:** 68–74.

Shtivelman, E., Lifshitz, B., Gale, R.P., and Canaani, E. 1985. Fused transcript of *abl* and *bcr* genes in chronic myelogenous leukaemia. *Nature* **315:** 550–554.

Sjoblom, T., Jones, S., Wood, L.D., Parsons, D.W., Lin, J., Barber, T.D., Mandelker, D., Leary, R.J., Ptak, J., Silliman, N., et al. 2006. The consensus coding sequences of human breast and colorectal cancers. *Science* **314:** 268–274.

Sorlie, T., Tibshirani, R., Parker, J., Hastie, T., Marron, J.S., Nobel, A., Deng, S., Johnsen, H., Pesich, R., Geisler, S., et al. 2003. Repeated observation of breast tumor subtypes in independent gene expression data sets. *Proc. Natl. Acad. Sci.* **100:** 8418–8423.

Speicher, M.R. and Ward, D.C. 1996. The coloring of cytogenetics. *Nat. Med.* **2:** 1046–1048.

Speicher, M.R., Gwyn Ballard, S., and Ward, D.C. 1996. Karyotyping human chromosomes by combinatorial multi-fluor FISH. *Nat. Genet.* **12:** 368–375.

Stacey, S.N., Gudbjartsson, D.F., Sulem, P., Bergthorsson, J.T., Kumar, R., Thorleifsson, G., Sigurdsson, A., Jakobsdottir, M., Sigurgeirsson, B., Benediktsdottir, K.R., et al. 2008. Common variants on 1p36 and 1q42 are associated with cutaneous basal cell carcinoma but not with melanoma or pigmentation traits. *Nat. Genet.* **40:** 1313–1318.

Stephens, P., Hunter, C., Bignell, G., Edkins, S., Davies, H., Teague, J., Stevens, C., O'Meara, S., Smith, R., Parker, A., et al. 2004. Lung cancer: Intragenic ERBB2 kinase mutations in tumors. *Nature* **431:** 525–526.

Stransky, N., Vallot, C., Reyal, F., Bernard-Pierrot, I., de Medina, S.G., Segraves, R., de Rycke, Y., Elvin, P., Cassidy, A., Spraggon, C., et al. 2006. Regional copy number-independent deregulation of transcrip-

tion in cancer. *Nat. Genet.* **38:** 1386–1396.

Strauss, B.S. 2007. Limits to the Human Cancer Genome Project? *Science* **315:** 762–764; author reply 764–765.

Suzuki, H., Gabrielson, E., Chen, W., Anbazhagan, R., van Engeland, M., Weijenberg, M.P., Herman, J.G., and Baylin, S.B. 2002. A genomic screen for genes upregulated by demethylation and histone deacetylase inhibition in human colorectal cancer. *Nat. Genet.* **31:** 141–149.

Tabin, C.J., Bradley, S.M., Bargmann, C.I., Weinberg, R.A., Papageorge, A.G., Scolnick, E.M., Dhar, R., Lowy, D.R., and Chang, E.H. 1982. Mechanism of activation of a human oncogene. *Nature* **300:** 143–149.

Tanay, A., Sharan, R., Kupiec, M., and Shamir, R. 2004. Revealing modularity and organization in the yeast molecular network by integrated analysis of highly heterogeneous genomewide data. *Proc. Natl. Acad. Sci.* **101:** 2981–2986.

Tanke, H.J., Wiegant, J., van Gijlswijk, R.P., Bezrookove, V., Pattenier, H., Heetebrij, R.J., Talman, E.G., Raap, A.K., and Vrolijk, J. 1999. New strategy for multi-colour fluorescence in situ hybridisation: COBRA: COmbined Binary RAtio labelling. *Eur. J. Hum. Genet.* **7:** 2–11.

Tapper, W., Hammond, V., Gerty, S., Ennis, S., Simmonds, P., Collins, A., and Eccles, D. 2008. The influence of genetic variation in 30 selected genes on the clinical characteristics of early onset breast cancer. *Breast Cancer Res.* **10:** R108.

Tenesa, A., Farrington, S.M., Prendergast, J.G., Porteous, M.E., Walker, M., Haq, N., Barnetson, R.A., Theodoratou, E., Cetnarskyj, R., Cartwright, N., et al. 2008. Genome-wide association scan identifies a colorectal cancer susceptibility locus on 11q23 and replicates risk loci at 8q24 and 18q21. *Nat. Genet.* **40:** 631–637.

Thomas, G., Jacobs, K.B., Kraft, P., Yeager, M., Wacholder, S., Cox, D.G., Hankinson, S.E., Hutchinson, A., Wang, Z., Yu, K., et al. 2009. A multistage genome-wide association study in breast cancer identifies two new risk alleles at 1p11.2 and 14q24.1 (RAD51L1). *Nat. Genet.* **41:** 579–584.

Thomas, G., Jacobs, K.B., Yeager, M., Kraft, P., Wacholder, S., Orr, N., Yu, K., Chatterjee, N., Welch, R., Hutchinson, A., et al. 2008. Multiple loci identified in a genome-wide association study of prostate cancer. *Nat. Genet.* **40:** 310–315.

Tomlins, S.A., Mehra, R., Rhodes, D.R., Cao, X., Wang, L., Dhanasekaran, S.M., Kalyana-Sundaram, S., Wei, J.T., Rubin, M.A., Pienta, K.J., et al. 2007. Integrative molecular concept modeling of prostate cancer progression. *Nat. Genet.* **39:** 41–51.

Tomlins, S.A., Mehra, R., Rhodes, D.R., Smith, L.R., Roulston, D., Helgeson, B.E., Cao, X., Wei, J.T., Rubin, M.A., Shah, R.B., et al. 2006. *TMPRSS2:ETV4* gene fusions define a third molecular subtype of prostate cancer. *Cancer Res.* **66:** 3396–3400.

Tomlins, S.A., Rhodes, D.R., Perner, S., Dhanasekaran, S.M., Mehra, R., Sun, X.W., Varambally, S., Cao, X., Tchinda, J., Kuefer, R., et al. 2005. Recurrent fusion of TMPRSS2 and ETS transcription factor genes in prostate cancer. *Science* **310:** 644–648.

van de Vijver, M.J., He, Y.D., van't Veer, L.J., Dai, H., Hart, A.A., Voskuil, D.W., Schreiber, G.J., Peterse, J.L., Roberts, C., Marton, M.J., et al. 2002. A gene-expression signature as a predictor of survival in breast cancer. *N. Engl. J. Med.* **347:** 1999–2009.

van't Veer, L.J., Dai, H., van de Vijver, M.J., He, Y.D., Hart, A.A., Mao, M., Peterse, H.L., van der Kooy, K., Marton, M.J., Witteveen, A.T., et al. 2002. Gene expression profiling predicts clinical outcome of breast cancer. *Nature* **415:** 530–536.

Vire, E., Brenner, C., Deplus, R., Blanchon, L., Fraga, M., Didelot, C., Morey, L., Van Eynde, A., Bernard, D., Vanderwinden, J.M., et al. 2006. The Polycomb group protein EZH2 directly controls DNA methylation. *Nature* **439:** 871–874.

Vogelstein, B. and Kinzler, K.W. 1993. The multistep nature of cancer. *Trends Genet.* **9:** 138–141.

Vogelstein, B. and Kinzler, K.W. 2004. Cancer genes and the pathways they control. *Nat. Med.* **10:** 789–799.

Vogelstein, B., Lane, D., and Levine, A.J. 2000. Surfing the p53 network. *Nature* **408:** 307–310.

Volik, S., Raphael, B.J., Huang, G., Stratton, M.R., Bignel, G., Murnane, J., Brebner, J.H., Bajsarowicz, K., Paris, P.L., Tao, Q., et al. 2006. Decoding the fine-scale structure of a breast cancer genome and transcriptome. *Genome Res.* **16:** 394–404.

Volik, S., Zhao, S., Chin, K., Brebner, J.H., Herndon, D.R., Tao, Q., Kowbel, D., Huang, G., Lapuk, A., Kuo, W.L., et al. 2003. End-sequence profiling: Sequence-based analysis of aberrant genomes. *Proc. Natl. Acad. Sci.* **100:** 7696–7701.

Volinia, S., Calin, G.A., Liu, C.G., Ambs, S., Cimmino, A., Petrocca, F., Visone, R., Iorio, M., Roldo, C., Ferracin, M., et al. 2006. A microRNA expression signature of human solid tumors defines cancer gene targets. *Proc. Natl. Acad. Sci.* **103:** 2257–2261.

Wan, P.T., Garnett, M.J., Roe, S.M., Lee, S., Niculescu-Duvaz, D., Good, V.M., Jones, C.M., Marshall, C.J., Springer, C.J., Barford, D., et al. 2004. Mechanism of activation of the RAF-ERK signaling pathway by oncogenic mutations of B-RAF. *Cell* **116:** 855–867.

Wang, Y., Broderick, P., Webb, E., Wu, X., Vijayakrishnan, J., Matakidou, A., Qureshi, M., Dong, Q., Gu, X., Chen, W.V., et al. 2008. Common 5p15.33 and 6p21.33 variants influence lung cancer risk. *Nat. Genet.* **40:** 1407–1409.

Wang, Z., Shen, D., Parsons, D.W., Bardelli, A., Sager, J., Szabo, S., Ptak, J., Silliman, N., Peters, B.A., van der Heijden, M.S., et al. 2004. Mutational analysis of the tyrosine phosphatome in colorectal cancers. *Science* **304:** 1164–1166.

Wang, Z., Zang, C., Rosenfeld, J.A., Schones, D.E., Barski, A., Cuddapah, S., Cui, K., Roh, T.Y., Peng, W., Zhang, M.Q., et al. 2008. Combinatorial patterns of histone acetylations and methylations in the human genome. *Nat. Genet.* **40:** 897–903.

Weber, M., Davies, J.J., Wittig, D., Oakeley, E.J., Haase, M., Lam, W.L., and Schubeler, D. 2005. Chromosome-wide and promoter-specific analyses identify sites of differential DNA methylation in normal and transformed human cells. *Nat. Genet.* **37:** 853–862.

Weir, B.A., Woo, M.S., Getz, G., Perner, S., Ding, L., Beroukhim, R., Lin, W.M., Province, M.A., Kraja, A., Johnson, L.A., et al. 2007. Characterizing the cancer genome in lung adenocarcinoma. *Nature* **450:** 893–898.

Wood, L.D., Parsons, D.W., Jones, S., Lin, J., Sjoblom, T., Leary, R.J., Shen, D., Boca, S.M., Barber, T., Ptak, J., et al. 2007. The genomic landscapes of human breast and colorectal cancers. *Science* **318:** 1108–1113.

Yamashita, K., Upadhyay, S., Osada, M., Hoque, M.O., Xiao, Y., Mori, M., Sato, F., Meltzer, S.J., and Sidransky, D. 2002. Pharmacologic unmasking of epigenetically silenced tumor suppressor genes in esophageal squamous cell carcinoma. *Cancer Cell* **2:** 485–495.

Yeager, M., Orr, N., Hayes, R.B., Jacobs, K.B., Kraft, P., Wacholder, S., Minichiello, M.J., Fearnhead, P., Yu, K., Chatterjee, N., et al. 2007. Genome-wide association study of prostate cancer identifies a second risk locus at 8q24. *Nat. Genet.* **39:** 645–649.

Zender, L., Spector, M.S., Xue, W., Flemming, P., Cordon-Cardo, C., Silke, J., Fan, S.T., Luk, J.M., Wigler, M., Hannon, G.J., et al. 2006. Identification and validation of oncogenes in liver cancer using an integrative oncogenomic approach. *Cell* **125:** 1253–1267.

Zhang, L., Volinia, S., Bonome, T., Calin, G.A., Greshock, J., Yang, N., Liu, C.G., Giannakakis, A., Alexiou, P., Hasegawa, K., et al. 2008. Genomic and epigenetic alterations deregulate microRNA expression in human epithelial ovarian cancer., *Proc. Natl. Acad. Sci.* **105:** 7004–7009.

Zheng, W., Long, J., Gao, Y.T., Li, C., Zheng, Y., Xiang, Y.B., Wen, W., Levy, S., Deming, S.L., Haines, J.L., et al. 2009. Genome-wide association study identifies a new breast cancer susceptibility locus at 6q25.1. *Nat. Genet.* **41:** 324–328.

WEB RESOURCES

http://cancergenome.nih.gov Cancer Genome Atlas Research Network.

http://cgap.nci.nih.gov/Chromosomes/Mitelman Mitelman Database of Chromosome Aberrations in Cancer (now part of the NCI/NCBI Cancer Chromosomes database).

http://www.ebi.ac.uk/microarray-as/ae Parkinson et al. 2007. European Bioinformatics Institute's ArrayExpress.

http://www.epigenome.org/ International Human Epigenome Project 2008. International Human Epigenome Project.

http://www.hapmap.org Consortium 2005. International HapMap Project.

http://www.icgc.org/ International Cancer Genome Consortium.

http:// www.methdb.de Grunau et al. 2001. DNA Methylation Database (MethDB).

http://methycancer.psych.ac.cn He et al. 2008. MethyCancer.

http://mit.lifescience.ntu.edu.tw/index.html Fang et al. 2008. MeInfoText.

http://www.ncbi. nlm.nih.giv/projects/SNP dbSNP Single Nucleotide Polymorphism Database.

http://www.ncbi.nlm.nih.gov/geo Barrett et al. 2005. NCBI Gene Expression Omnibus.

http://www.ncbi.nlm.nih.gov/RefSeq Wood et al. 2007. RefSeq genes.

http://www.ncbi.nlm.nih.gov/sites/entrez?db=cancerchromosomes NCBI/NCI's Cancer Chromosomes.

http://www.oncomine.org Rhodes et al. 2007. Oncomine.

http://www-p53.iarc.fr/ Olivier et al. 2002. p53 Database.

http://www.progenetix.net/progenetix Progenetix.

http://www.pubmeth.org Ongenaert et al. 2008. PubMeth.

http://research.nhgri.nih.gov/bic/ Breast Cancer Mutations Database.

http://www.sanger.ac.uk/genetics/CGP/ Cancer Genome Project.

http://www.sanger.ac.uk/genetics/CGP/Census/ Cancer Gene Census.

http://www. sanger.ac.uk/genetics/CGP/cosmic Catalogue of Somatic Mutations in Cancer. COSMIC.

http://www.somaticmutations-egfr.org/ EGFR Mutations Database.

http://smd.stanford.edu Marinelli et al. 2008. Stanford Microarray Database.

15

When the Genetic Code Is Not Enough—How Sequence Variations Can Affect Pre-mRNA Splicing and Cause (Complex) Disease

Brage Storstein Andresen[1] and Adrian R. Krainer[2]

[1]Department of Biochemistry and Molecular Biology, University of Southern Denmark, Campusvej 55, 5230 Odense M., Denmark; [2]Cold Spring Harbor Laboratory, Cold Spring Harbor, New York 11724

INTRODUCTION

Pre-mRNA splicing is a fundamental step in the expression of the >90% of eukaryotic genes that have introns interrupting their coding sequences. Although the pre-mRNA splicing process has been known for three decades, its enormous potential for creating diversity in gene expression and its involvement in nearly all disease processes has only recently become widely appreciated.

For a long time it was assumed that the complexity of an organism is reflected in the number of genes in its genome. But recently this simple assumption has been challenged by the steadily decreasing estimates for the total number of genes in the human genome, which appears not to be significantly higher than that of many less complex organisms. Fortunately for us, it seems that human genes exhibit a higher degree of alternative splicing than, for instance, mice, *Drosophila*, and *Arabidopsis thaliana* (Kim et al. 2007). Thus, we can produce more different proteins than these lower species, despite a comparable number of genes. In humans, an average of two to three protein isoforms per gene is produced by alternative splicing, and some genes can potentially express more than 1000 different isoforms (Tabuchi et al. 2002). Moreover, between 65% and 70%, if not more, of all human protein-coding genes undergo alternative splicing, and a large proportion show tissue-specific differences (Johnson et al. 2003; Castle et al. 2008). A corollary of all this diversity is that it requires a high level of flexibility in the splicing process, which makes both constitutive and alternative splicing very complex and critically dependent on numerous splicing-regulatory elements (SREs) in coding and noncoding regions throughout our genes. In some constitutively spliced exons, the balance between all these SREs is so finely tuned that changes at most exonic positions may potentially affect splicing (Singh et al. 2004; Pagani et al. 2005; Hua et al. 2007). Moreover, intronic sequences harbor many more SREs than previously believed, meaning that sequence variation in all parts of our introns may potentially affect splicing (Hua et al. 2008).

As a result, our view of missplicing as a disease mechanism has changed from a situation in which this possibility was considered to be rare and only relevant when one out of a few critical nucleotides at the exon–intron borders are changed, to a situation in which missplicing may be caused by nearly all sequence variation in a gene. This is of relevance to complex disease because, owing to the inherent dynamics of splicing, many sequence variations may cause only a partial and conditional change in the splicing pattern of a candidate gene. Furthermore, the severity of such a change in splicing may be dependent on environmental stimuli that modulate the activity of critical splicing-regulatory factors and/or be influenced by the simultaneous presence of genetic variants in general or specific splicing factors. Thus, it is not difficult to imagine that genetic variants like single-nucleotide polymorphisms (SNPs) may cause conditional splicing defects, which together with environmental factors could be a significant contributing factor to complex disease. In this chapter, we will describe the mechanism of pre-mRNA splicing and how it is regulated, as well as describe some interesting examples of how sequence variation may affect splicing and cause disease.

HOW SPLICING OCCURS

Pre-mRNA splicing is a highly complex process that relies on the interplay between numerous regulatory *cis* elements in the pre-mRNA transcript and the spliceosome (Fig. 15.1), a dynamic multiprotein complex responsible for catalyzing the splicing reaction. In addition, splicing of each pre-mRNA is also dependent on the activity of a wide variety of auxiliary splicing-regulatory proteins. Five different small nuclear RNAs (snRNAs)—U1, U2, U4, U5, U6—and more than 150 different proteins (Jurica and Moore 2003) constitute the spliceosome, which is responsible for splicing of pre-mRNAs with U2-type introns (which usually begin with GT or GC and end with AG). The vast majority of human introns are of the U2 type; the other types (Patel and Steitz 2003) are very rare and will not be discussed in this chapter.

The basic process of pre-mRNA splicing consists of recognition of the correct splice sites, precise cleavage at the 5'-splice site (5'-ss) and the 3'-splice site (3'-ss) at each end of the intron, and joining of the two flanking exons. The chemistry proceeds via two transesterification reactions. In the first reaction, the 5'-exon is cleaved by a nucleophilic attack from the 2'-OH of the branchpoint (BP) adenosine, and the 5'-end of the intron is joined to the BP via a 2'-5' phosphodiester bond, thereby creating an intermediate intron-3'-exon lariat structure and an upstream exon with a free 3'-end. The second reaction occurs when the 3'-OH of the free 5'-exon performs a nucleophilic attack at the downstream 3'-ss, joining the two exons and releasing the intron lariat.

Although this sounds relatively simple, it is a very complicated process in which spliceosome components assemble and disassemble sequentially. Moreover, extensive remodeling of the RNA–RNA and RNA–protein interactions within the spliceosome and between the spliceosomal components and the pre-mRNA takes place during these steps.

In the initial step (complex E, Fig. 15.1), the 5'-ss is recognized by base pairing with the U1 snRNA of the U1 small nuclear ribonucleoprotein (snRNP). The 3'-ss is also recognized in the initial step of splicing. The 3'-ss consists of three distinct elements: the BP sequence (BPS) is usually located 20–40 nucleotides upstream of the AG dinucleotide at the 3'-end of the intron; there is a polypyrimidine tract (PPT) sequence between the BP and the terminal AG. First, the BPS is recognized by Splicing Factor 1 (SF1). This factor then interacts cooperatively with the large subunit (U2AF65) of the U2 auxiliary factor (U2AF) heterodimer, which binds the PPT with its RNA-binding domain. The smaller subunit (U2AF35) recognizes the first AG dinucleotide downstream of the PPT in a sequence-specific manner. This first step is thought to commit the pre-mRNA for splicing and is necessary for further assembly of the spliceosome.

Next U2AF65 recruits the U2 snRNP through a protein–protein interaction, and SF1 is replaced at the BPS by the U2 snRNP, such that the U2 snRNA base pairs with the BPS and thereby causes the nearly invariant adenosine of the BPS to bulge out and expose its 2'-OH for the first nucleophilic attack in catalysis. The complex generated when both splice sites are recognized in this way is referred to as the pre-spliceosome complex (complex A). Then the last three snRNPs (U4, U5, and U6), which are assembled into a trimeric complex, are recruited (complex B). The U5 snRNP interacts with both the 5'- and the 3'-exon borders, allowing the spliceosome to "tether" the "free" 5'-exon from the first transesterification reaction and align it with the 3'-exon in the second transesterification reaction. The catalytically active spliceosome complex (complex C) is created by conformational alterations of the snRNPs, where, for instance, the base pairing of the U1 snRNA to the 5'-ss is destabilized and

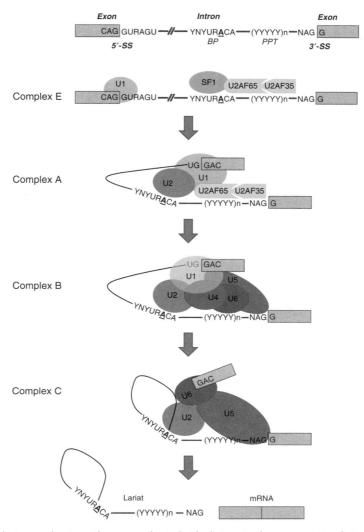

FIGURE 15.1 Splicing mechanism. Shown are the individual steps in the pre-mRNA splicing process, including the consensus sequences for important elements, such as the 5'-ss, branchpoint (BP), and 3'-ss.

replaced by the U6 snRNA. A new interaction between the U6 snRNP and U2 snRNP also occurs. When the exons have been spliced, the mRNA is released from the spliceosome.

Although the splice-site sequences are the strongest determinants for correct splicing, their consensus sequences (determined by comparing splice sites from annotated genes) show a high degree of degeneracy (see Fig. 15.1), and many functional splice sites do not conform well to the consensus sequence. For instance, only ~2% of human U2-type 5'-ss perfectly match the consensus sequence CAG/GTRAGT (Sheth et al. 2006).

Splicing Enhancers and Silencers Are Needed to Define Splice Sites

It is clear that the consensus motifs are by themselves insufficient to completely define the splice sites—other elements also need to be present. In fact, nonfunctional splice-site sequences that match the consensus (known as pseudo-splice sites) occur more frequently in most genes than the real splice sites (Sun and Chasin 2000). Thus, negative SREs may be needed to suppress recognition of nonfunctional splice-site sequences, and, conversely, positive SREs may be needed to assist in recognition of the real splice sites, especially when these are weak and conform poorly to the consensus motifs.

The existence of SREs became apparent in 1987, when a study showed that an 81-bp sequence in an exon from the fibronectin gene is essential for alternative splicing regulation (Mardon et al. 1987). The SREs can be divided into exonic splicing enhancers (ESEs), exonic splicing silencers (ESSs), intronic splicing enhancers (ISEs), and intronic splicing silencers (ISSs), according to their function and location in the pre-mRNA. Although SREs were initially described as regulators of alternative splicing, it is now generally accepted that they are also fundamental for constitutive splicing. Global analyses have shown that splicing enhancers are overrepresented in spliced exons and underrepresented in introns and pseudoexons, whereas splicing silencers show the opposite pattern (Wang et al. 2004; Zhang and Chasin 2004; Wang et al. 2005).

ESEs usually function by binding of serine/arginine-rich (SR) proteins. The SR proteins constitute a family of splicing factors characterized by an amino-terminal RNA-binding domain with one or two RNA-recognition motifs (RRMs) and a carboxy-terminal arginine/serine-rich (RS) splicing-activation domain with alternating serine and arginine residues (Graveley et al. 2000). SR proteins are important for both constitutive and alternative splicing (Krainer et al. 1990). The RRM(s) of SR proteins recognize short, degenerate sequence motifs, which can function as ESEs responsive to the corresponding SR protein (Liu et al. 1998). The RS domain functions through protein–protein interactions with other RS-domain-possessing proteins, like the 70-kDa subunit of the U1 snRNP and the 35-kDa subunit of U2AF (U2AF35). ESS and ISS elements often function by binding heterogeneous nuclear ribonucleoprotein (hnRNP) proteins, which typically repress splicing. The hnRNP proteins belong to a large protein superfamily (Dreyfuss et al. 2002). Many hnRNP proteins have important regulatory roles in pre-mRNA splicing and they each regulate distinct targets (Venables et al. 2008). The hnRNP proteins have one or more amino-terminal RNA-binding domains (usually RRM or hnRNP K-homology domains) and they also harbor carboxy-terminal repressive domains (e.g., a glycine-rich domain in hnRNP A1), which enable protein–protein interactions with other hnRNPs or other splicing-regulatory proteins.

It is important to realize that SR proteins may in some cases function as negative regulators of splicing, and hnRNP proteins may function as positive regulators. In addition, numerous other splicing-regulatory proteins also bind to SREs and are involved in splicing regulation. Coactivators and corepressors (e.g., SRm160/300, Raver, etc.) can also function in conjunction with the above factors, without contacting the SREs directly (Fig. 15.2A).

There are many ways in which SREs function in splicing regulation. In Figure 15.2 we show some simple, general mechanisms. In the simplest model for ESE or ISE function (Fig. 15.2A), the binding of an SR protein may help to recruit U2AF to a "weak" PPT of a 3'-ss or to recruit the U1 snRNP to a suboptimal 5'-ss (Wu and Maniatis 1993; Zuo and Maniatis 1996). Alternatively, binding of an SR protein may sterically hinder binding of negatively acting splicing-regulatory proteins (like the hnRNPs) to a flanking ESS or ISS (Fig. 15.2B). This mechanism does not require the RS domain (Zhu et al. 2001; Shaw et al. 2007). Another interesting mechanism for ISE function is to effectively decrease intron length by forming dimers between hnRNP molecules bound to ISE elements at the ends of long introns (Fig. 15.2C), thereby looping out the intervening sequence and bringing distant splice sites into closer proximity and thus stimulating splicing (Martinez-Contreras et al. 2006).

In a related but opposite way, hnRNPs bound to ISS and/or ESS elements could cause exon skipping by looping out the entire exon (Fig. 15.2D) or a splice site into an inaccessible conformation (Blanchette and Chabot 1999; Amir-Ahmady et al. 2005). A simpler mechanism of ISS and ESS function is to directly block access to flanking or overlapping splice sites (Fig. 15.2E) (Tange et al. 2001) or to ESE/ISE elements through binding of hnRNP proteins. Alternatively, ESS and ISS elements may inhibit splicing by providing a high-affinity binding site for hnRNP proteins, followed by multimerization and spreading along the pre-mRNA, thereby masking other important SREs or the splice sites (Fig. 15.2F) (Zhu et al. 2001).

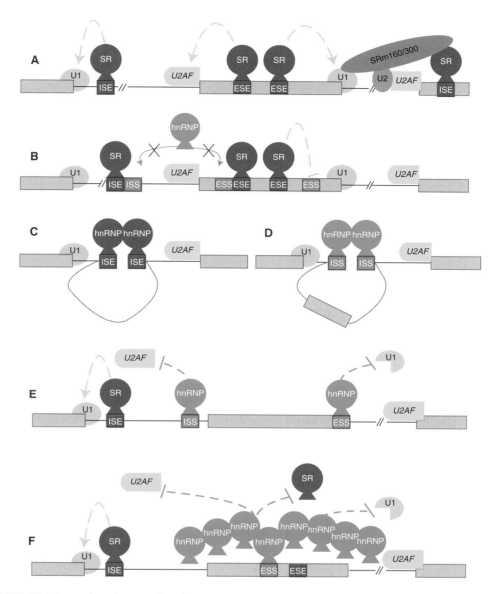

FIGURE 15.2 General mechanisms for splicing regulation by SREs. (*A*) Simple recruitment via serine/arginine-rich-domain interaction, with or without coactivator involvement. (*B*) Antagonistic splicing enhancer function by steric hindrance. (*C* and *D*) "Looping." (*E*) Splicing silencer function by direct blockage. (*F*) Multimerization of splicing inhibitory proteins initiated from a high-affinity splicing silencer.

Although the traditional view, as described above, indicates that splicing-regulatory proteins usually function at the earliest steps of spliceosome assembly (the E complex) by either enhancing or inhibiting the initial recognition of the splice sites, recent studies have indicated that splicing modulation may also take place at later steps. Splicing-regulatory proteins bound to a silencer may stall splicing at the A step by promoting a locked exon-bridging complex through protein–protein interactions between the prespliceosomal complexes, with U1 and U2 snRNPs bound at each end of the exon. This has been proposed for regulation of CD45 exon 4 and Fas receptor exon 6 alternative splicing (House and Lynch 2006; Bonnal et al. 2008; Topp et al. 2008). Finally, a recent study has suggested that proteins bound to splicing silencers may inhibit interaction between the U1 snRNP bound at the 5′-ss and the U2 snRNP bound at the BPS by altering the conformation of the bound U1 snRNP or by shielding a surface of the U1 snRNP needed for the interaction with the U2 snRNP (Yu et al. 2008).

Apart from SREs, other parameters may modulate splicing. Because pre-mRNA splicing usually occurs cotranscriptionally, the transcription elongation rate by RNA polymerase II may influence splicing by determining how many splice sites are available simultaneously. A fast elongation rate may, for instance, result in a stronger downstream splice site becoming available before splicing to a weaker upstream splice site has occurred, resulting in direct competition between the two splice sites (Kornblihtt et al. 2004). Thus, it is conceivable that some SREs function at the DNA level, modulating transcription elongation by forcing RNA polymerase II to pause. It is also becoming increasingly clear that intramolecular base pairing in the pre-mRNA can lead to formation of secondary structures that influence splicing choices by either exposing or hiding important elements, like the splice sites and splicing enhancers/silencers (Buratti et al. 2004; Hiller et al. 2007; Shepard and Hertel 2008).

Taking all of these factors and mechanisms into consideration, it is obvious that there are numerous ways in which sequence variation in our genes may affect splicing. An additional complication is the fact that these alternative ways are not mutually exclusive but may act in concert.

DISREGULATION OF SPLICING CAN CAUSE DISEASE

Disregulation of splicing may be caused by mutations in splicing-regulatory sequences in a gene, by changes in the abundance or activity of splicing-regulatory proteins, or by a combination of these mechanisms. Moreover, in complex diseases, it is possible that disease results from a combination of such changes in several genes and also that environmental triggering factors are involved.

Mutations that directly affect splice sites, by either creating them or abolishing them, are frequent in most human genes. They constitute close to 10% of the more than 80,000 different mutations that are currently known in more than 3000 different genes, as compiled in the Human Genome Mutation Database (HGMD; Professional release 2008.2, August 2008). As many as 6% of all mutations map to the GT/AG dinucleotides at the splice sites, reflecting the relative importance of these positions. The overrepresentation of mutations in the GT/AG dinucleotides may also be influenced by underreporting of other mutations at the splice sites, as their effect is more difficult to predict. Whereas mutation of the GT/AG dinucleotides almost always results in complete missplicing, mutation of the other positions may result in a variable degree of missplicing, depending in part on the intrinsic strength of the splice site, which in turn is determined by the nature of the nucleotides at the other positions (Roca et al. 2008).

An illustrative example is a disease-causing +3A→G mutation in the 5′-ss of exon 3 from the human *SBCAD* gene, causing short/branched-chain acyl-CoA dehydrogenase deficiency (Madsen et al. 2006). This mutation causes complete exon skipping, although it is well known that a G at position +3 matches the consensus and it is well tolerated in many other splice sites. However, this 5′-ss is weak because of mismatches at the −3, +4, and +5 positions and a weaker wobble base pair (G:U) at the −2 position, causing suboptimal base pairing with U1 snRNA. Therefore, the additional loss of stable base pairing with U1 snRNA caused by the subtle change from +3A→G, which only slightly destabilizes the hydrogen bonding by creating a (G:U) wobble base pair, results in a dramatic effect on splicing. Disease-causing +3A→G mutations in other genes show the same dependence on mismatches at other positions in the 5′-ss (Ohno et al. 1999; Madsen et al. 2006; Roca et al. 2008). In this rather simple example, there is a direct correlation between the degree of complementarity to the U1 snRNA and the amount of missplicing. This is further exemplified in Figure

15.3A, where the effect of mutating at another position in this splice site using a minigene system is shown.

It is important to realize that complementarity to U1 snRNA is not the only factor that determines the effect of splice-site mutations. Alternative splicing of exon 10 from the *MAPT* gene requires that a very precise ratio between skipping and inclusion be maintained, because disruption of the precise ratio between the resulting τ-3R and τ-4R isoforms causes disease (for review, see

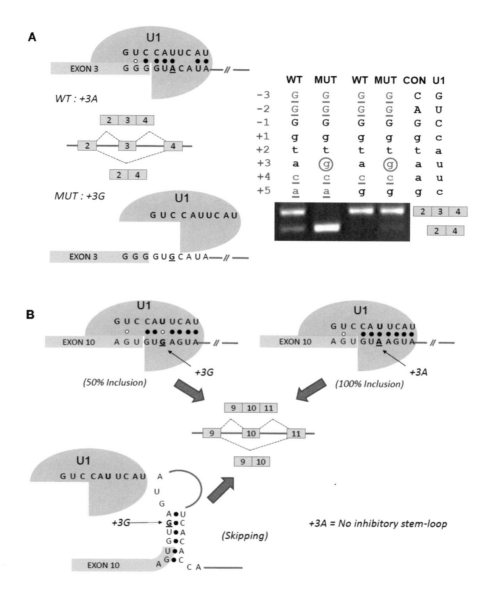

FIGURE 15.3 Splice-site mutations. (*A*) A +3A→G mutation in the *SBCAD* gene causes exon 3 skipping because of decreased base-pairing stability to U1 snRNA (Madsen et al. 2006). Results from compensatory mutations at position +5 using a *SBCAD* minigene are shown. (*B*) A +3G→A mutation in the *MAPT* gene increases complementarity to U1 snRNA and abolishes an inhibitory stem–loop structure at the 5′ splice site of exon 10. This leads to increased exon 10 inclusion and disruption of the correct ratio between the encoded protein isoforms (for review, see Kar et al. 2005). Stable base pairs are shown as *solid circles*, whereas less stable wobble base pairs are shown as *open circles*.

Kar et al. 2005). A disease-causing +3G→A mutation in the 5'-ss of *MAPT* exon 10 increases the complementarity to U1 snRNA and results, as expected, in increased inclusion. However, here the situation is more complex, owing to a hairpin loop, which is crucial for correct splicing regulation by decreasing the access of U1 snRNA (Fig. 15.3B). The +3G→A mutation disrupts this structure, and a number of other disease-causing mutations, which likewise destabilize it but without affecting complementarity to U1 snRNA, also increase exon 10 inclusion. Therefore, it is clear that the effect of the +3G→A mutation is also caused by disruption of the hairpin structure.

Splice-site mutations may also have additional negative effects by directly creating competing binding sites for splicing-inhibitory proteins, or, alternatively, the decreased complementarity to the snRNA portion of U1 or U2 snRNPs may be further aggravated if a binding site for a splicing-inhibitory protein overlaps with their binding sites. Disease-causing mutations in the 5'-ss of exon 3 of the neurofibromin 1 gene (*NF1*) and in the exon 2 5'-ss of the thyrotropin β-subunit (*TSHB*) gene both disrupt complementarity to U1 snRNA, but some of the negative effect can be attributed to an increase in binding of hnRNP H to an overlapping ISS motif (Buratti et al. 2004).

Finally, it was recently demonstrated that certain 5'-ss may function in a "shifted mode," which means that what may appear to be a weak 5'-ss with a poor match to U1 snRNA, may in fact function efficiently, because base pairing instead occurs in a shifted register (Roca and Krainer 2009). In turn, this observation can explain why a +5A→G mutation in intron 2 of the *RARS2* gene, associated with pontocerebellar hypoplasia, causes missplicing, even though G is the consensus nucleotide at this position (Edvarson et al. 2007); because this atypical 5'-ss is recognized by U1 in a shifted register, the mutation results in weaker U1 binding (Roca and Krainer 2009).

These examples illustrate that the effects of mutations in the 5'-ss are critically dependent on how the resulting changes affect the affinity to the U1 snRNP, mainly determined by the complementarity to U1 snRNA. In addition, the effects may be highly context-specific, so that even identical mutations at a particular position of the 5'-ss do not necessarily have the same effect in different introns. Although we only described examples of the effects of mutations in 5'-ss, similar principles apply to mutations in 3'-ss.

Exon skipping is the most frequently observed outcome of a splice-site mutation; however, other outcomes are also frequently observed. The basic underlying mechanism is that a splice-site mutation shifts the balance in the relative strength between a normal (authentic) splice site and a competing splice site, which may be either a cryptic splice site that is not used in the wild-type context, a new splice site created by the mutation, or a natural splice site from a flanking exon. In the latter case, the outcome is typically exon skipping, but use of a cryptic or de novo splice site may lead to inclusion of part of the intron or deletion of part of the exon, depending on the location of the activated splice site. The least frequent scenario is intron retention, which only occurs when the mutant splice site and the splice site at the opposite end of the intron are not recognized at all. It is at present very difficult to accurately predict which scenario will take place, and often more than one of the alternatives are active simultaneously and with different efficiencies. A particularly interesting outcome is inclusion of part of an intronic sequence in the form of a pseudoexon, although relatively few examples of this have been described (Buratti et al. 2006). Activation of pseudoexons may be caused by an intronic mutation, which creates a new splice site or strengthens a preexisting pseudo-splice site, which may then function together with a previously unmatched splice site at the other end of the pseudoexon (Buratti et al. 2006).

As discussed in more detail below, mutations may also affect SREs and thereby influence splicing, and there are a few reports of an intronic mutation in an SRE increasing recognition of a pseudo-splice site, leading to pseudoexon inclusion (Pagani et al. 2002; Davis et al. 2009). The fact that introns typically make up >90% of the sequence of a gene, together with the high frequency of pseudo-splice sites in introns (Sun and Chasin 2000), means that activation of pseudoexons is probably heavily underreported. Moreover, disease-causing mutations may be located deep within introns, where they are missed by routine mutation-detection techniques.

SPLICING MUTATIONS NOT DIRECTLY INVOLVING SPLICE-SITE SEQUENCES

Exons that are alternatively spliced typically have weak splice sites (Garg and Green 2007), making their recognition highly dependent on the balance between the positive and negative SREs. Splicing thus becomes dynamic and determined by the relative amounts/activity of the splicing-regulatory proteins that bind to the various SREs in a given set of conditions. In constitutive splicing, regulation is pushed as far as possible in the direction of only one product being formed, and this is typically achieved by the presence of strong splice sites, thereby making splice-site recognition less dependent or nearly independent on the SREs. However, there are many constitutively spliced exons with weak splice sites that are therefore highly dependent on the correct balance between positive and negative SREs. It is likely that such exons are more vulnerable to perturbations, because this fine balance between the SREs may be easily disturbed by sequence variations/mutations. Three well-studied examples of this are provided below (Fig. 15.4).

SRE Mutations in *MCAD*: The Importance of Haplotype

Mutations in the medium-chain acyl-CoA dehydrogenase (*MCAD*) gene cause the most frequent inborn error of fatty-acid oxidation. Several patients with MCAD deficiency have a disease-causing mutation, c.362C→T, in the middle of exon 5, which causes exon skipping (Nielsen et al. 2007). Because exon 5 has a weak 3′-ss, it is dependent on an ESE in order to recruit U2AF. The c.362C→T mutation decreases the activity of this ESE, but the splicing defect can be corrected by overexpression of SF2/ASF, because the mutation only decreases the affinity for SF2/ASF but does not completely abrogate the site (Fig. 15.4A,B). This illustrates that a mutation may result in a conditional splicing defect, in which the extent of missplicing is dependent on the activity or levels of a splicing-regulatory protein. An intriguing finding from the *MCAD* exon 5 study is that a synonymous SNP (c.351A/C), which flanks the ESE, determines if this element is needed (Fig. 15.4C,D). The c.351A variant creates an ESS, which binds hnRNP A1, whereas the c.351C variant abolishes hnRNP A1 binding and ESS function. Therefore, the c.362C→T mutation is only deleterious in the context of the c.351A variant, whereas alleles with the c.351C variant are immune. Thus, presumed neutral SNPs may have an important role in determining splicing, and this function may only be unmasked by mutations elsewhere. This example illustrates the complexity of splicing regulation of vulnerable exons and the need for evaluating sequence variations in the context of the correct haplotype. When the suboptimal 3′-ss was optimized in the context of a minigene, the need for a functional ESE was also alleviated, showing that it is the weak splice site that makes exon 5 vulnerable.

SRE Mutations and Missplicing in *CFTR* Exon 9

Another interesting example is exon 9 of the cystic fibrosis transmembrane conductance regulator (*CFTR*) gene, in which mutations in SREs cause missplicing (Fig. 15.4E,F). In this case, some of the SREs in exon 9 are organized into a complex element (termed CERES, or composite regulatory element of splicing) with both positive and negative SREs (Pagani et al. 2003). The *CFTR* exon 9 has a weak 5′-ss, but a flanking ISE in intron 9 stimulates its recognition through binding of the splicing-regulatory protein TIA-1. This ISE is counteracted by a flanking ISS, which is somewhat atypical in that inhibition is mediated by SF2/ASF and SRp40 binding. These SR proteins usually have a stimulatory effect, but here they inhibit splicing, possibly by stimulating recognition of a nonfunctional decoy splice site (Buratti et al. 2007). An interesting finding in *CFTR* is that polymorphic variation in a TG repeat and variation in the length of a run of Ts in the polypyrimidine tract of intron 8 together determine 3′-ss strength and thereby influence the risk for disease. In individuals with a short polypyrimidine tract (five Ts), the risk for nonclassical cystic fibrosis and congenital bilateral absence of the vas deferens is dependent on the number of upstream TG dinucleotides (Groman et al. 2004). Apparently, the short polypyrimidine tract causes the 3′-ss to be weak, and a high number of upstream

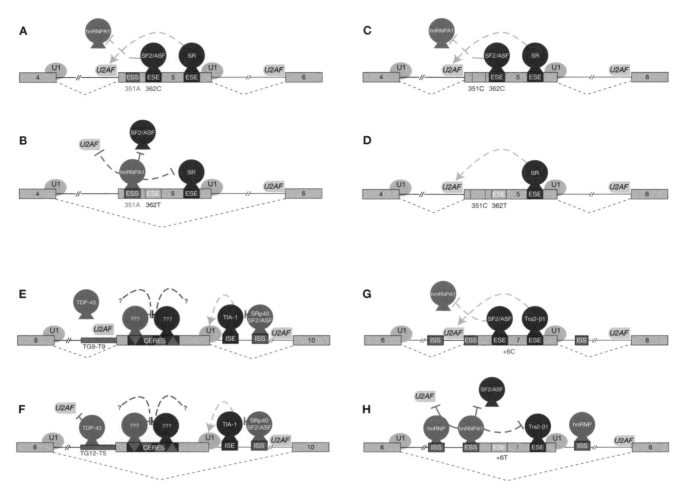

FIGURE 15.4 Models for splicing regulation in MCAD exon 5, CFTR exon 9, and SMN1/2 exon 7. (*A–D*) The effect of the different combinations (haplotypes) of the c.351A/C polymorphism and the wild-type c.362C or mutant c.362T variations in exon 5 of the *MCAD* gene (Nielsen et al. 2007). c.351A creates an ESS that binds hnRNP A1 and this is abolished by the c.351C variant. The c.362C→T mutation disrupts an ESE, so that it has a reduced affinity for SF2/ASF. The c.362C→T mutation only causes skipping when the c.351A ESS is present. (*E* and *F*) The effect of the different combinations (haplotypes) of the TG-Tn polymorphic variation in intron 8 of the *CFTR* gene (Buratti et al. 2004, 2007; Pagani et al. 2003). Long TG repeats (TG12) with a short T stretch (T5) cause exon 9 skipping as a result of binding of the inhibitory TDP43 protein to the TG repeat. Other important elements described in the text are also depicted. (*G* and *H*) One model for splicing inhibition by the +6C→T variation in *SMN2* is illustrated (Cartegni et al. 2006). Similar to the c.362C→T mutation in *MCAD*, the +6C→T variation in *SMN2* disrupts an SF2/ASF-specific ESE. This ESE may be important in antagonizing hnRNP A1 binding to splicing silencers and functions together with a Tra2-1-specific ESE to assist in U2AF binding to the weak 3'-ss. Several other elements are also involved.

TGs (12 or 13) further inhibits recognition of the 3'-ss by increasing binding of the splicing-regulatory protein TDP43, which sterically represses U2AF binding (Buratti et al. 2004).

The studies of *CFTR* exon 9 splicing further illustrate that even weak constitutive exons are dependent on an extensive and delicately balanced interplay between many different SREs. A unifying model that incorporates the complex interplay among all these elements is still missing. Importantly, *CFTR* exon 9 further emphasizes the importance of polymorphic variation in determining splicing efficiency.

SMN1/2, Splicing, and Spinal Muscular Atrophy

Perhaps the most extensively studied example of disregulated splicing as a result of exonic sequence variation is that in a gene associated with spinal muscular atrophy (SMA), a degenerative neuromuscular disease. The most frequent cause of SMA is deletion of the survival of motor neuron 1 (*SMN1*) gene. However, a nearly identical gene named *SMN2*, which encodes a protein with an identical amino acid sequence, is located next to *SMN1*, as a result of recent gene duplication. However, *SMN2* cannot fully compensate for the loss of *SMN1*, because a +6C→T silent substitution in exon 7 results in exon skipping (Monani et al. 1999) (Fig. 15.4G,H).

Initially, a simple explanation for this was proposed, according to which an ESE, which binds SF2/ASF, is disrupted by the +6C→T mutation in *SMN2*, and this ESE, together with another ESE, which binds hTra2β1 (Lorson and Androphy 2000), functions by facilitating recruitment of U2AF to the weak intron 6 3′-ss (Cartegni and Krainer 2002). Other studies have suggested that the +6C→T change may instead create an ESS, with a binding site for the splicing repressor protein hnRNP A1 (Kashima and Manley 2003). Several other SREs are located in intron 6, exon 7, and intron 7 (Miyajima et al. 2002; Singh et al. 2004; Cartegni et al. 2006; Singh et al. 2006; Hua et al. 2007, 2008; Kashima et al. 2007). Interestingly, the 5′-portion of exon 7 from the *SMN1/2* genes and the sequence of *MCAD* exon 5 that harbors both an ESE and an ESS are nearly identical, with similar positioning of ESE and ESS elements. This indicates that in different genes, some SREs may be organized into similar regulatory units with flanking and perhaps overlapping positive and negative SREs arranged into a more general architecture.

The detailed studies of *SMN1/2* splicing regulation are not only important for contributing to our general knowledge about the complexity of splicing regulation, but they are also crucial for the development of a treatment for SMA. A therapy based on splicing modulation is particularly appealing, because a method that increases exon 7 inclusion from the *SMN2* gene could serve as a treatment strategy applicable to nearly all SMA patients as they all have the endogenous *SMN2* gene available for manipulation. Different studies have shown that *SMN2* splicing can be dramatically improved by modulation of some of the SREs, either through changing the level of the corresponding splicing-regulatory proteins or by hybridizing oligonucleotides, which block access of repressor proteins (Miyajima et al. 2002; Cartegni and Krainer 2003; Singh et al. 2006; Hua et al. 2007, 2008).

Exon Skipping and Vulnerability

A common theme from these examples and others is that the affected exons are all dependent on multiple SREs and have weak splice sites. Therefore, such exons may represent a group of particularly vulnerable exons, present in most or all genes, which are difficult to splice because they are already on the verge of not being recognized by the splicing machinery. Consistent with this notion, significant levels of *CFTR* exon 9 (Bremer et al. 1992) and *MCAD* exon 5 (Gregersen et al. 1991) skipping can be observed from the wild-type alleles in control cells. The same is true for *MCAD* exon 2 (Gregersen et al. 1991), which has a weak 5′-ss, and for which the vulnerability concept is supported by the finding of a silent mutation that inactivates and ESE and causes exon skipping (B.S. Andresen, unpubl.). Finally, *SBCAD* also exhibits detectable levels of exon 3 skipping from the normal allele (Madsen et al. 2006). Thus, low levels of exon skipping may be a common characteristic of vulnerable exons. In all the examples discussed above and in many others, strengthening the weak splice sites by mutagenesis always overcomes the effect of mutations in the SREs. Not only does this observation confirm that splice-site strength is usually the major determinant for splicing, but it also indicates that exons with strong splice sites are not vulnerable and may readily tolerate mutations that abolish an ESE motif or create an ESS without any observable effect. This should be kept in mind when evaluating the possibility of a splicing defect resulting from an exonic sequence variation.

COMMON GENETIC VARIATION IN SPLICING-REGULATORY PROTEINS AND COMPONENTS OF THE SPLICEOSOME CAN RESULT IN SUBOPTIMAL SPLICING

The examples from missplicing of *CFTR* exon 9 and *MCAD* exon 5 nicely illustrate that polymorphic variation (e.g., SNPs) may be important for splicing efficiency. It is thus conceivable that there are variants (SNPs) in disease genes that compromise splicing-regulatory sequences, so that they either increase the dependence on the activity of splicing factors or result in lower splicing efficiency. Such alleles can be regarded as conditional splicing alleles, which, depending on the efficiency of the general splicing apparatus or of certain splicing factors (e.g., in different cell types), may have different levels of correct splicing. Depending on the conditions, they may have a level of correct splicing that is either above or below a certain threshold for disease manifestation. Given the estimates that the human genome contains more than six million SNPs (http://www.ncbi.nlm.nih.gov/SNP/snp_summary.cgi), it is likely that a significant proportion of SNPs affect the splicing pattern and/or efficiency of disease genes. In fact, a recent study (Nemgaware et al. 2008) identified many thousands of SNPs in splice sites and potential SREs. In addition, there are an increasing number of known SNPs that directly affect splicing and determine the risk for disease. For example, a SNP in the 3′-ss of exon 7 of the *OAS1* gene influences splicing efficiency and thereby the enzyme activity of the antiviral 2′,5′-oligo-adenylate synthetase enzyme (Bonnevie-Nielsen et al. 2005). Another example is a SNP in exon 4 of the *CYP2B6*6* gene that causes exon skipping. This SNP is presumed to inactivate an ESE and may have clinical relevance for the use of several drugs for human immunodeficiency virus (HIV) treatment by affecting the drug-metabolizing cytochrome P450 enzyme (Hofmann et al. 2008). Recently, genome-wide searches for differences in splicing associated with SNPs have indicated that the majority of such differences are tissue-specific (Heinzen et al. 2008). This suggests that a number of the SNPs that influence splicing may only manifest their pathogenic effect in a tissue-specific manner.

In addition to *cis* effects, in which SNPs or combinations of SNPs in a certain haplotype result in an allele with a (conditional) splicing defect, we can speculate that common genetic variation in some of the numerous components of the spliceosome, or in splicing-regulatory factors, results in suboptimal splicing activity. There are, in fact, numerous SNPs in splicing-regulatory proteins and components of the splicesosome. dbSNP (the Single-Nucleotide Polymorphism Database) reveals 16 missense, 39 synonymous, 11 frameshift, and more than 350 intronic SNPs in 12 representative splicing factors (SF2/ASF, SC35, SRp55, SRp40, hnRNP A1, hnRNP A2/B1, hnRNP H1, hnRNP I [PTB], U2AF35, U2AF65, U1-70K, U1C). The vast majority of these SNPs are probably neutral, but some of them may influence splicing efficiency. In fact, a recent study showed interindividual differences in the splicing of transcripts encoding several proteins (SRp40, SRp30c, HTra2β1, U2AF35, hnRNP A0, and hnRNPF) that are themselves important for splicing (Zhang et al. 2009). In mice, a mutation in a 5′-ss in a sodium channel gene, *Scn8a*, results in different disease severity in two mouse strains. This difference in phenotype depends on a mutation in the sodium channel modifier 1 (*Scnm1*) gene (Buchner et al 2003: Howell et al. 2007). Scnm1 is an auxiliary splicing factor and an accessory component of the U1 snRNP complex, and it was suggested that it might influence the efficiency of recognition of noncanonical 5′-ss in multiple genes, although splicing-microarray analysis failed to identify other targets (Howell et al. 2008). Interestingly, a polymorphic missense mutation (rs.1115, dbSNP) is present in ~1% of alleles of the human *Scnm1* homolog, and we can speculate that this may result in lower splicing efficiency in humans.

ENVIRONMENT INFLUENCES SPLICING

There are numerous ways in which the environment can influence splicing and thereby affect the onset and development of complex disease from susceptibility alleles. Extracellular stimuli may affect the activity, amounts, and localization of splicing-regulatory proteins and thus the splicing patterns of target genes. One common regulatory pathway involves phosphorylation/dephosphorylation cycles, because the activity of both hnRNP and SR proteins is controlled by their phosphorylation status, which affects their subcellular localization, as well as their participation in both protein–protein interactions and binding to pre-mRNA (Stamm 2008). There are many examples of extracellular stimuli that affect splicing through phosphorylation/dephosphorylation cascades. Osmotic stress can influence the cellular localization of hnRNP A1 through changes in phosphorylation status, and the consequent reduction in the nuclear levels of this splicing repressor affects alternative splicing (Allemand et al. 2005). Changes in insulin levels in response to metabolic changes may also affect splicing, because insulin levels influence the phosphorylation status of SR proteins, such as SRp40 and SF2/ASF (Patel et al. 2005). Thus, metabolic changes could modulate splicing of vulnerable alleles of disease genes through an insulin-mediated effect on splicing factors like SRp40.

Variation in low-density lipoprotein cholesterol (LDL-C) is a complex inherited trait influenced by multiple different genes (e.g., *LDLr* and *HMGCR*) and is influenced by environmental factors (Kathiresan et al. 2008). A SNP in the middle of exon 12 of the *LDLr* gene causes increased exon skipping and is presumed to abolish an ESE needed for recognition of the suboptimal 5′-ss (Zhu et al. 2007). This change in splicing is associated with increased LDL-C, mainly in premenopausal women, so the splicing defect is therefore sex-specific. The variant that is associated with exon skipping presumably affects SRp40 binding, and it is likely that this is the reason for the sex-specific difference in splicing. Indeed, SRp40 expression and activity has been shown to be regulated in a sex-specific way (for instance, in the myometrium during pregnancy [Tyson-Capper et al. 2005]), presumably in response to hormonal regulation.

An intronic SNP in the *HMGCR* gene, which is associated with higher LDL-C, determines the level of exon 13 skipping (Burkhardt et al 2008; Medina et al. 2008). The minor allele results in more exon 13 inclusion and thus increased *HMGCR* activity and consequently higher LDL-C. *HMGCR* exon 13 has a suboptimal 5′-ss and the SNP is located 47 nucleotides downstream, where the minor allele creates an ISE, which is presumed to bind SF2/ASF and SRp40.

It is interesting that in both the *LDLr* and *HMGCR* genes the SNPs affect potential SRp40-binding sites, and thus variation in SRp40 levels and activity may simultaneously affect splicing from both susceptibility alleles. This would lead to an additive effect on LDL-C levels in individuals who possess both SNPs. This illustrates that a combination of conditional splicing susceptibility alleles, in interplay with environmental factors, such as hormones, may be involved in determining a complex trait. In addition, the *LDLr* SNP is also a risk factor for Alzheimer's disease (Zhou et al. 2008), presumably because LDLr is the major receptor in the brain for apolipoprotein E (ApoE), and ApoE is itself a risk factor for Alzheimer's disease. This illustrates how SNPs that confer conditional splicing defects may be involved in more than one complex disease trait and also that the deleterious effects of these SNPs may only be pathogenic in certain tissues.

Obesity is another trait with complex inheritance involving several disease genes and environmental factors. A large number of SNPs have been associated with extreme body mass, and many of these SNPs are located in exons with weak splice sites and affect putative SREs. Functional testing confirmed that some of these SNPs cause altered splicing, indicating that they could each be contributing an additive risk for obesity (Goren et al. 2008).

One of the most general parameters that may affect splicing is temperature. Shifts within the range of physiological temperatures may modulate both 5′-ss choice and 3′-ss choice in different human disease genes (Weil et al. 1989; Kralovicova et al. 2004; Madsen et al. 2006). The simplest explanation for this modulation is a direct effect caused by increased/decreased stability of the base pairing between RNAs during spliceosome assembly, such as base pairing between U1 snRNA and a suboptimal 5′-ss or between U2 snRNA and a suboptimal BP. Consistent with this idea, a +5G→A disease-causing mutation in the 5′-ss of the human pro-α2(I) collagen gene (*COL1A2*) causes partial exon 6 skipping at 37°C in cell culture. Lowering of the temperature to 31°C increases exon inclusion, whereas skipping increases further at 39°C (Weill et al. 1989). The same pattern was observed with the normal *SBCAD* exon 3 (Madsen et al. 2006), which was described above and shown in Figure 15.3A; it is intrinsically weak, with mismatches at positions –2, +4, and +5 of the 5′-ss. This illustrates that even normal alleles with weak splice sites may be responsive to changes in temperature, and it emphasizes that even modest changes in splice-site strength caused by SNPs may result in thermosensitivity.

Temperature can also modulate splicing through other mechanisms. A polymorphic variation in the intron 3 BP sequence of the human *DQB1* gene, which causes decreased complementarity to U2 snRNA, results in increased exon 4 skipping when the temperature is lowered. This indicates that in this case improved hydrogen bonding between U2 snRNA and the BP sequence does not improve 3′-ss recognition, but instead another mechanism results in the opposite effect (Kralovicova et al. 2004).

From these examples, it is not difficult to imagine that temperature-dependent susceptibility alleles exist in a large number of disease genes. Such alleles could be variants in which a SNP results in a suboptimal splice site or weak exon definition, which in turn may cause a temperature-dependent change in splicing efficiency. In the normal situation, missplicing is at a sufficiently low level to avoid clinically manifest disease, but changes in temperature could contribute to increased missplicing and thus disease manifestation. Fever resulting from common infections is therefore an obvious risk factor, and it is well known that in inherited metabolic diseases, like SBCAD deficiency and MCAD deficiency, feverish illness is a frequent triggering factor. Conversely, exposure to low temperatures is known to be involved in diseases like rheumatoid arthritis, which is associated with the *DQB1* alleles that show increased exon 4 skipping at low temperature (Kralovicova et al. 2004).

IN SILICO EVALUATION TOOLS CAN BE USED TO EVALUATE SPLICING EFFECTS

Because any sequence variant in a gene may potentially affect splicing, and functional testing remains a cumbersome process, there is a very strong need for in silico tools to evaluate their potentially deleterious effects. A comprehensive description of all currently available methods is beyond the scope of this review; instead, we briefly describe some of the most frequently used below; the corresponding URLs are listed in Table 15.1.

Programs for prediction of splice-site strength either employ position weight matrices (Shapiro and Senapathy 1987), use neural networks (Brunak et al. 1991), or calculate scores according to interdependences between adjacent or more distant positions of the splicing consensus sequences (Yeo and Burge 2004; Roca et al. 2008). There are also many methods for prediction and scoring of positive and negative SREs. Most of them (RESCUE-ESE, Fairbrother et al. 2002; PESX, Zhang and Chasin 2004; FAS-ESS, Wang et al. 2004) are based on identification of hexamer or octamer motifs,

TABLE 15.1 In silico tools for analysis of splicing-regulatory elements (SREs)

Program	Type	Web address	Reference
Maximum Entropy	Splice sites	http://genes.mit.edu/burgelab/maxent/Xmaxentscan_scoreseq.html	Yeo and Burge 2004
SpliceRack	Splice sites	http://katahdin.cshl.edu:9331/splice/index.cgi?database=spliceNew	Roca et al. 2008
Neural network	Splice sites	http://www.cbs.dtu.dk/services/NetGene2/	Brunak et al. 1991
ESEfinder	SREs	http://rulai.cshl.edu/cgi-bin/tools/ESE3/esefinder.cgi?process=home	Cartegni et al. 2003
RESCUE-ESE	SREs	http://genes.mit.edu/burgelab/rescue-ese/	Fairbrother et al. 2002
PESX	SREs	http://cubweb.biology.columbia.edu/pesx/	Zhang and Chasin 2004
FAS-ESS	SREs	http://genes.mit.edu/fas-ess/	Wang et al. 2004
mfold	mRNA folding	http://mfold.bioinfo.rpi.edu/cgi-bin/rna-form1.cgi	Zuker 2003

which are overrepresented or underrepresented in different types of spliced or unspliced segments (introns vs. exons, pseudoexons vs. functional exons, etc.) of the transcriptome. An experimental approach, termed functional SELEX (Systematic Evolution of Ligands by Exponential Enrichment), in which random sequences are iteratively enriched for their ability to drive in vitro splicing of a reporter pre-mRNA in an extract complemented with individual SR proteins, has been used to identify functional recognition motifs for the SR proteins SF2/ASF, SRp40, SRp55, and SC35. The identified motifs were used to generate 6- to 7-nucleotide-long scoring matrices, which can be used to search for presumed ESEs and to calculate changes in the score resulting from mutations (ESE-finder, Cartegni et al. 2003). RNA secondary structure effects can be determined by the mfold program (Zuker 2003).

The main problem at present is that the outputs from the individual programs may lead to opposite conclusions, and it is currently impossible to decide which program is most suitable in a given situation and also to effectively combine information from different types of programs. There are, of course, many examples of correct and rather unambiguous predictions, but it is important to remember that the binding motifs for many of the splicing-regulatory proteins are degenerate and overlapping, and, in addition, that splicing regulation is context-specific and influenced by many different factors, each of which contributes to the final choice. The major task in the future will be to generate prediction algorithms in which the individual effects on all important parameters (splice sites, SREs, intron and exon size, RNA secondary structure, etc.) are predicted, given correct weighting, and combined into a conclusive result.

CONCLUSION

Splicing is a highly complex process that is dependent on the concerted action of numerous sequence elements in a gene to determine which parts of the pre-mRNA are to be included in the mature mRNA. The principal determinant for splicing is the relative strength of the splice sites, which is mainly dependent on their match to the consensus sequences, but also to a large extent on the context. Positive and negative SREs, which may be located far away or may even overlap with the splice sites, may be essential for their recognition or may modulate the efficiency of this recognition. In addition, the activity of splicing-regulatory factors, which bind and determine the function of the SREs, may vary developmentally, between tissues, or in response to extracellular stimuli, such as hormones. Moreover, factors such as temperature or genetic variation in the splicing-regulatory factors or components of the spliceosome could influence splicing. The combined effect of all these parameters determines if a mutation or polymorphism affects splicing and causes disease. It is also important that the effect of a given sequence variation be evaluated in the context of the relevant haplotype, because its effect may depend on linked modifying polymorphisms.

At present, any sequence variant that may be involved in a disease should be considered as potentially affecting splicing, irrespective of its expected effect according to the genetic code. Although computer analysis may provide some clues to evaluate the potential effect that a sequence variant might have on splicing, we are still at the stage where functional testing is required.

REFERENCES

Allemand, E., Guil, S., Myers, M., Moscat, J., Cáceres, J.F., and Krainer, A.R. 2005. Regulation of heterogeneous nuclear ribonucleoprotein A1 transport by phosphorylation in cells stressed by osmotic shock. *Proc. Natl. Acad. Sci.* **102:** 3605–3610.

Amir-Ahmady, B., Boutz, P.L., Markovtsov, V., Phillips, M.L., and

Black, D.L. 2005. Exon repression by polypyrimidine tract binding protein. *RNA* **11:** 699–716.

Blanchette, M. and Chabot, B. 1999. Modulation of exon skipping by high-affinity hnRNPA1 binding sites and by intronic elements that repress splice site utilization. *EMBO J.* **18:** 1939–1952.

Bonnal, S., Martínez, C., Förch, P., Bachi, A., Wilm, M., and Valcárcel, J. 2008. RBM5/Luca-15/H37 regulates Fas alternative splice site pairing after exon definition. *Mol. Cell* **32:** 81–95.

Bonnevie-Nielsen, V., Field, L.L., Lu, S., Zheng, D.J., Li, M., Martensen, P.M., Nielsen, T.B., Beck-Nielsen, H., Lau, Y.L., and Pociot, F. 2005. Variation in antiviral 2′,5′-oligoadenylate synthetase (2′5′AS) enzyme activity is controlled by a single-nucleotide polymorphism at a splice-acceptor site in the *OAS1* gene. *Am. J. Hum. Genet.* **76:** 623–633.

Bremer, S., Hoof, T., Wilke, M., Busche, R., Scholte, B., Riordan, J.R., Maass, G., and Tümmler, B. 1992. Quantitative expression patterns of multidrug-resistance P-glycoprotein (MDR1) and differentially spliced cystic-fibrosis transmembrane-conductance regulator mRNA transcripts in human epithelia. *Eur. J. Biochem.* **206:** 137–149.

Brunak, S., Engelbrecht, J., and Knudsen, S. 1991. Prediction of human mRNA donor and acceptor sites from the DNA sequence. *J. Mol. Biol.* **220:** 49–65.

Buchner, D.A., Trudeau, M., and Meisler, M.H. 2003. SCNM1, a putative RNA splicing factor that modifies disease severity in mice. *Science* **301:** 967–969.

Buratti, E., Brindisi, A., Pagani, F., and Baralle, F.E. 2004. Nuclear factor TDP-43 binds to the polymorphic TG repeats in CFTR intron 8 and causes skipping of exon 9: A functional link with disease penetrance. *Am. J. Hum. Genet.* **74:** 1322–1325.

Buratti, E., Baralle, M., and Baralle, F.E. 2006. Defective splicing, disease and therapy: Searching for master checkpoints in exon definition. *Nucleic Acids Res.* **34:** 3494–510.

Buratti, E., Chivers, M., Královicová, J., Romano, M., Baralle, M., Krainer, A.R., and Vorechovsky, I. 2007. Aberrant 5′ splice sites in human disease genes: Mutation pattern, nucleotide structure and comparison of computational tools that predict their utilization. *Nucleic Acids Res.* **35:** 4250–4263.

Burkhardt, R., Kenny, E.E., Lowe, J.K., Birkeland, A., Josowitz, R., Noel, M., Salit, J., Maller, J.B., Pe'er, I., Daly, M.J., et al. 2008. Common SNPs in HMGCR in Micronesians and whites associated with LDL-cholesterol levels affect alternative splicing of exon13. *Arterioscler. Thromb. Vasc. Biol.* **28:** 2078–2084.

Cartegni, L. and Krainer, A.R. 2002. Disruption of an SF2/ASF-dependent exonic splicing enhancer in SMN1 causes spinal muscular atrophy in the absence of SMN1. *Nat. Genet.* **30:** 377–384.

Cartegni, L. and Krainer, A.R. 2003. Correction of disease-associated exon skipping by synthetic exon-specific activators. *Nat. Struct. Biol.* **10:** 120–125.

Cartegni, L., Wang, J., Zhu, Z., Zhang, M.Q., and Krainer, A.R. 2003. ESEfinder: A web resource to identify exonic splicing enhancers. *Nucleic Acid Res.* **31:** 3568–3571.

Cartegni, L., Hastings, M.L., Calarco, J.A., de Stanchina, E., and Krainer, A.R. 2006. Determinants of exon 7 splicing in the spinal muscular atrophy genes, *SMN1* and *SMN2. Am. J. Hum. Genet.* **78:** 63–77.

Castle, J.C., Zhang, C., Shah, J.K., Kulkarni, A.V., Kalsotra, A., Cooper, T.A., and Johnson, J.M. 2008. Expression of 24,426 human alternative splicing events and predicted *cis* regulation in 48 tissues and cell lines. *Nat. Genet.* **40:** 1416–1425.

Davis, R.L., Homer, V.M., George, P.M., and Brennan, S.O. 2009. A deep intronic mutation in FGB creates a consensus exonic splicing enhancer motif that results in afibrinogenemia caused by aberrant mRNA splicing, which can be corrected in vitro with antisense oligonucleotide treatment. *Hum. Mutat.* **30:** 221–227.

Dreyfuss, G., Kim, V.N., and Kataoka, N. 2002. Messenger-RNA-binding proteins and the messages they carry. *Nat. Rev. Mol. Cell. Biol.* **3:** 195–205.

Edvardson, S., Shaag, A., Kolesnikova, O., Gomori, J.M., Tarassov, I., Einbinder, T., Saada, A., and Elpeleg, O. 2007. Deleterious mutation in the mitochondrial arginyl-transfer RNA synthetase gene is associated with pontocerebellar hypoplasia. *Am. J. Hum. Genet.* **81:** 857–862.

Fairbrother, W.G., Yeh, R.F., Sharp, P.A., and Burge, C.B. 2002. Predictive identification of exonic splicing enhancers in human genes. *Science* **297:** 1007–1013.

Garg, K. and Green, P. 2007. Differing patterns of selection in alternative and constitutive splice sites. *Genome Res.* **17:** 1015–1022.

Goren, A., Kim, E., Amit, M., Bochner, R., Lev-Maor, G., Ahituv, N., and Ast, G. 2008. Alternative approach to a heavy weight problem. *Genome Res.* **18:** 214–220.

Graveley, B.R. 2000. Sorting out the complexity of SR protein functions. *RNA* **6:** 1197–1211.

Gregersen, N., Andresen, B.S., Bross, P., Winter, V., Rüdiger, N., Engst, S., Christensen, E., Kelly, D., Strauss, A., Kølvraa, S., et al. 1991. Molecular characterization of medium-chain acyl-CoA dehydrogenase (MCAD) deficiency: Identification of a lys329 to glu mutation in the *MCAD* gene, and expression of inactive mutant protein in *E. coli. Hum. Genet.* **86:** 545–551.

Groman, J.D., Hefferon, T.W., Casals, T., Bassas, L., Estivill, X., Des Georges, M., Guittard, C., Koudova, M., Fallin, M.D., Nemeth, K., et al. 2004. Variation in a repeat sequence determines whether a common variant of the cystic fibrosis transmembrane conductance regulator gene is pathogenic or benign. *Am. J. Hum. Genet.* **74:** 176–179.

Heinzen, E.L., Ge, D., Cronin, K.D., Maia, J.M., Shianna, K.V., Gabriel, W.N., Welsh-Bohmer, K.A., Hulette, C.M., Denny, T.N., and Goldstein, D.B. 2008. Tissue-specific genetic control of splicing: Implications for the study of complex traits. *Plos Biol.* **6:** e1000001.

Hiller, M., Zhang, Z., Backofen, R., and Stamm, S. 2007. Pre-mRNA secondary structures influence exon recognition. *PLoS Genet.* **3:** e204.

Hofmann, M.H., Blievernicht, J.K., Klein, K., Saussele, T., Schaeffeler, E., Schwab, M., and Zanger, U.M. 2008. Aberrant splicing caused by single nucleotide polymorphism c.516G>T [Q172H], a marker of CYP2B6*6, is responsible for decreased expression and activity of CYP2B6 in liver. *J. Pharmacol. Exp. Ther.* **325:** 284–292.

House, A.E. and Lynch, K.W. 2006. An exonic splicing silencer represses spliceosome assembly after ATP-dependent exon recognition. *Nat. Struct. Mol. Biol.* **13:** 937–944.

Howell, V.M., Jones, J.M., Bergren, S.K., Li, L., Billi, A.C., Avenarius, M.R., and Meisler, M.H. 2007. Evidence for a direct role of the disease modifier SCNM1 in splicing. *Hum. Mol. Genet.* **16:** 3506–3516.

Howell, V.M., de Haan, G., Bergren, S., Jones, J.M., Culiat, C.T., Michaud, E.J., Frankel, W.N., and Meisler, M.H. 2008. A targeted deleterious allele of the splicing factor SCNM1 in the mouse. *Genetics* **180:** 1419–1427.

Hua, Y., Vickers, T.A., Okunola, H.L., Bennett, C.F., and Krainer, A.R. 2008. Antisense masking of an hnRNP A1/A2 intronic splicing silencer corrects SMN2 splicing in transgenic mice. *Am. J. Hum. Genet.* **82:** 834–848.

Hua, Y., Vickers, T.A., Baker, B.F., Bennett, C.F., and Krainer, A.R. 2007. Enhancement of SMN2 exon 7 inclusion by antisense oligonucleotides targeting the exon. *PLoS Biol.* **5:** e73.

Johnson, J.M., Castle, J., Garrett-Engele, P., Kan, Z., Loerch, P.M., Armour, C.D., Santos, R., Schadt, E.E., Stoughton, R., and Shoemaker, D.D. 2003. Genome-wide survey of human alternative pre-mRNA splicing with exon junction microarrays. *Science* **302:** 2141–2144.

Jurica, M.S. and Moore, M.J. 2003. Pre-mRNA splicing: Awash in a sea of proteins. *Mol. Cell* **12:** 5–14.

Kar, A., Kuo, D., He, R., Zhou, J., and Wu, J.Y. 2005. Tau alternative splicing and frontotemporal dementia. *Alzheimer. Dis. Assoc. Disord.* (suppl. 1) **19:** S29–S36.

Kashima, T. and Manley, J.L. 2003. A negative element in SMN2 exon 7 inhibits splicing in spinal muscular atrophy. *Nat. Genet.* **34:** 460–463.

Kashima, T., Rao, N., and Manley, J.L. 2007. An intronic element contributes to splicing repression in spinal muscular atrophy. *Proc. Natl. Acad. Sci.* **104:** 3426–3431.

Kathiresan, S., Melander, O., Guiducci, C., Surti, A., Burtt, N.P., Rieder, M.J., Cooper, G.M., Roos, C., Voight, B.F., Havulinna, A.S., et al. 2008. Six new loci associated with blood low-density lipoprotein cholesterol, high-density lipoprotein cholesterol or triglycerides in humans. *Nat. Genet.* **40:** 189–197.

Kim, E., Magen, A., and Ast, G. 2007. Different levels of alternative splicing among eukaryotes. *Nucleic Acids Res.* **35:** 125–131.

Kornblihtt, A.R., de la Mata, M., Fededa, J.P., Munoz, M.J., and Nogues, G. 2004. Multiple links between transcription and splicing. *RNA* **10:** 1489–1498.

Krainer, A.R., Conway, G.C., and Kozak, D. 1990. Purification and characterization of pre-mRNA splicing factor SF2 from HeLa cells. *Genes. Dev.* **4:** 1158–1171.

Královicová, J., Houngninou-Molango, S., Krämer, A., and Vorechovsky, I. 2004. Branch-site haplotypes that control alternative splicing. *Hum. Mol. Genet.* **13:** 3189–3202.

Liu, H.X., Zhang, M., and Krainer, A.R. 1998. Identification of functional splicing enhancer motifs recognized by individual SR proteins. *Genes Dev.* **12:** 1998–2012.

Lorson, C.L. and Androphy, E.J. 2000. An exonic enhancer is required for inclusion of an essential exon in the SMA-determining gene SMN. *Hum. Mol. Genet.* **9:** 259–265.

Madsen, P.P., Kibæk, M., Roca, X., Sachidanandam, R., Krainer, A.R., Christensen, E., Steiner, R., Gibson, K.M., Corydon, T.J., Knudsen, I., et al. 2006. Short/branched-chain acyl-CoA dehydrogenase deficiency due to an IVS3+3A→G mutation that causes exon skipping. *Hum. Genet.* **118:** 680–690.

Mardon, H.J., Sebastio, G., and Baralle, F.E. 1987. A role for exon sequences in alternative splicing of the human fibronectin gene. *Nucleic Acids Res.* **15:** 7725–7733.

Martinez-Contreras, R., Fisette, J.F., Nasim, F.H., Madden, R., Cordeau, M., and Chabot, B. 2006. Intronic binding sites for hnRNP A/B and hnRNP F/H proteins stimulate pre-mRNA splicing. *PLoS Biol.* **4:** e21.

Medina, M.W., Gao, F., Ruan, W., Rotter, J.I., and Krauss, R.M. 2008. Alternative splicing of 3-hydroxy-3-methylglutaryl coenzyme A reductase is associated with plasma low-density lipoprotein cholesterol response to simvastatin. *Circulation* **118:** 355–362.

Miyajima, H., Miyaso, H., Okumura, M., Kurisu, J., and Imaizumi, K. 2002. Identification of a *cis*-acting element for the regulation of SMN exon 7 splicing. *J. Biol. Chem.* **277:** 23271–23277.

Monani, U.R., Lorson, C.L., Parsons, D.W., Prior, T.W., Androphy, E.J., Burghes, A.H., and McPherson, J.D. 1999. A single nucleotide difference that alters splicing patterns distinguishes the SMA gene *SMN1* from the copy gene *SMN2*. *Hum. Mol. Genet.* **8:** 1177–1183.

Nembaware, V., Lupindo, B., Schouest, K., Spillane, C., Scheffler, K., and Seoighe, C. 2008. Genome-wide survey of allele-specific splicing in humans. *BMC Genomics* **9:** 265.

Nielsen, K.B., Sørensen, S., Cartegni, L., Corydon, T.J., Doktor, T.K., Schroeder, L.D., Reinert, L.S., Elpeleg, O.N., Krainer, A.R., Gregersen, N., et al. 2007. Seemingly neutral polymorphic variants may confer immunity to splicing inactivating mutations. *Am. J. Hum. Genet.* **80:** 416–432.

Ohno, K., Brengman, J.M., Felice, K.J., Cornblath, D.R., and Engel, A.G. 1999. Congenital end-plate acetylcholinesterase deficiency caused by a nonsense mutation and an A→G splice-donor-site mutation at position +3 of the collagenlike-tail-subunit gene (*COLQ*): How does G at position +3 result in aberrant splicing? *Am. J. Hum. Genet.* **65:** 635–644.

Pagani, F., Buratti, E., Stuani, C., Bendix, R., Dörk, T., and Baralle, F.E. 2002. A new type of mutation causes a splicing defect in ATM. *Nat. Genet.* **30:** 426–429.

Pagani, F., Buratti, E., Stuani, C., and Baralle, F.E. 2003. Missense, nonsense, and neutral mutations define juxtaposed regulatory elements of splicing in cystic fibrosis transmembrane regulator exon

9. *J. Biol. Chem.* **278:** 26580–26588.

Pagani, F., Raponi, M., and Baralle, F.E. 2005. Synonymous mutations in CFTR exon 12 affect splicing and are not neutral in evolution. *Proc. Natl. Acad. Sci.* **102:** 6368–6372.

Patel, A.A. and Steitz, J.A. 2003. Splicing double: Insights from the second spliceosome. *Nat. Rev. Mol. Cell. Biol.* **4:** 960–970.

Patel, N.A., Kaneko, S., Apostolatos, H.S., Bae, S.S., Watson, J.E., Davidowitz, K., Chappell, D.S., Birnbaum, M.J., Cheng, J.Q., and Cooper, D.R. 2005. Molecular and genetic studies imply Akt-mediated signaling promotes protein kinase CbetaII alternative splicing via phosphorylation of serine/arginine-rich splicing factor SRp40. *J. Biol. Chem.* **280:** 14302–14309.

Roca, X. and Krainer, A.R. 2009. Recognition of atypical 5′ splice sites by shifted base-pairing to U1 snRNA. *Nat. Struct. Mol. Biol.* **16:** 176–182.

Roca, X., Olson, A.J., Rao, A.R., Enerly, E., Kristensen, V.N., Borresen-Dale, A.L., Andresen, B.S., Krainer, A.R., and Sachidanandam, R. 2008. Disease-causing mutations at 5′ splice sites and comparative genomics help identify features determining splice-site efficiency. *Genome Res.* **18:** 77–87.

Shapiro, M.B. and Senapathy, P. 1987. RNA splice junctions of different classes of eukaryotes: Sequence statistics and functional implications in gene expression. *Nucleic Acids Res.* **15:** 7155–7174.

Shaw, S.D., Chakrabarti, S., Ghosh, G., and Krainer, A.R. 2007. Deletion of the N-terminus of SF2/ASF permits RS-domain-independent pre-mRNA splicing. *PLoS ONE* **2:** e854.

Shepard, P.J. and Hertel, K.J. 2008. Conserved RNA secondary structures promote alternative splicing. *RNA* **14:** 1463–1469.

Sheth, N., Roca, X., Hastings, M.L., Roeder, T., Krainer, A.R., and Sachidanandam, R. 2006. Comprehensive splice-site analysis using comparative genomics. *Nucleic Acids Res.* **34:** 3955–3967.

Singh, N.N., Androphy, E.J., and Singh, R.N. 2004. In vivo selection reveals combinatorial controls that define a critical exon in the spinal muscular atrophy genes. *RNA* **10:** 1291–1305.

Singh, N.K., Singh, N.N., Androphy, E.J., and Singh, R.N. 2006. Splicing of a critical exon of human survival motor neuron is regulated by a unique silencer element located in the last intron. *Mol. Cell. Biol.* **26:** 1333–1346.

Stamm, S. 2008. Regulation of alternative splicing by reversible protein phosphorylation. *J. Biol. Chem.* **283:** 1223–1227.

Sun, H. and Chasin, L.A. 2000. Multiple splicing defects in an intronic false exon. *Mol. Cell. Biol.* **20:** 6414–6425.

Tabuchi, K. and Südhof, T.C. 2002. Structure and evolution of neurexin genes: Insight into the mechanism of alternative splicing. *Genomics* **79:** 849–859.

Tange, T.O., Damgaard, C.K., Guth, S., Valcárcel, J., and Kjems, J. 2001. The hnRNP A1 protein regulates HIV-1 tat splicing via a novel intron silencer element. *EMBO J.* **20:** 5748–5458.

Topp, J.D., Jackson, J., Melton, A.A., and Lynch, K.W. 2008. A cell-based screen for splicing regulators identifies hnRNP LL as a distinct signal-induced repressor of CD45 variable exon 4. *RNA* **14:** 2038–2049.

Tyson-Capper, A.J., Bailey, J., Krainer, A.R., Robson, S.C., and Europe-Finner, G.N. 2005. The switch in alternative splicing of cyclic AMP response element modulator protein CREM 2 (activator) to CREM (repressor) in human myometrial cells is mediated by srp40. *J. Biol. Chem.* **280:** 34521–34529.

Venables, J.P., Koh, C.S., Froehlich, U., Lapointe, E., Couture, S., Inkel, L., Bramard, A., Paquet, E.R., Watier, V., Durand, M., et al. 2008. Multiple and specific mRNA processing targets for the major human hnRNP proteins. *Mol. Cell. Biol.* **28:** 6033–6043.

Wang, Z., Rolish, M.E., Yeo, G., Tung, V., Mawson, M., and Burge, C.B. 2004. Systematic identification and analysis of exonic splicing silencers. *Cell* **119:** 831–845.

Wang, J., Smith, P.J., Krainer, A.R., and Zhang, M.Q. 2005. Distribution of SR protein exonic splicing enhancer motifs in human protein-coding genes. *Nucleic Acids Res.* **33**: 5053–5062.

Weil, D., D'Alessio, M., Ramirez, F., Steinmann, B., Wirtz, M.K., Glanville, R.W., and Hollister, D.W. 1989. Temperature-dependent expression of a collagen splicing defect in the fibroblasts of a patient with Ehlers–Danlos syndrome type VII. *J. Biol. Chem.* **264**: 16804–16809.

Wu, J.Y. and Maniatis, T. 1993. Specific interactions between proteins implicated in splice site selection and regulated alternative splicing. *Cell* **75**: 1061–1070.

Yeo, G. and Burge, C.B. 2004. Maximum entropy modeling of short sequence motifs with applications to RNA splicing signals. *J. Comput. Biol.* **11**: 377–394.

Yu, Y., Maroney, P.A., Denker, J.A., Zhang, X.H., Dybkov, O., Lührmann, R., Jankowsky, E., Chasin, L.A., and Nilsen, T.W. 2008. Dynamic regulation of alternative splicing by silencers that modulate 5′ splice site competition. *Cell* **135**: 1224–1236.

Zhang, W., Duan, S., Bleibel, W.K., Wisel, S.A., Huang, R.S., Wu, X., He, L., Clark, T.A., Chen, T.X., Schweitzer, A.C., et al. 2009. Identification of common genetic variants that account for transcript isoform variation between human populations. *Hum. Genet.* **125**: 81–93.

Zhang, X.H. and Chasin, L.A. 2004. Computational definition of sequence motifs governing constitutive exon splicing. *Genes Dev.* **18**: 1241–1250.

Zou, F., Gopalraj, R.K., Lok, J., Zhu, H., Ling, I.F., Simpson, J.F., Tucker, H.M., Kelly, J.F., Younkin, S.G., Dickson, D.W., et al. 2008. Sex-dependent association of a common low-density lipoprotein receptor polymorphism with RNA splicing efficiency in the brain and Alzheimer's disease. *Hum. Mol. Genet.* **17**: 929–935.

Zhu, J., Mayeda, A., and Krainer, A.R. 2001. Exon identity established through differential antagonism between exonic splicing silencer-bound hnRNPA1 and enhancer-bound SR proteins. *Mol. Cell.* **8**: 1351–1361.

Zhu, H., Tucker, H.M., Grear, K.E., Simpson, J.F., Manning, A.K., Cupples, L.A., and Estus, S. 2007. A common polymorphism decreases low-density lipoprotein receptor exon 12 splicing efficiency and associates with increased cholesterol. *Hum. Mol. Genet.* **16**: 1765–1772.

Zuker, M. 2003. Mfold web server for nucleic acid folding and hybridization prediction. *Nucleic Acids Res.* **31**: 3406–3415.

Zuo, P. and Maniatis, T. 1996. The splicing factor U2AF35 mediates critical protein–protein interactions in constitutive and enhancer-dependent splicing. *Genes Dev.* **10**: 1356–1368.

WEB RESOURCES

http://www.hgmd.cf.ac.uk/ac/index.php Human Genome Mutation Database (HGMD).

http://www.ncbi.nlm.nih.gov/SNP/snp_summary.cgi dbSNP.

16 Laboratory Methods for High-Throughput Genotyping

Howard J. Edenberg[1] and Yunlong Liu[2]

[1]Department of Biochemistry and Molecular Biology and Medical and Molecular Genetics, and Center for Medical Genomics, Indiana University School of Medicine, Indianapolis, Indiana 46202; [2]Division of Biostatistics, Department of Medicine and Center for Computational Biology and Bioinformatics, Indiana University School of Medicine, Indianapolis, Indiana 46202

INTRODUCTION

The genetics of complex diseases has been given a tremendous boost in recent years by the introduction of high-throughput laboratory methods that allow us to approach larger questions in larger populations and to cover the genome more comprehensively. The ability to determine genotypes of many individuals accurately and efficiently has allowed genetic studies that cover more of the variation within individual genes, instead of focusing only on one or a few coding variants, and to do so in study samples of reasonable power. Chip-based genotyping assays, combined with knowledge of the patterns of coinheritance of markers (linkage disequilibrium, LD) developed through the HapMap Project (http://www. hapmap.org), have stimulated genome-wide association studies (GWAS) of complex diseases. These are being encouraged and supported by the National Institutes of Health (NIH) and other groups, notably The Wellcome Trust. Recent successes of GWAS in identifying specific genes that affect risk for common diseases are dramatic illustrations of how improved technology can lead to scientific breakthroughs. Rapid developments in high-throughput sequencing may enable new kinds of studies.

A key issue in high-throughput genotyping is to choose the appropriate technology for your goals and for the stage of your experiment, being cognizant of your sample numbers and resources. This chapter introduces some of the commonly used methods of high-throughput single-nucleotide polymorphism (SNP) genotyping for different stages of genetic studies and briefly reviews some of the high-throughput sequencing methods just coming into use. We will also note some recent developments in "next-generation" sequencing that will enable other kinds of studies. We cannot be comprehensive, and technology in this area is rapidly changing, so our comments should be taken as a starting point for further investigation.

For simplicity, we will discuss three main types of studies: candidate genes, linkage studies and their follow-up, and GWAS and their follow-up. There are choices of genotyping technologies suited to each of these types of studies (Fig. 16.1). Throughput, cost per SNP genotype, and costs per sample can be very different for different technologies. Some technologies, which we call "serial," allow testing of small to modest numbers of SNPs on many subjects in each reaction and are easy to customize. Others, called "parallel" methods, test up to a million SNPs on each subject at one time in fixed panels. The cost per SNP for a serial method is much larger than for a parallel method, but the cost per subject is much less.

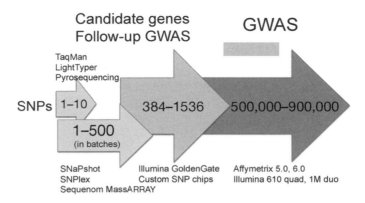

FIGURE 16.1 Different technologies are appropriate for different types of projects and scales of SNPs to be genotyped.

WHY ARE SNPS USED?

At this time, most SNPs do not have known effects on gene expression or function; they are used as markers for genetic differences in their vicinity. Some SNPs are known to cause differences in gene expression or function. The most obvious functional SNPs are those that alter the amino acid encoded at a particular position in a protein or terminate translation leading to a shortened (and often rapidly degraded) polypeptide. Historically, much of the attention to "functional" SNPs has been restricted to these nonsynonymous coding SNPs. More recently, increasing attention has been paid to SNPs that potentially alter splicing, transcription, or mRNA stability. These are generally located in or near a gene and can include synonymous coding SNPs that alter binding sites for the splicing machinery. Such SNPs are harder to recognize and to distinguish from those that do not affect gene expression. It is even harder to tell whether a variation located at some distance from any known gene might have a function, although that is more likely if it lies within a region that is highly conserved among distantly related species.

The binary nature of SNPs has made them the marker of choice in most current work, particularly high-throughput studies. There are about 7 million validated human SNPs in the dbSNP database (http://www.ncbi.nlm.nih.gov/SNP/), with many more listed. Genotypic data on about 4 million different SNPs for three major continental groups are available from the HapMap Project, with more coming. SNPs are also available as markers for model organisms. New approaches allow high-throughput SNP genotyping in multiplex reactions. Copy-number variations (CNVs, described in Chapter 13) are increasingly recognized as important, and many of the SNP genotyping platforms also provide information on CNVs (although standardization and interpretation are more difficult).

SNPs can, obviously, be detected by direct sequencing at high accuracy. This is the primary method of SNP discovery, and it demands sequencing of a sufficient number of individuals at an accuracy and coverage that distinguishes real SNPs from sequencing artifacts. Databases still contain many SNPs that are likely to be artifacts. The extraordinary efforts of the HapMap Project (The International HapMap Consortium 2005; Frazer et al. 2007) to date have been very valuable in identifying common SNPs, primarily those with minor allele frequencies (MAFs) greater than 5%. Efforts are under way to examine a wider variety of populations to discover additional SNPs and broaden our understanding of genetic diversity (e.g., 1000 Genomes, A Deep Catalog of Human Genetic Variation, at http://www.1000genomes.org). However, when one is studying a particular disease and has found an association between SNPs and the disease, resequencing in the group with the disease generally leads to discovery of many more SNPs in that gene, often including ones that are common within the study group, although not in the larger population. For example, sequencing just 16 individuals (eight with a high-risk haplotype and eight with a low-risk haplotype for alcohol dependence) in the exons and proximal 5′ and 3′ regions of *OPRK1* revealed seven new SNPs and an 830-bp insertion/deletion (Xuei et al. 2006; Edenberg et al. 2008).

However, sequencing is not at this time an efficient way to genotype SNPs, although as sequencing technology progresses toward the much sought "$1000 genome" it will become the method of choice. The most efficient methods at this time involve single-base extension (in effect, microsequencing) with readouts that include mass differences (e.g., Sequenom MassARRAY), light flashes (pyrosequencing), or hybridization, with readouts that include measures of the amount of oligonucleotide hybridized (e.g., microarrays), cleavage of hybridized oligonucleotides (e.g., Taq-Man, Applied Biosystems), or melting curves (e.g., LightTyper, Roche Applied Science). There are platforms that work best for targeted SNP genotyping and others that are aimed at whole genomes. Each has advantages for particular studies.

CAN CANDIDATE GENES BE USED?

Many studies focus on candidate genes that are chosen on the basis of physiology or suggestive results from other studies. One problem is that many candidate genes are weak candidates, with low prior probabilities. This, plus some publication bias toward positive findings, might explain the large number of candidate-gene association studies that have not been replicated. As association findings are reported, many groups will want to test the significant genes in their own study sample. This is a good strategy for confirmation studies and for some with too small a subject population to withstand the multiple testing corrections required for GWAS. With a small sample, the necessity to correct for the very high levels of multiple testing in a GWAS make it very difficult to detect effects of modest size with genome-wide significance (see Chapter 6). Even samples that are large in terms of our ability to do sophisticated and reproducible phenotyping (1000–2000 cases and controls) are proving to be underpowered in studies of complex traits in which the contribution of any one genetic variant is small. Small samples can, however, be very useful in replicating results. In this case, the prior probability is much higher and the likelihood of confirming a finding is increased.

Many early candidate gene studies tested only a single SNP, which carries very limited information about the overall variation within that gene. This approach can be successful if there is a known, strong functional SNP (e.g., Thomasson et al. 1991), but it often leads to false-negative results. Even in simple situations (Mendelian disorders) in which a single change in one gene can lead to disease, there is often allelic heterogeneity, and testing only one SNP can miss the one(s) that are functional in that family or population. It is generally better to cover a larger fraction of the variation by genotyping multiple SNPs chosen based on LD in addition to hypothesized functional SNPs, although cost and time (and concerns about multiple testing) can preclude full coverage.

HapMap data (The International HapMap Consortium 2005; Frazer et al. 2007) can be used to select the SNPs that best report on common variation within a gene or region ("tag" the region). It is useful first to visualize the LD structure in the region of interest, using Haploview (Barrett et al. 2005) (which can be performed within the HapMap website). A program such as Tagger (de Bakker et al. 2005) can aid in the selection of SNPs; it can be run from a server (through HapMap or directly at http://www.broad.mit.edu/mpg/tagger). Parameters such as minimum MAF of SNPs to be "tagged" and the degree of correlation (r^2) to be accepted as adequate can be adjusted so that a set of SNPs reasonable for sample size, technology, and budget can be selected.

The genotyping technologies most appropriate for candidate-gene studies are what we call "serial" technologies; that is, they test from 1 to 48 SNPs at a time on each sample. A set of assays can be designed, run, and analyzed; then, if needed, a follow-up set of assays can be run. These generally allow testing of a large number of samples for a modest number of targeted SNPs at modest cost. There are many technologies that are good choices for candidate-gene studies (Tsuchihashi and Dracopoli 2002).

TaqMan SNP Genotyping Assays (Applied Biosystems) measure individual SNPs using a 5′ nuclease assay (De la Vega et al. 2005). Many assays (>4.5 million) have been predesigned and it is relatively straightforward to design new assays. Single SNPs are generally run on sets of 96 or 384 samples at a time. Moving up to low levels of multiplexing, the SNaPshot Multiplex System (Applied Biosystems) is a primer extension-based method with detection by capillary electrophoresis; it allows multiplexing up to 10 SNPs starting from as little as 3 ng DNA per sample. The SNPlex Genotyping System from Applied Biosystems (De la Vega et al. 2005) uses the oligonucleotide ligation assay (Nickerson et al. 1990) to discriminate SNPs, followed by PCR and capillary electrophoresis; it allows genotyping of up to 48 SNPs at a time. Both SNaPshot and SNPlex use capillary sequencing instruments, which might already be available in many laboratories. The LightTyper system (Roche Applied Science) uses melting curve analysis to discriminate individual SNPs. These methods use fluorescently labeled oligonucleotides for detection of the SNPs (Bennett et al. 2003).

Pyrosequencing detects SNPs by a synthesis reaction with detection based on flashes of light when a nucleotide is incorporated (Ahmadian et al. 2000; Pourmand et al. 2002). The Invader assay (Third Wave Technologies, Inc., now Hologic) involves formation and then cleavage of a flap created by hybridization of two oligonucleotides to the target sequence; the signal from this initially cleaved flap is amplified in a second fluorescence resonance energy transfer reaction (Lyamichev et al. 1999).

The Sequenom MassARRAY system (Jurinke et al. 2001; Jurinke et al. 2002) can measure up to 36 SNPs on 384 samples per assay. A region is amplified by polymerase chain reaction (PCR) and then a single-base primer extension is performed using modified deoxyribonucleoside triphosphates that increase the resolution with which a mass spectrometer can distinguish the four possible nucleotides added. An advantage of this is that unmodified oligonucleotides can be used, reducing the initial cost.

These genotyping techniques are good for testing candidate genes and small regions. One can test a modest number of SNPs, analyze them, and then genotype additional SNPs to whatever depth of coverage is desired in genes or regions that remain of interest. The investigator has nearly complete freedom to design assays for any SNPs desired for the particular project.

LINKAGE REGIONS CAN BE FINE-MAPPED BY HIGH-THROUGHPUT SNP GENOTYPING

Although new availability of commercial platforms has stimulated interest in GWAS, there are many projects that performed linkage studies and identified broad regions likely to contain genes in which variations affect risk for a disease or a related phenotype. Linkage studies have relatively low resolution, so one needs a way to follow up such studies with fine-mapping. High-throughput SNP genotyping offers an attractive approach to this task. One can either use a parallel approach, such as a custom microarray of thousands or tens of thousands of SNPs or an Illumina GoldenGate assay (Fan et al. 2006) to test from 384 to 1536 SNPs, or a serial approach in which SNPs are tested in smaller numbers at a time.

If there are particularly good candidate genes within the linkage region, the serial approach of testing a limited number of such candidates may be optimum because it limits costs and also limits multiple testing. An example is targeting a set of four genes that encode subunits of the γ-aminobutyric acid A receptor in a region of chromosome 4 linked to both alcohol dependence and an electrophysiological phenotype (Edenberg et al. 2004). These made excellent candidate genes on the basis of both physiological knowledge and location in the center of the linkage peak. A serial approach was taken, first testing five to six SNPs in each gene and then covering the gene in which multiple SNPs were associated with alcohol dependence with additional SNPs in an attempt to further localize the key SNPs. The Sequenom MassARRAY system was used. *GABRA2* was shown to be associated with alcoholism, with a large LD block extending from intron 3 past the 3′ end of the gene (Edenberg et al. 2004); this finding has since been replicated by many groups.

In a case in which there are no strong candidate genes or the candidate genes tested did not prove to be associated, a parallel approach may be called for. Illumina offers GoldenGate Custom Panels that can measure 384–1536 assays per reaction. These can be valuable for testing many sites across a linkage region, or setting up panels to follow up the best "hits" from an earlier study (either a GWAS or a compilation of candidate genes). The Center for Inherited Disease Research at Johns Hopkins University is a resource for getting such genotyping performed, if it is approved by their advisory panel.

One can also make custom-designed genotyping microarrays. An example of this approach was the design of a panel of 1536 SNPs (using the Illumina GoldenGate assay) that used LD information to capture data (at $r^2 > 0.8$) on >4000 SNPs with MAF > 0.10, along with a small number of nonsynonymous coding SNPs, in a linkage region on chromosome 7q22 covering about 18 Mb (Dick et al. 2007). Of these, eight SNPs were found to be associated with alcohol dependence at $p < 0.01$, four of which clustered in a single gene. This gene was followed up by genotyping 16 additional SNPs, using the Sequenom MassARRAY assay; 12 SNPs in that gene were nominally significant, with eight remaining significant when corrected for multiple testing (Dick et al. 2007).

GWAS AND FOLLOW-UP BY ADDITIONAL GENOTYPING

Parallel genotyping tests many SNPs on a single sample (or two samples) at one time, using an array-based format. The number of SNPs per array has dramatically increased in the past few years, from about 10,000 to 100,000, 300,000, 600,000, and now more than 1 million. Currently, the Affymetrix Genome Wide Human SNP Array 6.0 has more than 906,600 SNPs and more than 946,000 probes for the detection of copy number variation, and the Illumina Human1M-Duo Bead-Chip assesses more than 1.1 million loci per sample.

Both of these array-based methods are widely used. There are differences between these two platforms in design, SNP selection, and biochemistry, but overall data quality and coverage appear similar. About 480,000 of the SNPs on the Affymetrix Genome Wide Human SNP Array 6.0 were selected based on what worked in a "complexity-reduction" step that involved selection of restriction fragments between about 200 and 1100 bp. These were supplemented with 424,000 tag SNPs, plus the 946,000 monomorphic sites of which about 202,000 are in known CNV regions and the rest were chosen by spacing to detect CNV. SNP detection is by differential hybridization to 25-mers designed to match both alleles at each site. Illumina probes are larger (about 50 nucleotides) and were chosen to cover the genome based on HapMap LD data plus nonsynonymous coding SNPs. Both platforms give excellent genome-wide coverage in European populations and very good coverage even in African populations. It should be noted that there are significant gaps in both. For example, an analysis of SNPs (MAF ≥ 5%) in about 900 genes relating to addiction, many were not well tagged ($r^2 \leq 0.8$) by either platform (Saccone et al. 2009). Therefore, follow-up of initial results with additional genotyping is valuable.

Initially, the high cost of the genome-wide microarray assays made individual genotyping of large samples prohibitive for many groups. Pooling of samples can reduce costs (Bansal et al. 2002; Sham et al. 2002). Pooling requires very careful measurement of DNA concentrations to equalize the contributions of each individual, but when performed carefully this method can measure relative allele frequencies to within a few percent. Pooling has provided interesting leads that can be followed up by individual genotyping of the most significant findings. However, pooling does not provide nearly as much information as individual genotyping. It requires either a dichotomous phenotype or the decision to make a pseudo-dichotomous phenotype—for example, by contrasting the two ends of a distribution. Only the phenotype by which the pools were created can be analyzed; information on endophenotypes or related phenotypes is lost, although one could theoretically make smaller pools matched on more than one phenotype. Pooling approaches have been effective in studies of alcohol dependence and bipolar disorder genetics (e.g., Johnson et al. 2006; Baum et al. 2008).

As the density of SNPs on arrays has increased and the costs gone down, microarrays are increasingly used for GWAS on individual samples. This approach is much more powerful and allows analysis of multiple phenotypes and endophenotypes at once, as well as analysis of quantitative traits. This approach has been successful in studies of macular degeneration, height, and type 2 diabetes (e.g., Klein et al. 2005; Weedon et al. 2007; Zeggini et al. 2008; see also Chapter 18).

GWAS must be followed up. One can attempt replication either of the leading SNPs in another population, using one of the serial genotyping methods described previously (see "Can Candidate Genes Be Used?") or one can perform follow-up GWAS on another population. In some cases, the strongest signal from an analysis or meta-analysis might be an imputed SNP (i.e., an SNP not actually genotyped but rather predicted based on the genotypes and the known LD in the region; see Chapter 10) (e.g., Ferreira et al. 2008). Such imputed SNPs should be genotyped directly in the initial population, using one of the serial methods, because imputation is not exact. Finally, denser genotyping and resequencing of the significantly associated genes in a GWAS is often performed in search of potentially functional polymorphisms.

QUALITY CONTROL OF SNP DATA IS REQUIRED

Quality control (QC) of the SNP data is required before analysis. This is not a simple task. Although the details of QC could be a chapter themselves, some basic approaches are relatively standard. QC is usually performed in two cycles: (1) Remove problem samples based on the sample QC metrics and (2) recalculate the SNP metrics on the remaining samples before SNP QC metrics are applied.

It is useful to test DNA quality before running the whole-genome chips. First, a spectrum from 220 to 350 nm is better than just examining A_{260}/A_{280}, because sometimes there are contaminants that absorb in the 230–270-nm region. These contaminants can lead to large overestimates of the amount of DNA actually present and can also inhibit genotyping or sequencing reactions. Use of a dye selective for double-stranded DNA (e.g., PicoGreen, Molecular Probes, Invitrogen) can give a better measurement of DNA quantity in the presence of RNA or free nucleotides, but it does not reveal the presence of contaminants. A combination of the two approaches is best. Then, evaluation of fragment size should be performed on agarose gels. Poor-quality samples (contaminated or too fragmented) generally give poor results. The average fragment size needed depends on the platform. The Affymetrix Genome Wide Human SNP Array 6.0 uses a size selection of fragments from 200 to 1100 bp, so fragmented DNA does not work well. The Sequenom MassARRAY generally amplifies regions of about 100 bp and therefore is less affected by small fragment size.

For any genotyping methods, samples that show too many dropouts or no-calls are suspect and are generally removed from analysis if <97%–98% of the SNPs give genotypes. Samples with heterozygosity that is too high or too low compared with others in the population are also suspect and could reflect mixed samples containing DNA from more than one individual. Probes on the X and Y chromosome can be used to confirm sex. Chromosomal loss and duplication, particularly of the X chromosome, are often observed in immortalized cell lines; if the other data appear to be okay, one can choose to selectively remove the data from the X chromosome and retain the rest. With genome-wide data, one can identify cryptic relatives (i.e., individuals in the data set who are related to each other) based on allele sharing. When samples from multiple studies are analyzed jointly, this might include relatives or even the same subject participating in more than one study.

After sample QC, SNP QC is needed. Duplicate samples should give the same calls. It is a good idea to include one or more HapMap samples to compare the genotype call with that in the database. In many studies, SNPs with MAF < 0.01 are not analyzed, because calling rare genotypes is subject to more errors. Although many SNPs may be lost by this filter, the overall cost to the project power is not great because even in studies with 1000 cases and 1000 controls, the power to detect effects with rare genotypes is very limited. SNPs for which many samples do not give genotypes (dropouts) are suspect and are usually omitted from analysis. The acceptable call rate depends on the nature of the study, but usually is set in the region of 95% or better. Tests of Hardy–Weinberg equilibrium (HWE) are useful in flagging SNPs with nonrandom dropping out of a genotype, but the problem of multiple testing limits how strictly they can be screened and some SNPs with bias might still pass this filter. When measuring 1 million SNPs in a GWAS, deviations from HWE must generally be more significant than $p < 0.000001$ to remove an SNP. Differential dropouts of one homozygote (usually the minor allele) or over- or underrepresentation of heterozygotes should raise a flag and lead to a close look at the raw data from that SNP (see the following).

Examining the clustering of the three genotypes, using either a Cartesian or polar plot of the intensity of alleles, is useful (Fig. 16.2 shows examples from Sequenom data, but the basic ideas and method are general). Good-quality SNPs show three clearly defined and tight clusters, with the ho-

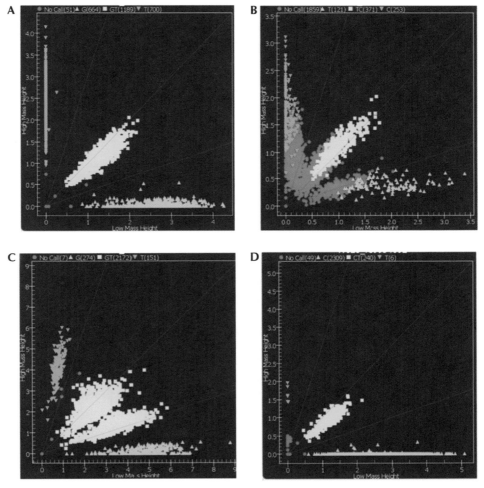

FIGURE 16.2 Quality control of SNP genotyping. One method of examining the quality of the genotype calls is to plot the intensity of the signal for one allele (green) versus the intensity of the signal of the other (blue) with heterozygotes shown in yellow, and uncalled alleles in red. Data shown are from the Sequenom MassARRAY reaction, but in principle could be from many technologies including the whole-genome scale microarrays. (*A*) Good-quality assay, with good clustering and separation of the alleles. (*B*) Bad assay, with overlap of homozygotes and heterozygotes such that many samples are not called. (*C*) Questionable assay; the two clusters of heterozygote intensities might suggest copy number problems. (*D*) Questionable assay; although clustering looks tight and heterozygotes are well balanced, close inspection suggests that many of the unknowns might represent homozygotes for the minor allele. Differential loss of one of the homozygous genotypes will bias the analysis.

mozygotes falling along the vertical or horizontal axis and the heterozygotes at a 45° angle (Fig. 16.2A). One can often detect a problem such as overlap of the clusters from one allele and from the heterozygotes (Fig. 16.2B). One might also detect more than three clusters or a split cluster (Fig. 16.2C), suggesting the possibility of a relatively frequent CNV. The hardest potential error to detect is the selective dropping out of the minor allele (Fig. 16.2D); such SNPs will clearly give incorrect results. The huge numbers of SNPs in a GWAS precludes looking at intensity plots of all of the SNPs, but one should look at the intensity plots of those deemed significant. It is valuable to regenotype significant SNPs by a different technique to insure that the finding reflects the real genetics rather than a technical artifact.

Most QC issues are the same for GWAS as for the more targeted studies, but the vast amount of data makes some additional checks both possible and necessary. In GWAS the biochemistry is typically performed in an automated or semiautomated manner on sets of 48 or 96 samples (plates), so an additional analysis is usually performed to detect situations in which one of the plates differs from the others, perhaps because of some aspect of its processing. This "plate effect" is detected by analyzing the allele frequency of the SNP on that plate against the allele frequency on the sum of the other plates by a chi-square test, and discarding data if a plate effect is seen at the level of 10^{-8}, or multiple plates show effects at 10^{-4} or worse. For best results, cases and controls should be evenly distributed on each plate; otherwise, subtle differences in the biochemistry can lead to biases in the allele calling. These subtle biases will be an issue when, as is happening increasingly, data from controls genotyped at one time in one study are used with data from samples genotyped at a different time and perhaps in a different laboratory to increase power in another study.

NEXT-GENERATION SEQUENCING WILL REVOLUTIONIZE GENOTYPING

The rapid development of high-throughput "next-generation sequencing" technology, by which we mean "massively parallel" sequencing, offers great potential to revolutionize genotyping (Mardis 2008). It is already playing a role in identifying novel SNPs (Hodges et al. 2007; Van Tassell et al. 2008; Wheeler et al. 2008) and other structural variants such as insertion/deletion (indels) and CNVs (Campbell et al. 2008). An early study reported that at least 13-fold coverage is needed to identify 99% of the heterozygous SNPs (Wheeler et al. 2008).

Next-generation sequencing technology is capable of sequencing from hundreds of thousands to hundred millions of DNA (or cDNA) fragments in a single instrument run in a massively parallel fashion. Such high-throughput sequencing technology has been used in applications including de novo sequencing, resequencing to detect SNPs and other variants, transcriptome sequencing, immunoprecipitation-based protein–DNA or protein–RNA interaction mapping, and DNA methylation using bisulfite-mediated cytosine conversion. When the cost of sequencing a human genome approaches $1000, the current target, sequencing will probably replace genotyping for GWAS.

So far, three major platforms are commercially available: the Roche GS FLX Sequencer (454 technology, 454 Life Sciences, Roche), the Illumina Genome Analyzer (Solexa), and the Applied Biosystems SOLiD sequencer (SOLiD 3 System). The 454 Sequencer produces longer reads of >250 bp per fragment, in contrast to 35- to 50-bp reads of the other two platforms (Solexa and SOLiD). The short-sequence instruments produce many more reads for each instrument run. The capacities of these instruments are increasing rapidly with new development in the chemistry and physics of the techniques, so detailed figures would be out of date before this chapter is published, but some already claim >20 Gb of sequence per run. Therefore, we will restrict our discussion to some general issues.

High-throughput sequencing offers the potential to identify de novo SNPs as well as previously reported SNPs. Complete genome-wide resequencing of an individual has been published for two individuals, James D. Watson (Wheeler et al. 2008) and Craig Venter (Levy et al. 2007). These sequences showed many new variants, particularly insertions, deletions, and translocations. Given present costs, however, a major current application is detection of variations in a focused genomic region, usually to follow up significant association studies by identifying potentially functional variation.

There are several methods available to select focused genomic regions from individual samples. One approach is to amplify the region through multiplex PCRs (Porreca et al. 2007), but this can be expensive and time-consuming, and PCR artifacts are possible. One can also produce reduced representation libraries by constructing cDNA libraries of a temporally and spatially specific transcriptome (Bainbridge et al. 2006; Barbazuk et al. 2007), which focuses on variations within transcripts, or by selecting DNA fragments of a specific size after complete restriction endonuclease digestion (Barbazuk et al. 2007; Van Tassell et al. 2008).

Recently, microarrays have been used to capture genomic regions of interest (Albert et al. 2007; Hodges et al. 2007). The NimbleGen Sequence Capture Array (Roche) can be used to isolate up to 5 Mb of DNA using probes of 50–80 nucleotides tiled across the region(s) of interest. Similar technology has also been implemented in another study, in which a customized array was designed that contains 55,000 100-mer oligonucleotides (Porreca et al. 2007). The selected regions can be eluted and sequenced. One concern is whether the capture efficiency in different samples is comparable, given the presence of SNPs and other variations; platforms with longer probes are less sensitive to this.

This technology is quite good at identifying structural variants such as deletions, insertions, duplications, and inversions. Sequencing short reads from both ends of millions of DNA fragments of known size (hundreds to thousands of base pairs) allows one to determine if the spacing and orientation of the paired sequences matches that in the genome (Campbell et al. 2008), and thereby detect many insertions and deletions.

SUMMARY AND CONCLUSIONS

High-throughput genotyping technology has enabled GWAS that have great promise for identifying genes that contribute to complex diseases and phenotypes and for large-scale follow-up of the top candidate genes from such studies. There are a range of techniques from whole-genome scale to individual SNPs. As with all techniques, appropriate choice of the approach for different projects or phases of projects is important. Careful attention to quality control is also essential. In the future, whole-genome sequencing may overtake the current large-scale technologies.

ACKNOWLEDGMENTS

We thank Dr. Xiaoling Xuei for help in selecting figures and for helpful comments on the manuscript, and Dr. Jeanette McClintick for helpful comments on the manuscript. Related work in the investigators' laboratories has been funded by grants AA008401, AA006460, AA07611 from NIAAA, and MH078151 from the National Institute of Mental Health and the Indiana Genomics Initiative (INGEN, which is partially funded by The Lilly Endowment, Inc.).

REFERENCES

Ahmadian, A., Gharizadeh, B., Gustafsson, A.C., Sterky, F., Nyren, P., Uhlen, M., and Lundeberg, J. 2000. Single-nucleotide polymorphism analysis by pyrosequencing. *Anal. Biochem.* **280:** 103–110.

Albert, T.J., Molla, M.N., Muzny, D.M., Nazareth, L., Wheeler, D., Song, X., Richmond, T.A., Middle, C.M., Rodesch, M.J., Packard, C.J., et al. 2007. Direct selection of human genomic loci by microarray hybridization. *Nat. Methods* **4:** 903–905.

Bainbridge, M.N., Warren, R.L., Hirst, M., Romanuik, T., Zeng, T., Go, A., Delaney, A., Griffith, M., Hickenbotham, M., Magrini, V., et al. 2006. Analysis of the prostate cancer cell line LNCaP transcriptome using a sequencing-by-synthesis approach. *BMC Genomics* **7:** 246.

Bansal, A., van den Boom, D., Kammerer, S., Honisch, C., Adam, G., Cantor, C.R., Kleyn, P., and Braun, A. 2002. Association testing by DNA pooling: An effective initial screen. *Proc. Natl. Acad. Sci.* **99:** 16871–16874.

Barbazuk, W.B., Emrich, S.J., Chen, H.D., Li, L., and Schnable, P.S. 2007. SNP discovery via 454 transcriptome sequencing. *Plant J.* **51:** 910–918.

Barrett, J.C., Fry, B., Maller, J., and Daly, M.J. 2005. Haploview: Analysis and visualization of LD and haplotype maps. *Bioinformatics* **21:** 263–265.

Baum, A.E., Akula, N., Cabanero, M., Cardona, I., Corona, W., Klemens, B., Schulze, T.G., Cichon, S., Rietschel, M., Nothen, M.M., et al. 2008. A genome-wide association study implicates diacylglycerol kinase eta (DGKH) and several other genes in the etiology of bipolar disorder. *Mol. Psychiatry* **13:** 197–207.

Bennett, C.D., Campbell, M.N., Book, C.J., Eyre, D.J., Nay, L.M., Nielsen, D.R., Rasmussen, R.P., and Bernard, P.S. 2003. The Light-Typer: High-throughput genotyping using fluorescent melting curve analysis. *Biotechniques* **34:** 1288–1292, 1294–1295.

Campbell, P.J., Stephens, P.J., Pleasance, E.D., O'Meara, S., Li, H., Santarius, T., Stebbings, L.A., Leroy, C., Edkins, S., Hardy, C., et al. 2008. Identification of somatically acquired rearrangements in cancer using genome-wide massively parallel paired-end sequencing. *Nat. Genet.* **40:** 722–729.

de Bakker, P.I., Yelensky, R., Pe'er, I., Gabriel, S.B., Daly, M.J., and Altshuler, D. 2005. Efficiency and power in genetic association studies. *Nat. Genet.* **37:** 1217–1223.

De la Vega, F.M., Lazaruk, K.D., Rhodes, M.D., and Wenz, M.H. 2005. Assessment of two flexible and compatible SNP genotyping platforms: TaqMan SNP Genotyping Assays and the SNPlex Genotyping System. *Mutat. Res.* **573:** 111–135.

Dick, D.M., Aliev, F., Wang, J.C., Saccone, S., Hinrichs, A., Bertelsen, S., Budde, J., Saccone, N., Foroud, T., Nurnberger Jr., J., et al. 2007. A systematic single nucleotide polymorphism screen to fine-map alcohol dependence genes on chromosome 7 identifies association with a novel susceptibility gene *ACN9*. *Biol. Psychiatry* **63:** 1047–1053.

Edenberg, H.J., Dick, D.M., Xuei, X., Tian, H., Almasy, L., Bauer, L.O., Crowe, R.R., Goate, A., Hesselbrock, V., Jones, K., et al. 2004. Variations in *GABRA2*, encoding the α2 subunit of the GABA$_A$ receptor, are associated with alcohol dependence and with brain oscillations. *Am. J. Hum. Gen.* **74:** 705–714.

Edenberg, H.J., Wang, J., Tian, H., Pochareddy, S., Xuei, X., Wetherill, L., Goate, A., Hinrichs, T., Kuperman, S., Nurnberger Jr., J.I., et al. 2008. A regulatory variation in *OPRK1*, the gene encoding the κ-opioid receptor, is associated with alcohol dependence. *Hum. Mol. Genet.* **17:** 1783–1789.

Fan, J.B., Chee, M.S., and Gunderson, K.L. 2006. Highly parallel genomic assays. *Nat. Rev. Genet.* **7:** 632–644.

Ferreira, M.A., O'Donovan, M.C., Meng, Y.A., Jones, I.R., Ruderfer, D.M., Jones, L., Fan, J., Kirov, G., Perlis, R.H., Green, E.K., et al. 2008. Collaborative genome-wide association analysis supports a role for *ANK3* and *CACNA1C* in bipolar disorder. *Nat. Genet.* **40:** 1056–1058.

Frazer, K.A., Ballinger, D.G., Cox, D.R., Hinds, D.A., Stuve, L.L., Gibbs, R.A., Belmont, J.W., Boudreau, A., Hardenbol, P., Leal, S.M., et al. 2007. A second generation human haplotype map of over 3.1 million SNPs. *Nature* **449:** 851–861.

Hodges, E., Xuan, Z., Balija, V., Kramer, M., Molla, M.N., Smith, S.W., Middle, C.M., Rodesch, M.J., Albert, T.J., Hannon, G.J., and McCombie, W.R. 2007. Genome-wide in situ exon capture for selective resequencing. *Nat. Genet.* **39:** 1522–1527.

The International HapMap Consortium. 2005. A haplotype map of the human genome. *Nature* **437:** 1299–1320.

Johnson, C., Drgon, T., Liu, Q.R., Walther, D., Edenberg, H., Rice, J.,

Foroud, T., and Uhl, G.R. 2006. Pooled association genome scanning for alcohol dependence using 104,268 SNPs: Validation and use to identify alcoholism vulnerability loci in unrelated individuals from the collaborative study on the genetics of alcoholism. *Am. J. Med. Genet. B Neuropsychiatr. Genet.* **141B:** 844–853.

Jurinke, C., van den Boom, D., Cantor, C.R., and Koster, H. 2001. Automated genotyping using the DNA MassARRAY technology. *Methods Mol. Biol.* **170:** 103–116.

Jurinke, C., van den Boom, D., Cantor, C.R., and Koster, H. 2002. Automated genotyping using the DNA MassARRAY technology. *Methods Mol. Biol.* **187:** 179–192.

Klein, R.J., Zeiss, C., Chew, E.Y., Tsai, J.Y., Sackler, R.S., Haynes, C., Henning, A.K., SanGiovanni, J.P., Mane, S.M., Mayne, S.T., et al. 2005. Complement factor H polymorphism in age-related macular degeneration. *Science* **308:** 385–389.

Levy, S., Sutton, G., Ng, P.C., Feuk, L., Halpern, A.L., Walenz, B.P., Axelrod, N., Huang, J., Kirkness, E.F., Denisov, G., et al. 2007. The diploid genome sequence of an individual human. *PLoS Biol.* **5:** e254.

Lyamichev, V., Mast, A.L., Hall, J.G., Prudent, J.R., Kaiser, M.W., Takova, T., Kwiatkowski, R.W., Sander, T.J., de Arruda, M., Arco, D.A., et al. 1999. Polymorphism identification and quantitative detection of genomic DNA by invasive cleavage of oligonucleotide probes. *Nat. Biotechnol.* **17:** 292–296.

Mardis, E.R. 2008. The impact of next-generation sequencing technology on genetics. *Trends Genet.* **24:** 133–141.

Nickerson, D.A., Kaiser, R., Lappin, S., Stewart, J., Hood, L., and Landegren, U. 1990. Automated DNA diagnostics using an ELISA-based oligonucleotide ligation assay. *Proc. Natl. Acad. Sci.* **87:** 8923–8927.

Porreca, G.J., Zhang, K., Li, J.B., Xie, B., Austin, D., Vassallo, S.L., LeProust, E.M., Peck, B.J., Emig, C.J., Dahl, F., et al. 2007. Multiplex amplification of large sets of human exons. *Nat. Methods* **4:** 931–936.

Pourmand, N., Elahi, E., Davis, R.W., and Ronaghi, M. 2002. Multiplex pyrosequencing. *Nucleic Acids Res.* **30:** e31.

Saccone, S.F., Bierut, L.B., Chesler, E.J., Kalivas, P.W., Lerman, C., Saccone, N.L., Uhl, G.R., Li, C.-Y., Philip, V.M., Edenberg, H.J., et al. 2009. Supplementing high-density SNP microarrays for additional coverage of disease-related genes: Addiction as a paradigm. *PLoS ONE* **4:** e5225.

Sham, P., Bader, J.S., Craig, I., O'Donovan, M., and Owen, M. 2002. DNA Pooling: A tool for large-scale association studies. *Nat. Rev. Genet.* **3:** 862–871.

Thomasson, H.R., Edenberg, H.J., Crabb, D.W., Mai, X.L., Jerome, R.E., Li, T.K., Wang, S.P., Lin, Y.T., Lu, R.B., and Yin, S.J. 1991. Alcohol and aldehyde dehydrogenase genotypes and alcoholism in Chinese men. *Am. J. Hum. Genet.* **48:** 677–681.

Tsuchihashi, Z. and Dracopoli, N.C. 2002. Progress in high throughput SNP genotyping methods. *Pharmacogenomics J.* **2:** 103–110.

Van Tassell, C.P., Smith, T.P., Matukumalli, L.K., Taylor, J.F., Schnabel, R.D., Lawley, C.T., Haudenschild, C.D., Moore, S.S., Warren, W.C., and Sonstegard, T.S. 2008. SNP discovery and allele frequency estimation by deep sequencing of reduced representation libraries. *Nat. Methods* **5:** 247–252.

Weedon, M.N., Lettre, G., Freathy, R.M., Lindgren, C.M., Voight, B.F., Perry, J.R., Elliott, K.S., Hackett, R., Guiducci, C., Shields, B., et al. 2007. A common variant of *HMGA2* is associated with adult and childhood height in the general population. *Nat. Genet.* **39:** 1245–1250.

Wheeler, D.A., Srinivasan, M., Egholm, M., Shen, Y., Chen, L., McGuire, A., He, W., Chen, Y.J., Makhijani, V., Roth, G.T., et al. 2008. The complete genome of an individual by massively parallel

DNA sequencing. *Nature* **452:** 872–876.

Xuei, X., Dick, D., Flury-Wetherill, L., Tian, H.J., Agrawal, A., Bierut, L., Goate, A., Bucholz, K., Schuckit, M., Nurnberger Jr., J., et al. 2006. Association of the κ-opioid system with alcohol dependence. *Mol. Psychiatry* **11:** 1016–1024.

Zeggini, E., Scott, L.J., Saxena, R., Voight, B.F., Marchini, J.L., Hu, T., de Bakker, P.I., Abecasis, G.R., Almgren, P., Andersen, G., et al. 2008. Meta-analysis of genome-wide association data and large-scale replication identifies additional susceptibility loci for type 2 diabetes. *Nat. Genet.* **40:** 638–645.

WWW RESOURCES

http://www.1000genomes.org 1000 genomes, a deep catalog of human genetic variation.

http://www.broad.mit.edu/mpg/tagger de Bakker et al. 2005. Tagger.

http://www.broad.mit.edu/mpg/haploview Barrett et al. 2005. Haploview.

http://www.cidr.jhmi.edu Center for Inherited Disease Research, Johns Hopkins University.

http://www.hapmap.org International HapMap Project.

http://www.ncbi.nlm.nih.gov/SNP dbSNP, Single Nucleotide Polymorphism Database.

17 Gene Set Analysis and Network Analysis for Genome-Wide Association Studies

Inti Pedroso and Gerome Breen

Medical Research Council Social, Genetic, and Developmental Psychiatry Centre, Institute of Psychiatry, King's College London, De Crespigny Park, London, SE5 8AF, United Kingdom and National Institute for Health Research Biomedical Research Centre for Mental Health, South London, and Maudsley National Health Services Foundation Trust and Institute of Psychiatry, King's College London, De Crespigny Park, London, SE5 8AF, United Kingdom

INTRODUCTION

The application of high-throughput genotyping in humans has yielded numerous insights into the genetic basis of human phenotypes and an unprecedented amount of genetic data. By April 23, 2009, the Catalog of Published Genome-Wide Association Studies (http://www.genome.gov/gwastudies) registered 304 publications with 232 single-nucleotide polymorphisms (SNPs) reaching the widely agreed upon genome-wide significance level of 7.2×10^{-8} (Dudbridge and Gusnanto 2008). However, each of these variants explains only a tiny proportion of the estimated genetic contribution to phenotype variation, suggesting that the current approach lacks the necessary power to detect the bulk of risk variants (McCarthy and Hirschhorn 2008). In addition to a lack of statistical power, there are several possible explanations for this: (1) The coverage of relevant genetic variants may still be low, particularly if the variant is rare; (2) phenotypic heterogeneity may be more pervasive than previously thought; and (3) environmental heterogeneity between cohorts combined for genome-wide association studies (GWAS) may act to mask true signals. This may be particularly problematic in psychiatric genetics, where tests for physical symptoms are not available or are poorly defined, and environment is thought to have a key role (Caspi et al. 2002). The current approach for GWAS is based on the common disease/common variant (CDCV) hypothesis, making it less successful for phenotypes with a different genetic architecture (e.g., in schizophrenia, some variants identified have odds ratios of 3–10, but their frequency tends to be <0.1% in case populations and <0.01% in controls; Stefansson et al. 2008).

A notable exception is the APOE4 allele (Pericak-Vance et al. 1991), which may account for more than half of the genetic variance in risk for Alzheimer's disease (AD). Although this allele was discovered by linkage analysis in the pre-GWAS era, it is the only replicated locus for AD in GWAS (as yet). Thus far, its discovery has not led to new treatments for AD directly, although the APOE4 allele is showing the potential to influence clinical practice and treatment on the basis of subclassification of the disease into early- and late-onset forms (Roses et al. 2007). Nevertheless, APOE and other examples of replicated loci with poor power in diagnostics settings have led to strong criticism of the current GWAS strategy (Goldstein 2009).

It is important to remember that the technology and sophistication of GWAS analyses are advancing at a dramatic rate and that some or all of these limitations may be overcome because new technology will provide assays with better coverage. Such examples are a dense whole-genome array (>1 million variants) complemented by exon sequencing to identify important rare variants. Finally, total whole-genome sequencing will allow for deeper exploration of genetic variation. However, additional limitations require new methods for interpreting and testing the genotype–phenotype correlations; these methods probably will come in the shapes of diverse statistical and data-mining techniques. Examples of these limitations include the following: (1) After applying the stringent statistical thresholds necessary for multiple testing correction, very few significant variants remain (many times none!), representing a very unclear picture of the underlying biology of phenotypic variation. (2) There is a lack of formal methods for interpreting genetic association (e.g., identifying which variants are causative of a given association). (3) The statistical tests for genotype–phenotype correlation may not capture the real underlying effect of genetic variation, such as epistatic effects. (4) We now appreciate that some replicated loci do not reach genome-wide significance in individual studies (e.g., the peroxisome proliferative activated receptor γ gene *PPARG* in type 2 diabetes; Altshuler et al. 2000). (5) As yet, there are no systematic statistical methods available to extract information from numerous true hits immersed in the noise of multiple testing. Although combining data sets (e.g., by meta-analysis) has been a successful approach, it faces similar problems with genetic and phenotypic heterogeneity.

The main strength of GWAS, as for other "-omics" technologies, may be to provide a data-driven approach to reveal new (sometimes unsuspected) avenues for studying the biological underpinnings of trait variation (e.g., the relationship between Crohn's disease and autophagy; Zhang et al. 2008), some of which may be valuable for clinical practice and disease treatment (Roses et al. 2007). The many thousands of marginal significant results could hold the "higher-hanging fruit" of real but non-genome-wide significant associations because of either their low effect size or a lack of power caused by a low allele frequency and the limited sample sizes used in GWAS to date. To uncover these associations, we need tools to interpret genetic associations in the context of the genes and the biological processes that they may influence. The aims of this approach are to (1) identify biological processes or gene networks that are drivers or that confer susceptibility to the phenotype under study, (2) improve the interpretation of results, and (3) achieve increased power to detect genes with small effect sizes that cluster on common biological processes, which may in turn prove to be a fruitful target for treatment.

This chapter summarizes the progress made in the development of gene set analysis (GSA) and network analysis for GWAS as a way to identify the underlying molecular processes of human phenotypes, highlights some promising findings, and indicates future directions that may greatly enhance the analysis and interpretation of GWAS (Fig. 17.1).

GENE SET ANALYSIS AND NETWORK ANALYSIS

In contrast to gene-centric approaches, such as assessing expression differences of individual genes or gene products, the goal of GSA and network analysis is to capture the collective activity of sets of molecules (e.g., gene products) by testing a group instead of the individual components in isolation. (Note that GSA was originally developed to analyze gene expression data, and we have used terminology and examples from that literature. Other than small differences, the general ideas can be translated to GWAS.) Although large expression differences in one molecule provide supporting evidence for its perturbed state, these differences provide little information for answering relevant questions such as (1) What are the biological processes that explain this large differential expression? (2) How do the molecules showing large differences relate to one another? (3) Are any small expression changes likely to be biologically significant? GSA aims to address some of these questions.

There are several variants of GSA, but basically all test the enrichment of significant results in predefined groups of genes or gene sets. Gene sets are genes grouped by biological criteria, such as co-expression or presence in a defined metabolic pathway (Nam and Kim 2008). Gene set definitions (see Table 17.1 for some commonly used resources) can be constructed via literature review or compilation of results from high-throughput experiments, allowing the incorporation of previous knowledge

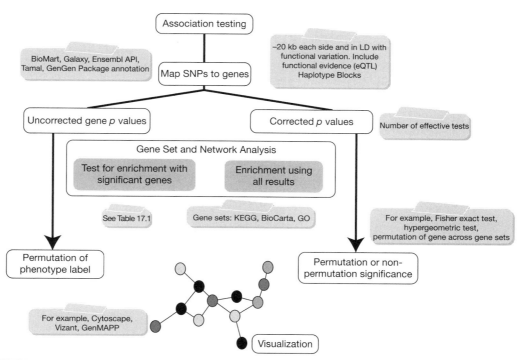

FIGURE 17.1 Gene set analysis and network analysis. The flow chart proposes key steps to perform GSA for GWAS. We highlight some useful resources throughout the chapter.

(e.g., tissue-specific expression profiles or manually curated gene–gene interaction information) and the evaluation of convergent evidence toward a specific biological process (Lehner and Lee 2008). A drawback of GSA is its strong dependence on the availability and quality of gene sets. Better annotation is to be expected for well-studied processes (e.g., cancer-related pathways) compared with biological processes that are difficult to study (such as neurotransmission) or are the focus of limited research.

Network analysis makes direct use of interaction information among biological molecules and can be used to research the network properties of phenotype-associated genes (Goh et al. 2007; Ideker and Sharan 2008) as well as to identify subnetworks harboring more associations than would be expected by chance (Chuang et al. 2007; Dittrich et al. 2008). As for GSA, network analysis presents its own limitations, including (1) finding significant subnetworks is a combinatorial NP-difficult problem, meaning that it is very difficult, if possible at all, to find an optional solution, and is usually tackled with computationally intensive and stochastic algorithms, and (2) the exponential nature of the interaction information makes its acquisition slow and biased against interactions of molecules more difficult to study, such as transmembrane proteins or interactions involving sugars or RNA molecules (just recently starting to be tested with high-throughput experiments) (Venkatesan et al. 2009). It is possible to overcome this problem by inferring a network using functional data, which can help to overcome coverage as well as to capture some of the tissue- or cell-specific nature of the interactions (Webster et al. 2009). Nevertheless, GSA and network analysis have been successfully used to characterize, prioritize, and shed new light on the molecular processes underlying phenotypes in human and model organisms (Fuller et al. 2007; Chen et al. 2008; Emilsson et al. 2008; Ideker and Sharan 2008; Presson et al. 2008; Zhu et al. 2008), strongly supporting their development and application in GWAS.

Before commenting on the current results of GSA and network analyses in GWAS, we need to point out some caveats. GSA was developed originally for and has been applied extensively to gene expression data. Gene expression data can be highly autocorrelated, making the identification of significant gene sets easier because gene expression changes occur in a concerted manner. Fur-

TABLE 17.1 Tools for exploring genome-wide association studies (GWAS) using gene set analysis (GSA)

Mapping genetic variation to genes	
Biomart	http://www.ensembl.org/biomart/martview/
BrainArray	http://brainarray.mbni.med.umich.edu/Brainarray/
Ensembl Perl API	http://www.ensembl.org/info/docs/api/index.html
Galaxy	http://main.g2.bx.psu.edu/
PLINK	http://pngu.mgh.harvard.edu/~purcell/plink/ (access Tamal database)
SNAP	http://www.broad.mit.edu/mpg/snap/
Tamal	http://neoref.ils.unc.edu/tamal/
Expression quantitative trait locus (eQTL) resources	
GENEVAR	http://www.sanger.ac.uk/humgen/genevar/
eQTL web browser	http://eqtl.uchicago.edu
Lab functional neurogenomics	http://labs.med.miami.edu/myers/LFuN/LFuN.html
Pathway definition resources	
BioCarta	http://www.biocarta.com
Gene Ontology	http://www.geneontology.org
KEGG	http://www.genome.jp/kegg
MSigDB	http://www.broad.mit.edu/gsea/msigdb
Gene set analysis tools	
FatiGO	http://fatigo.bioinfo.cnio.es
GenGen Package	http://www.openbioinformatics.org/gengen
GSEA	http://www.broad.mit.edu/gsea
GSEA-SNP	http://www.nr.no/pages/samba/area_emr_smbi_gseasnp
Set of tools for gene ontology	http://www.geneontology.org/GO.tools.shtml

thermore, gene expression analyses capture changes that are both the cause and the consequence of the phenotype. In addition, changes in gene expression can be of several orders of magnitude. On the other hand, genotype–phenotype correlations of common variants with complex traits (the focus of current GWAS) have subtle effect sizes, are not correlated with one another (although epistasis is expected to have a role, it has been challenging to find this in GWAS), and the magnitude of change is much smaller than in gene expression studies.

These caveats have important practical consequences. In gene expression analyses, a strong GSA finding is supported by several genes, providing evidence that a significant portion of a biological process is deregulated. However, in a GWAS, one significant association in the same gene set may not be statistically significant in GSA, but its functional consequences may indeed perturb the entire biological process. Although these differences do not invalidate the application of these methods, they must be taken into account when interpreting the results. Statistical support from GSA or network analysis provides strong evidence for the biological processes contributing to phenotype variation; because each of the signals is independent of one another (insofar as the genetic variants are not in linkage disequilibrium [LD]), each represents additional evidence, and they are not a consequence of the phenotype.

Mapping Genetic Associations to Genes

The first necessary step in GSA for GWAS is to map genetic variation to genes present in gene sets. At the moment, no "gold standard" exists for this, and the decisions rely on the subjective experimenter's criteria. The first question may be to determine the kind of functional relationship that needs to be mapped. For example, one may want to map genetic variation affecting protein function (e.g., nonsynonymous variants) or to include variation affecting gene transcript expression. Both are not mutually exclusive, but the results can be quite different. In the first case, it may be sufficient to take associated variants that map directly to a gene, but a more sensitive approach also should map variants in LD with associated variants to genes.

"Mapping to a gene" should include not only the coding region but also the introns (the impacts on splicing can affect protein function as well) and untranslated regions (possibly impacting microRNA regulation). In the case of variants affecting transcript expression, it is necessary to define a distance limit, because the inclusion of genetic variation affecting possible enhancers (which can be hundreds of kilobases away) may introduce excessive noise in downstream analysis. A sensible approach may be to define the limits based on evidence from expression quantitative trait locus (eQTL) studies (see Table 17.1 for eQTL resources), which have shown that the 95% of common genetic variation affecting transcript levels is within 20 kb of the transcription start and end sites (Veyrieras et al. 2008). Examples of both approaches can be found in the literature (e.g., Wang et al. 2007; Moskvina et al. 2009), but no systematic comparison of the effect on downstream analysis has been done.

GSA and Network Analysis on GWAS: Fishing Deeper for Smaller Fishes

A theoretical background as well as some empirical evidence supports the use of gene set methods in GWAS (e.g., Gibson 2009; Wang et al. 2009). The main argument here is that the underlying cause of phenotypic variability is changes in biological processes at the cellular and organismal level, meaning that perturbations can arise from any element acting on these processes.

Wang et al. (2007) introduced GSA for GWAS with an implementation of the gene set enrichment analysis (GSEA) algorithm (Subramanian et al. 2005). Since then, an increasing number of studies have applied or proposed new GSA and network methods for GWAS, demonstrating the interest of the field for their development. Published articles range from the application of gene expression analysis techniques to new approaches exploiting additional information for genetic mapping, such as genetic interactions or integration with functional data such as gene expression. Table 17.2 provides a list of publications using GSA, and Table 17.3 compiles publications using network and epistasis analysis on GWAS.

TABLE 17.2 Published applications of GSA on GWAS

Reference	Phenotype	Method	Software
Askland et al. (2009)	BD	Exploratory visual analysis	Exploratory Visual Analysis (EVA)
Chen et al. (2009)	BD, CD, CAD, T2D, HT	Prioritizing risk pathway fusing SNPs and pathways	
Craddock et al. (2008)	BD	Logistics regression	
Elbers et al. (2009)	T2D	Mixed	Webgestalt, GATHER, DAVID, PANTHER, BioCarta
Holden et al. (2008)		GSEA-SNP	GSEA-SNP
Inada et al. (2008)	TD	Fisher exact test	Ingenuity Pathway Analysis
Iossifov et al. (2008)	BD, SZC, AU	Gene mixture generative model	
Lesnick et al. (2007)	PD	Regression methods	
Perry et al. (2009)	T2D	GSEA	GenGen Package
Srinivasan et al. (2009)	PD	Boosted decision trees	
Torkamani et al. (2008)	BD, CD, CAD, T2D, T1D, RA, HT	Hypergeometric test	MetaCore
Walsh et al. (2008)	SZC	Fisher exact test	Ingenuity Pathway Analysis
Wang et al. (2007)	PD, AMD	GSEA	GenGen Package
Wang et al. (2009)	CD	GSEA	GenGen Package
Yu et al. (2009)	CSB	Adaptive combination of p values	

(AMD) Age-related macular degeneration; (AU) autism; (BD) bipolar disorder; (CAD) coronary artery disease; (CD) Crohn's disease; (CSB) cigarette smoking behaviors; (HT) hypertension; (PD) Parkinson's disease; (RA) rheumatoid arthritis; (SZC) schizophrenia; (T1D) type 1 diabetes; (T2D) type 2 diabetes; (TD) treatment-resistant tardive dyskinesia.

TABLE 17.3 Published methods and application of network and epistasis analysis on GWAS

Reference	Phenotype	Method	Software
Baranzini et al. (2009)	MS	Network analysis	jActive Modules
Emily et al. (2009)	BD, CD, CAD, T2D, T1D, RA, HT	Logistic regression model	BiRC
Gayán et al. (2008)	PD	Mixed	HFCC
Jiang et al. (2009)	AMD	Random forest	
Li et al. (2008)	RA	Decision trees	
Ma et al. (2008)		Linear models and extended Kempthorne model	EPISNP and EPISNPmpi
Macgregor et al. (2006)		Logistic regression model	GAIA
Marchini et al. (2005)		Contingency table	
Mechanic et al. (2008)		Mixed	PIA
Nunkesser et al. (2007)	BC	Genetic algorithm	GPAS
Tang et al. (2009)	AMD, PD	Bayesian marker partition model and Gibbs sampling	epiMODE
Wan et al. (2009)	PD, RA	Gradient boosting of regression tree	MegaSNPHunter
Yang et al. (2009)	RA	L_2 penalized logistic regression models	SNPHarvester
Zhang and Liu (2007)	AMD	Bayesian epistasis association mapping	BEAM

(AMD) Age-related macular degeneration; (BC) breast cancer; (BD) bipolar disorder; (CAD) coronary artery disease; (CD) Crohn's disease; (HT) hypertension; (MS) multiple sclerosis; (PD) Parkinson's disease; (RA) rheumatoid arthritis; (T1D) type 1 diabetes; (T2D) type 2 diabetes.

Rather than providing a technical report here, we provide a summary of the published research using GSA and network analysis on GWAS, not pretending to be exhaustive, but rather to be instructive for the reader aiming to apply these methods.

Application of GSA

Probably the most relevant methodological finding has been the ability of GSA to provide successful replication at the pathway level in the presence of genetic heterogeneity (Baranzini et al. 2009; Wang et al. 2009). These examples open the door for the application of GSA in large-scale studies, which in the long term will offer the evidence to judge the relative contribution of GSA to genetic mapping.

In a pharmacogenetic context, GSA can identify molecular systems that explain differential drug responses, which may help to improve drug and clinical interventions. Inada et al. (2008) applied GSA to treatment-resistant tardive dyskinesia (TD), an involuntary movement disorder in patients undergoing long-term treatment with antipsychotic medications. The authors found significant enrichment in the γ-aminobutyric acid receptor (GABRA) signaling pathway. Dysfunction of the GABAergic system previously had been associated with TD, and the findings of these investigators suggest that genetic variation in this system confers susceptibility to TD. An interesting complement to the pharmacogenetic approach is the estimation of the contribution to risk of a given pathway (Liu et al. 2008; Chen et al. 2009; Srinivasan et al. 2009). Further development of these methods may prove to be useful for predicting response or identifying a patient subgroup that carries risk alleles in susceptibility pathways.

The identification of patient subgroups characterized by allelic patterns is important for refining diagnostic categories and improving disease treatment. Craddock et al. (2008) have explored this avenue by seeking a clinically relevant subgroup of patients driving an association signal observed in the whole sample. They observed that an association with bipolar disorder (BD) found by the Wellcome Trust Case Control Consortium at an SNP in the *GABRB1* gene was enriched in patients diagnosed with the subtype schizoaffective bipolar (SABP) disorder. Interestingly, associations with SNPs within the genes coding for other subunits of the GABRA A (*GABRA4*, *GABRB3*, *GABRA5*, and *GABRR1*) were also enriched in SABP disorder, providing a system-wide association p value of 6.6×10^{-5}.

Elbes et al. (2009) evaluated the effect of pathway size and compared the performance of several pathway analysis tools. They showed that if one does not control for the number of variants seen, a large number of false positives are observed in GSA. Additionally, they found that different pathway tools gave different results for the same data set, and results seemed to be biased toward well-characterized gene sets. We have observed that several authors did not correct for the number of variants in the gene or pathway (Torkamani et al. 2008; Walsh et al. 2008; Askland et al. 2009), suggesting that some results may be a product of pathway size bias.

Three GWAS have associated the axon-guidance pathway with Parkinson's disease. In the most recent, Srinivasan et al. (2009) presented their results and reanalyzed the Fung et al. (2006) data set (genotypes were not available; Lesnick et al. 2007). They found a significant association with the axon-guidance pathway, but they did not find such a result in their reanalysis of the Fung et al. (2006) GWAS. The authors explain this discrepancy on the basis of methodological differences: They corrected for pathway size, whereas Fung et al. did not.

Application of Network and Epistasis Analyses

Baranzini et al. (2009) overlaid results from GWAS of nine diseases on a protein–protein interaction network and performed network analysis to find network modules associated with one or more disorders. They found that, in general, each disorder is represented by a distinctive group of modules, but some modules are shared across disorders, such as human leukocyte antigen (HLA) genes across autoimmune disorders, or modules are shared between central nervous system disorders such as BD, AD, and multiple sclerosis (MS), suggesting that common genetic variation can be used not only to reveal associations with network modules but also to explore the molecular basis of similar, and often comorbid, disorders.

The incoming GWAS of a plethora of phenotypes will allow data mining to uncover genetic variation in biological processes conferring risk across diagnostic categories and, together with information on environmental exposure, will offer the possibility of tackling questions related to pleiotropy, penetrance, differential expressivity, and gene– and/or pathway–environment interactions. Quantitative genetic research has shown that nonadditive epistatic effects have an important role in observed variation of human traits (Sing and Davignon 1985; Zerba et al. 2000). Several approaches have been developed to search for epistatic interactions at the genome level (see Table 17.3). Some represent a brute-force strategy and test exhaustively, whereas others make use of machine learning techniques to extract a subset of variants on which to apply additional analysis (including brute force). Several proposed methods have interesting features, such as the possibility of analyzing more than one sample at a time and testing for replication of the interactions (Gayán et al. 2008), not relying on predefined interaction models (Li et al. 2008; Jiang et al. 2009; Tang et al. 2009; Wan et al. 2009; C. Yang et al. 2009), and allowing identification of interacting modules (an appealing concept in the context of associations with biological processes) (Tang et al. 2009), with feasible running time on a desktop computer (Emily et al. 2009) or parallel implementation for high-performance computing (Ma et al. 2008).

At the moment, results from epistasis analyses are far from explaining more phenotypic variation than the main effect variant and have many of the limitations commented on above. However, epistasis mapping may prove to be a useful tool for constructing genetic interaction networks, probably capturing information relevant in the tissue and for the phenotype under study. Replication of epistatic effects promises to be a challenging process for which understanding of the functional consequences of an interaction may be crucial.

Combining GWAS with Network and eQTL Analyses

Association between genetic variation and mRNA levels (eQTL) can shed light on which gene is expressing the functional effects (or some of them) of a genetic association (for review, see Dermitzakis 2008; Cookson et al. 2009). It is expected that many disease susceptibility loci will be eQTLs, and close inspection shows that 10%–15% of genome-wide significant findings can be associated with an eQTL (Cookson et al. 2009). However, it should be noted that eQTL studies (as with almost any GWAS) are underpowered for detecting both subtle and rare variant effects, and some associations are likely to be explained by functional effects different from changes in mRNA levels.

Webster et al. (2009) used mRNA levels from brain to identify eQTLs associated with AD. They exploited the correlation between mRNA levels to construct a coexpression network that enabled them to identify network modules enriched in disease susceptibility eQTLs and explore the consequences of these genetic variants at a higher level of cellular organization. Presson et al. (2008) used a similar approach to map eQTLs to coexpression networks associated with chronic fatigue syndrome. Both studies used Network Edge Orienting (NEO) software to establish the causal relationship between the genetic variants and changes in mRNA levels (http://www.genetics.ucla.edu/labs/horvath/aten/NEO/; Aten et al. 2008).

CHALLENGES AND PERSPECTIVES

GSA and network analysis are proving to be a good complement to conventional approaches for the analysis of GWAS. We note several areas in which these methods can be improved technically to extract the most from their results. The bulk of studies has used the minimum p value in the gene to represent association. However, recent evidence suggests that allelic heterogeneity is more common than expected (e.g., see Schulze et al. 2009). Therefore, a sensible strategy might be to combine the effect of multiple variants at the gene and pathway levels. This would allow for the merging of information from common and rare variants, which may not be subjected to association testing, to obtain a statistic representing a more comprehensive picture of the biological effect of a particular variant. To this end, it may be crucial to integrate information on the predicted functionality of genetic variants in order to find biologically relevant signals. As more functional information for the human genome is accumulated, this avenue may be useful for prioritizing and understanding the effect of identified variations (Curtis et al. 2007; Eskin 2008).

Comparative genomics is revealing a better picture of the evolution, location, and relative importance of functional elements in the genome. This information will be important to the understanding of genotypic and phenotypic variation at the population and species levels. Each type of genetic variation may contribute to phenotypic variation with different odds ratios and population frequencies, and an integrated analysis of rare and common variation will provide a better picture of the biology underpinning a phenotype.

Methods for mining genomic information in a system-oriented approach may uncover findings with a profound impact on our understanding of phenotype variation and evolution. Integration with functional genomic data will be crucial for identifying key biological molecules to monitor and manipulate biological processes and thus improve drug efficacy and disease diagnosis and treatment. Although there have been significant advances in this direction (Schadt et al. 2009; C. Yang et al. 2009), this is still an open field, with substantial development of data mining and statistical methods required to integrate and interrogate a combination of GWAS and functional genomic data sets for identifying and prioritizing biological systems of clinical and scientific relevance.

REFERENCES

Altshuler, D., Hirschhorn, J.N., Klannemark, M., Lindgren, C.M., Vohl, M.C., Nemesh J., Lane, C.R., Schaffner, S.F., Bolk, S., Brewer, C., et al. 2000. The common *PPARγ* Pro12Ala polymorphism is associated with decreased risk of type 2 diabetes. *Nat. Genet.* **26:** 76–80.

Askland, K., Read, C., and Moore J. 2009. Pathways-based analyses of whole-genome association study data in bipolar disorder reveal genes mediating ion channel activity and synaptic neurotransmission. *Hum. Genet.* **125:** 63–79.

Aten, J.E., Fuller, T.F., Lusis, A.J., and Horvath, S. 2008. Using genetic markers to orient the edges in quantitative trait networks: The NEO software. *BMC Syst. Biol.* **2:** 34.

Baranzini, S.E., Galwey, N.W., Wang, J., Khankhanian, P., Lindberg, R., Pelletier, D., Wu, W., Uitdehaag, B.M., Kappos, L., Polman, C.H., et al. 2009. Pathway and network-based analysis of genome-wide association studies in multiple sclerosis. *Hum. Mol. Genet.* **18:** 2078–2090.

Caspi, A., Sugden, K., Moffitt, T.E., Taylor A., Craig, I.W., Harrington, H., McClay, J., Mill, J., Martin, J., Braithwaite, A., et al. 2003. Influence of life stress on depression: Moderation by a polymorphism in the *5-HTT* gene. *Science* **301:** 386–389.

Chen, L., Zhang, L., Zhao, Y., Xu, L., Shang, Y., Wang, Q., Li, W., Wang, H., and Li, X. 2009. Prioritizing risk pathways: A novel association approach to searching for disease pathways fusing SNPs and pathways. *Bioinformatics* **25:** 237–242.

Chen, Y., Zhu, J., Lum, P.Y., Yang, X., Pinto, S., MacNeil, D.J., Zhang, C., Lamb, J., Edwards, S., Sieberts, S.K., et al. 2008. Variations in DNA elucidate molecular networks that cause disease. *Nature* **452:** 429–435.

Chuang, H.Y., Lee, E., Liu, Y.T., Lee, D., and Ideker, T. 2007. Network-based classification of breast cancer metastasis. *Mol. Syst. Biol.* **3:** 140.

Cookson, W., Liang, L., Abecasis, G., Moffatt, M., and Lathrop, M. 2009. Mapping complex disease traits with global gene expression. *Nat. Rev. Genet.* **10:** 184–194.

Craddock, N., Jones, L., Jones, I.R., Kirov, G., Green, E.K., Grozeva, D., Moskvina, V., Nikolov, I., Hamshere, M.L., Vukcevic, D., et al. 2008. Strong genetic evidence for a selective influence of GABA$_A$ receptors on a component of the bipolar disorder phenotype. *Mol. Psychiatry* (in press).

Curtis, D., Vine, A.E., and Knight, J. 2007. A pragmatic suggestion for dealing with results for candidate genes obtained from genome wide association studies. *BMC Genet.* **8:** 20.

Dermitzakis, E.T. 2008. From gene expression to disease risk. *Nat. Genet.* **40:** 492–493.

Dittrich, M.T., Klau, G.W., Rosenwald, A., Dandekar, T., and Muller, T. 2008. Identifying functional modules in protein-protein interaction networks: An integrated exact approach. *Bioinformatics* **24:** i223–i231.

Dudbridge, F. and Gusnanto, A. 2008. Estimation of significance thresholds for genomewide association scans. *Genet. Epidemiol.* **32:** 227–234.

Elbers, C.C., van Eijk, K.R., Franke, L., Mulder, F., van der Schouw, Y.T., Wijmenga, C., and Onland-Moret, N.C. 2009. Using genomewide pathway analysis to unravel the etiology of complex diseases. *Genet. Epidemiol.* **33:** 419–431.

Emilsson, V., Thorleifsson, G., Zhang, B., Leonardson, A.S., Zink, F., Zhu, J., Carlson, S., Helgason, A., Walters, G.B., Gunnarsdottir, S., et al. 2008. Genetics of gene expression and its effect on disease. *Nature* **452:** 423–428.

Emily, M., Mailund, T., Hein, J., Schauser, L., and Schierup, M.H. 2009. Using biological networks to search for interacting loci in genomewide association studies. *Eur. J. Hum. Genet.* (in press).

Eskin, E. 2008. Increasing power in association studies by using linkage disequilibrium structure and molecular function as prior informa-

tion. *Genome Res.* **18:** 653–660.

Fuller, T.F., Ghazalpour, A., Aten, J.E., Drake, T.A., Lusis, A.J., and Horvath, S. 2007. Weighted gene coexpression network analysis strategies applied to mouse weight. *Mamm. Genome* **18:** 463–472.

Fung, H.C., Scholz, S., Matarin, M., Simon-Sanchez, J., Hernandez, D., Britton, A., Gibbs, J.R., Langefeld, C., Stiegert, M.L., Schymick, J., et al. 2006. Genome-wide genotyping in Parkinson's disease and neurologically normal controls: First stage analysis and public release of data. *Lancet Neurol.* **5:** 911–916.

Gayán, J., González-Pérez, A., Bermudo, F., Sáez, M.E., Royo, J.L., Quintas, A., Galan, J.J., Morón, F.J., Ramirez-Lorca, R., Real, L.M., et al. 2008. A method for detecting epistasis in genome-wide studies using case-control multi-locus association analysis. *BMC Genomics* **9:** 360.

Gibson, G. 2009. Decanalization and the origin of complex disease. *Nat. Rev. Genet.* **10:** 134–140.

Goh, K.I., Cusick, M.E., Valle, D., Childs, B., Vidal, M., and Barabasi, A.L. 2007. The human disease network. *Proc. Natl. Acad. Sci.* **104:** 8685–8690.

Goldstein, D.B. 2009. Common genetic variation and human traits. *N. Engl. J. Med.* **360:** 1696–1698.

Holden, M., Deng, S., Wojnowski, L., and Kulle, B. 2008. GSEA-SNP: Applying gene set enrichment analysis to SNP data from genomewide association studies. *Bioinformatics* **24:** 2784–2785.

Ideker, T. and Sharan, R. 2008. Protein networks in disease. *Genome Res.* **18:** 644–652.

Inada, T., Koga, M., Ishiguro, H., Horiuchi, Y., Syu, A., Yoshio, T., Takahashi, N., Ozaki, N., and Arinami, T. 2008. Pathway-based association analysis of genome-wide screening data suggests that genes associated with the γ-aminobutyric acid receptor signaling pathway are involved in neuroleptic-induced, treatment-resistant tardive dyskinesia. *Pharmacogenet. Genomics* **18:** 317–323.

Iossifov, I., Zheng, T., Baron, M., Gilliam, T.C., and Rzhetsky, A. 2008. Genetic-linkage mapping of complex hereditary disorders to a wholegenome molecular-interaction network. *Genome Res.* **18:** 1150–1162.

Jiang, R., Tang, W., Wu, X., and Fu, W. 2009. A random forest approach to the detection of epistatic interactions in case-control studies. *BMC Bioinformatics* (suppl. 1) **10:** S65.

Lehner, B. and Lee, I. 2008. Network-guided genetic screening: Building, testing and using gene networks to predict gene function. *Brief Funct. Genomics Proteomics* **7:** 217–227.

Lesnick, T.G., Papapetropoulos, S., Mash, D.C., Ffrench-Mullen, J., Shehadeh, L., de Andrade, M., Henley, J.R., Rocca, W.A., Ahlskog, J.E., and Maraganore, D.M. 2007. A genomic pathway approach to a complex disease: Axon guidance and Parkinson disease. *PLoS Genet.* **3:** e98.

Li, C., Zhang, G., Li, X., Rao, S., Gong, B., Jiang W., Hao, D., Wu, P., Wu, C., Du, L., et al. 2008. A systematic method for mapping multiple loci: An application to construct a genetic network for rheumatoid arthritis. *Gene* **408:** 104–111.

Liu, D., Ghosh, D., and Lin, X. 2008. Estimation and testing for the effect of a genetic pathway on a disease outcome using logistic kernel machine regression via logistic mixed models. *BMC Bioinformatics* **9:** 292.

Ma, L., Runesha, H.B., Dvorkin, D., Garbe, J.R., and Da, Y. 2008. Parallel and serial computing tools for testing single-locus and epistatic SNP effects of quantitative traits in genome-wide association studies. *BMC Bioinformatics* **9:** 315.

Macgregor, S. and Khan, I.A. 2006. GAIA: An easy-to-use web-based application for interaction analysis of case-control data. *BMC Med. Genet.* **7:** 34.

Marchini, J., Donnelly, P., and Cardon, L.R. 2005. Genome-wide strategies for detecting multiple loci that influence complex diseases. *Nat.*

Genet. **37:** 413–417.

McCarthy, M.I. and Hirschhorn, J.N. 2008. Genome-wide association studies: Potential next steps on a genetic journey. *Hum. Mol. Genet.* **17:** R156–R165.

Mechanic, L.E., Luke, B.T., Goodman, J.E., Chanock. S.J., and Harris, C.C. 2008. Polymorphism interaction analysis (PIA): A method for investigating complex gene-gene interactions. *BMC Bioinform.* **9:** 146.

Moskvina, V., Craddock, N., Holmans, P., Nikolov, I., Pahwa, J.S., Green, E., Owen, M.J., and O'Donovan, M.C. 2009. Gene-wide analyses of genome-wide association data sets: Evidence for multiple common risk alleles for schizophrenia and bipolar disorder and for overlap in genetic risk. *Mol. Psychiatry* **14:** 252–260.

Nam, D. and Kim, S.Y. 2008. Gene-set approach for expression pattern analysis. *Brief Bioinform.* **9:** 189–197.

Nunkesser, R., Bernholt, T., Schwender, H., Ickstadt, K., and Wegener, I. 2007. Detecting high-order interactions of single nucleotide polymorphisms using genetic programming. *Bioinformatics* **23:** 3280–3288.

Pericak-Vance, M.A., Bebout, J.L., Gaskell, P.C.J., Yamaoka, L.H., Hung, W.Y., Alberts, M.J., Walker, A.P., Bartlett, R.J., Haynes, C.A., Welsh, K.A., et al. 1991. Linkage studies in familial Alzheimer disease: Evidence for chromosome 19 linkage. *Am. J. Hum. Genet.* **48:** 1034–1050.

Perry, J.R., McCarthy, M.I., Hattersley, A.T., Zeggini, E., Weedon, M.N., and Frayling, T.M. 2009. Interrogating type 2 diabetes genome-wide association data using a biological pathway-based approach. *Diabetes* **58:** 1463–1467.

Presson, A.P., Sobel, E.M., Papp, J.C., Suarez, C.J., Whistler, T., Rajeevan, M.S., Vernon, S.D., and Horvath, S. 2008. Integrated weighted gene co-expression network analysis with an application to chronic fatigue syndrome. *BMC Syst. Biol.* **2:** 95.

Roses, A.D., Saunders, A.M., Huang, Y., Strum, J., Weisgraber, K.H., and Mahley R.W. 2007. Complex disease-associated pharmacogenetics: Drug efficacy, drug safety, and confirmation of a pathogenetic hypothesis (Alzheimer's disease). *Pharmacogenomics J.* **7:** 10–28.

Schadt, E.E., Zhang, B., and Zhu, J. 2009. Advances in systems biology are enhancing our understanding of disease and moving us closer to novel disease treatments. *Genetica* **136:** 259–269.

Schulze, T.G., Detera-Wadleigh, S.D., Akula, N., Gupta, A., Kassem, L., Steele, J., Pearl, J., Strohmaier, J., Breuer, R., Schwarz, M., et al. 2009. Two variants in Ankyrin 3 (ANK3) are independent genetic risk factors for bipolar disorder. *Mol. Psychiatry.* **14:** 487–491.

Sing, C.F. and Davignon, J. 1985. Role of the apolipoprotein E polymorphism in determining normal plasma lipid and lipoprotein variation. *Am. J. Hum. Genet.* **37:** 268–285.

Srinivasan, B.S., Doostzadeh, J., Absalan, F., Mohandessi, S., Jalili, R., Bigdeli, S., Wang, J., Mahadevan, J., Lee, C.L., Davis, R.W., et al. 2009. Whole genome survey of coding SNPs reveals a reproducible pathway determinant of Parkinson disease. *Hum. Mutat.* **30:** 228–238.

Stefansson, H., Rujescu, D., Cichon, S., Pietilainen, O.P., Ingason, A., Steinberg, S., Fossdal, R., Sigurdsson, E., Sigmundsson, T., Buizer-Voskamp, J.E., et al. 2008. Large recurrent microdeletions associated with schizophrenia. *Nature* **455:** 232–236.

Subramanian, A., Tamayo, P., Mootha, V.K., Mukherjee, S., Ebert, B.L., Gillette, M.A., Paulovich, A., Pomeroy, S.L., Golub, T.R., Lander, E.S., et al. 2005. Gene set enrichment analysis: A knowledge-based approach for interpreting genome-wide expression profiles. *Proc. Natl. Acad. Sci.* **102:** 15545–15550.

Tang, W., Wu X., Jiang, R., and Li, Y. 2009. Epistatic module detection for case-control studies: A Bayesian model with a Gibbs sampling strategy. *PLoS Genet.* **5:** e1000464.

Torkamani, A., Topol, E.J., and Schork, N.J. 2008. Pathway analysis of seven common diseases assessed by genome-wide association. *Genomics* **92:** 265–272.

Venkatesan, K., Rual, J.F., Vazquez, A., Stelzl, U., Lemmens, I., Hirozane-Kishikawa, T., Hao, T., Zenkner, M., Xin, X., Goh, K.I., et al. 2009. An empirical framework for binary interactome mapping. *Nat. Methods* **6:** 83–90.

Veyrieras, J.B., Kudaravalli, S., Kim, S.Y., Dermitzakis, E.T., Gilad, Y., Stephens, M., and Pritchard, J.K. 2008. High-resolution mapping of expression-QTLs yields insight into human gene regulation. *PLoS Genet.* **4:** e1000214.

Walsh, T., McClellan, J.M., McCarthy, S.E., Addington, A.M., Pierce, S.B., Cooper, G.M., Nord, A.S., Kusenda, M., Malhotra, D., Bhandari, A., et al. 2008. Rare structural variants disrupt multiple genes in neurodevelopmental pathways in schizophrenia. *Science* **320:** 539–543.

Wan, X., Yang, C., Yang, Q., Xue, H., Tang, N.L., and Yu, W. 2009. MegaSNPHunter: A learning approach to detect disease predisposition SNPs and high level interactions in genome wide association study. *BMC Bioinformatics* **10:** 13.

Wang, K., Li, M., and Bucan, M. 2007. Pathway-based approaches for analysis of genomewide association studies. *Am. J. Hum. Genet.* **81:** 1278–1283.

Wang, K., Zhang, H., Kugathasan, S., Annese, V., Bradfield, J.P., Russell, R.K., Sleiman, P.M., Imielinski, M., Glessner, J., Hou, C., et al. 2009. Diverse genome-wide association studies associate the IL12/IL23 pathway with Crohn disease. *Am. J. Hum. Genet.* **84:** 399–405.

Webster, J.A., Gibbs, J.R., Clarke, J., Ray, M., Zhang, W., Holmans, P., Rohrer, K., Zhao A., Marlowe, L., Kaleem, M., et al. 2009. Genetic control of human brain transcript expression in Alzheimer disease. *Am. J. Hum. Genet.* **84:** 445–458.

Yang, C., He, Z., Wan, X., Yang, Q., Xue, H., and Yu, W. 2009. SNP-Harvester: A filtering-based approach for detecting epistatic interactions in genome-wide association studies. *Bioinformatics* **25:** 504–511.

Yang, X., Deignan, J.L., Qi, H., Zhu, J., Qian, S., Zhong, J., Torosyan, G., Majid, S., Falkard, B., Kleinhanz, R.R., et al. 2009. Validation of candidate causal genes for obesity that affect shared metabolic pathways and networks. *Nat. Genet.* **41:** 415–423.

Yu, K., Li, Q., Bergen, A.W., Pfeiffer, R.M., Rosenberg, P.S., Caporaso, N., Kraft, P., and Chatterjee, N. 2009. Pathway analysis by adaptive combination of *P*-values. *Genet. Epidemiol.* (in press).

Zerba, K.E., Ferrell, R.E., and Sing, C.F. 2000. Complex adaptive systems and human health: The influence of common genotypes of the apolipoprotein E (ApoE) gene polymorphism and age on the relational order within a field of lipid metabolism traits. *Hum. Genet.* **107:** 466–475.

Zhang, H., Massey, D., Tremelling, M., and Parkes, M. 2008. Genetics of inflammatory bowel disease: Clues to pathogenesis. *Br. Med. Bull.* **87:** 17–30.

Zhu, J., Zhang, B., Smith, E.N., Drees, B., Brem, R.B., Kruglyak, L., Bumgarner, R.E., and Schadt, E.E. 2008. Integrating large-scale functional genomic data to dissect the complexity of yeast regulatory networks. *Nat. Genet.* **40:** 854–861.

WWW RESOURCES

http://www.genetics.ucla.edu/labs/horvath/aten/NEO/ Network Edge Orienting software.

http://www.genome.gov/gwastudies Catalog of Published Genome-Wide Association Studies.

Index

A

ABO blood type
    association with ulcer susceptibility, 12–13
    determining frequency by affection status, 7
aCGH (array comparative genomic hybridization),
    131–132, 137, 145, 156
Additive genetic correlation, 44
Additive genetic effects, 38–39, 40
Additive variance, 37, 38–39
Affected family-based controls test (AFBAC), 121–122
Affected sib-pair (ASP) analysis
    nonparametric analysis, 23
    use in transmission disequilibrium test, 122, 123
Affymetrix arrays, 76, 96–97, 187, 188
Age-related macular degeneration, 84
Allele frequency
    in case–control study, 64
    Hardy–Weinberg equilibrium, 6–7, 67–68
    linkage disequilibrium (LD) and, 74–75
    population stratification of, 32
    sib transmission disequilibrium test (S-TDT), 124–125
    testing for difference in gene association studies, 64
Allelic heterogeneity, 24
Allelic spectra, 134
Alternative hypothesis
    description, 11
    effect size, 49
    examples, 12
    power and, 14
Alternative splicing, 165, 168, 173
Alzheimer's disease, 177, 195
American Cancer Prevention Study, 34
Amish, *FTO*-physical activity interaction in the Old
    Order, 109, 111
Aneuploidy, in tumor cells, 145–146
APL (association in the presence of linkage) test, 125,
    126, 127

APOE4 allele, 195
Apolipoprotein E (ApoE), 177
Array comparative genomic hybridization (aCGH),
    131–132, 137, 145, 156
ArrayExpress, 152, 154
Ascertainment correction, 43
Ascertainment scheme, used to select families for
    study, 42–43
Association, hypothesis testing for, 12–13
Association analysis for quantitative traits, 44–45
Association in the presence of linkage (APL) test, 125,
    126, 127
Association study. *See also* Case–control study;
    Cohort study
  conduct of, 27–31
    bias, 30
    changes in exposure, 31
    choice of design, 27–28
    governance, 27
    marker selection, 28–29
    quality control, 31
    random error, 30
    sample selection, 29–30
    sample size, 30
  copy-number analysis in, 135–138
  direct association, 62–65
    case–control study, 62–65
    example of statistical analysis, 64–65
    quantitative measures, use of, 65
  family-based genetic association tests, 119–128
  indirect association, 65–67
    haplotypes, analysis of, 67
    interactions between genes in disease risk, 67
    linkage disequilibrium, 65–66
    multiple markers, analysis of, 66
  interpretation, 31–33
    causality, 31–33
    confounding, 31–33
    direct association, 62–65
    effect modification, 32

Association study (*continued*)
    indirect association, 65–67
    of significant genetic association, 62
    large-scale data resources (LSDRs)
      biorepositories (biobanking), 33–34
      integrated fungible studies, 35
      remote studies, 34–35
    multiple testing in, 49–55
    overview, 61–62
    pitfalls and problems with studies, 69–70
      false-positive finding, 69–70
      heterogeneity across studies, 70
      heterogeneity between studies, 70
      replication study lacking power, 70
    power, statistical, 14, 30
      calculations, 55–59
      improving using multiple markers, 66
      in quantitative trait study, 65
      replication study lacking power, 70
    problems in analysis, addressing, 67–69
      Hardy–Weinberg equilibrium, 67–68
      missing genotypes, 68–69
      population stratification, 69
      quality control, 67
Assortative mating, 68
Autism, role of copy-number variants in, 136
Axon-guidance pathway, association with Parkinson's disease, 201

B

Bacterial artificial chromosomes (BACs), 132, 137
Bar codes, 77
Bateson, William, 2
Bayesian analysis, and multiple-testing correction, 51–52
Bayesian information criterion (BIC), 47
Bayes' theorem, 6
Bell-shaped curve, 8–9
Bias
    description, 30
    publication, 30, 117, 185
BIMBAM program, 96
Binomial distribution, 10
BioCarta, 198
Biomart, 198
Biorepositories (biobanks), 33–34
Bipolar disorder (BD), 200–201
Birdsuite, 85
Bisulfate treatment of DNA, 150, 151
Blood type
    association with ulcer susceptibility, 12–13
    determining frequency of ABO by affection status, 7
Bonferroni method of correction, 50–51, 70

BP sequence (BPS), 166, 177–178
BRAF kinase, 147
BrainArray, 198
Breast cancer, 149
Breast Cancer Mutations Database, 148, 152

C

Calculator, genetic power, 58
CaMP (cancer mutation prevalence) score, 148
Cancer. *See also* Oncogenomics
    candidate genes
      predicting, difficulties in, 148–149
      prioritization of, 154–155
    chromosomal rearrangements in, 146–147
    copy-number changes in, 145–146
    epigenomic alterations in, 149–151
    genetic basis of, 143–145
    somatic mutations in, 147–149
    transcriptomic changes in tumors, 151–164
    variants in, 149
Cancer Gene Census, 153, 154–155
CancerGenes, 153
Cancer Genome Atlas, 153, 157
Cancer Genome Atlas Research Network, 156
Cancer Genome Project, 153, 157
Cancer Genomics Browser, UCSC, 153
Cancer mutation prevalence (CaMP) score, 148
Cancer outlier profile analysis (COPA), 154
Candidate gene study
    gene–environment interaction, 115
    genotyping technologies appropriate for, 185–186
Carlson Greedy algorithm for tagSNP selection, 91–92
Case–control study
    changes in exposure, 31
    controls, selection of, 62
    copy-number variants (CNVs) and disease, 136
    description, 28
    family-based tests for association as alternatives to, 121–122
    gene–environment interaction, 112
    genome-wide association study
      data analysis methods, 81–82
      haplotype analysis, 82–84
      preparation for analysis, 77–81
    logarithm of odds ratio as measure of effect size, 49
    nesting within cohort studies, 28, 112
    population stratification in, 69, 119–120
    power calculation example, 56–57
    problems, 67–69
      genotyping errors, 67, 68–69
      population admixture, 120
      population stratification, 69, 119–120

random error, 30
quality control, 67
quantitative measures, use of, 65
sample selection, 29
statistical analysis of, 62–64
Case-only study, 112
Caspase 8, 149
Catalog of Published Genome-Wide Association Studies, 195
Catalogue of Somatic Mutations in Cancer (COSMIC), 148, 152
CATT (Cochran–Armitage test for trend), 63–64, 80–81
Causality, 31
CCDS (consensus coding sequences), mutational screens for, 148
Center for Inherited Disease Research at Johns Hopkins University, 187
Central dogma of molecular biology, 2
Central limit theorem, 9
CEPG (conditional exchangeable on parental genotypes), 127
CERES (composite regulatory element of splicing), 173
CFTR (cystic fibrosis transmembrane conductance regulator) gene, 173–174
CGH. *See* Comparative genomic hybridization (CGH)
CGPrio, 153, 155, 157–158
Chance, and departure from Hardy–Weinberg equilibrium, 68
Chip call quality measures, 77
Chi-square
    correction for imputation uncertainty, 97–99
    distribution, 9
    Pearson's, 62–63, 67–68
    statistic for allelic association, 81
    use in correction for population stratification, 80–81
    use in genome-wide association study (GWAS), 81
Chromatin immunoprecipitation (ChIP), 150–151
Chromosomal rearrangements, methods for detecting, 146–147
Chronic fatigue syndrome, 202
Cloning, positional, 144
CLR (conditional logistic regression), 128
CNV. *See* Copy-number variation (CNV)
COBRA (combined binary ratio labeling), 147
Cochran–Armitage test for trend (CATT), 63–64, 80–81
Cochran/Mantel–Haenszel test, 69, 81
Coefficient of relationship, 39
Cohort study
    case–control study nested in, 28, 112
    changes in exposure, 31

description, 28
gene–environment interaction, 112–113
sample selection, 29–30
COLIA2 gene, 177
Combinatorial partitioning method (CPM), 128
Combined binary ratio labeling (COBRA), 147
Common disease/common variant (CDCV)
        hypothesis, 81, 85, 134, 195
    testing, 73–74
Comparative genomic hybridization (CGH)
    array (aCGH), 131–132, 137, 145, 156
    oncogenomics, 144, 145–146
Complement factor H, 84
Complex disorders, description of term, 61
Conditional exchangeable on parental genotypes (CEPG), 127
Conditional logistic regression (CLR), 128
Conditional on parental genotypes (CPG), 127
Conditional probability, 6
Conditional splicing alleles, 165, 176
Confidence in a result, providing with hypothesis testing, 11
Confidence intervals, calculating on odds ratio by Woolf method, 64
Confounding
    causes of, 32–33
    description, 31–32
    of familial effects with genetic ones, 39
    by population of origin, 119
    population stratification, 69
Consensus coding sequences (CCDS), mutational screens for, 148
Consent, 27, 34
Contingency table, 62–63
Continuous data
    chi-square distribution, 9
    definition of, 8
    normal distribution, 8–9
Controls
    case–control study, selecting for, 62
    supernormal, 62
    universal control strategy, 84–85
COPA (cancer outlier profile analysis), 154
Copy-number variation (CNV), 131–138, 184, 187
    biallelic *versus* multiallelic, 134, 135
    in cancer, 145–146
    definition, 131–138
    detection methods, advances in, 137–138
    dynamic nature of, 132
    evolutionary drive responsible for, 133
    gene expression, impact of, 133
    genome-wide association studies (GWAS), 135–137
    mechanisms of structural variation, 133
    microassay analysis of, 131–132, 137–138, 184, 187

Copy-number variation (*continued*)
mutation rate, 134
population characteristics, 134
search for variants, 85
single-nucleotide polymorphisms (SNPs) compared, 134
whole-genome catalog in humans, 131–132
Correction constant, for nonnormality, 43
COSMIC (Catalogue of Somatic Mutations in Cancer), 148, 152
Covariance, in trait among family members, 38
Covariates
effect of selection on variance component analysis, 40–41
measured environmental factors used as, 40
CPG (conditional on parental genotypes), 127
CpG islands, 150, 151
CPM (combinatorial partitioning method), 128
Crick, Francis H.C., 2
Crohn's disease, 135
Cross-sectional case–control studies, 112
CYP2B6*6 gene, 176
Cystic fibrosis transmembrane conductance regulator *(CFTR)* gene, 173–174

D

Darwin, Charles, 1
Data analysis methods, in GWAS, 81–82
Database of Genotypes and Phenotypes (dbGAP), 137
Databases. *See specific applications; specific database names*
Data security, association study, 27
DataSHaPER, 34
Data storage, electronic, 31
dbGAP (Database of Genotypes and Phenotypes), 137
dbSNP (Single Nucleotide Polymorphism Database), 148, 176
Dependence, 53
Diabetes Genetics Initiative, 96, 101
Diabetes mellitus
gene associations with type 1, 85
gene–environment interaction in, 107, 109, 110
in Pima Native Americans, 107, 120
transient neonatal (TNDM), 127
transmission disequilibrium test (TDT), 122
Diabetes Prevention Program, 110
Diogenes Project, 114
Discrete data
binomial distribution, 10
uniform distribution, 8
Discrete traits, linkage analysis of, 17–25
Discrete uniform distribution, 8

Disease risk, interaction between genes in, 67
Distribution
binomial, 10
chi-square, 9
correction constant for nonnormality, 43
discrete uniform, 8
kurtosis of, 43
leptokurtosis of, 43
multinomial, 10
normal, 8–9
observed, 7, 8
skewness of, 43
t-distribution, 43
theoretical, 7, 8
DNA methylation, 149–151
DNA Methylation Database (MethDB), 151
DNA microarray genotyping platforms. *See* Microarray
DNA structure, 2
Dominance variance, 37
DQB1 gene, 178
DRD2 A1 allele, 120

E

Effect modifier, 32
EGFR Mutations Database, 152
EIGENSTRAT, 81
EIGENSTRAT/EIGENSOFT, 77
Electronic data storage, 31
Elston and Stewart algorithm, 19
E-M algorithm. *See* Expectation–maximization (E-M) algorithm
ENCODE projects, 91
End sequence pair (ESP) approaches, for copy-number variant detection, 132
Ensembl Perl API, 198
Environment. *See* Gene–environment interaction
gene–environment interaction, 28, 105–117
candidate gene study, 115
case-only study, 112
cross-sectional case–control study, 112
in diabetes, 107, 109, 110
epidemiological definition, 106
evidence for, in common disease, 106–107
family-based association tests, 128
genetic perspective, 107, 108
genome-wide study, 115–116
large-scale studies, 113, 114
meta-analysis, 115
in obesity, 106, 107
overview, 105, 116–117
power in studies, 113, 114
prospective cohort study, 112–113
public health perspective, 108, 109

reasons for study of, 109–111
replication of study results, 115
sample size in studies, 113–115
study design, 111–113
in variance component analysis framework, 45–46
influence on splicing, 176–178
using measured environmental factors as covariates, 40
Environmental variance, 37, 38, 39–40, 44, 45–46
EPIC (European Prospective Investigation into Cancer and Nutrition), 34, 114
Epidemiology
association studies
conduct, 27–31
interpretation, 31–33
design choices
case–control study, 28
cohort study, 28
gene–environment interaction, 106
large-scale data resources, 33–35
Epigenomic alterations in cancer, 149–151
databases, 151
methods for detecting, 150–151
types of changes, 150
Epistasis
as source of confounding, 33
in variance component analysis framework, 46
Epistasis analysis, application on GWAS, 200, 201
Epistatic variance, 37, 46
eQTL (expression quantitative trait locus) studies, 198, 199, 202
eQTL web browser, 198
Error
genotyping, 67, 68–69
random, 30
systematic, 30, 77
ESE-finder, 179
ESP (end sequence pair) approaches, for copy-number variant detection, 132
European Prospective Investigation into Cancer and Nutrition (EPIC), 34, 114
Evolutionary shadow, 73–74
Exonic splicing enhancers (ESEs), 168, 173, 174, 179
Exonic splicing silencers (ESSs), 168, 173, 174
Exons. *See* Splicing
Exon skipping, 170–175
Expectation–maximization (E-M) algorithm, 7, 67, 82–84, 127
Exposure, changes in during association study, 31
Expression analysis, 151, 154
Expression quantitative trait locus (eQTL) studies, 198, 199, 202
Extended tagging, approach for prediction of untyped loci, 92

F

False discovery rate (FDR), 12, 52–55
description, 52–53
example, 54
modification for dependent tests, 54
parametric methods, 55
π_0 estimation, 53–54
problems with methods, 55
q value, 53
weighted, 54
False negative
description, 11
in gene–environment interaction study, 117
from methods that control family-wise error rate, 51
power and probability, 14
in underpowered studies, 30
False positive
Bonferroni method of correction applied to *p* values, 50–51, 70
causes of in association study, 69–70
description, 11
in gene–environment interaction study, 116
in genome-wide association studies, 14
increase with population stratification, 69
kurtosis and, 43
multiple-testing correction for, 50–55
Bayesian analysis, 51–52
false discovery rate, 52–55
family-wise error rate, control of, 50–51
publication bias, 30
Family-Based Association Tests (FBAT) program, 125
Family-based genetic association tests, 119–128
as alternative to case–control study, 121–122
applications
tests for association and linkage in general nuclear families and pedigrees, 125
tests for association with continuous quantitative traits, 126
tests for gene–gene and gene–environment interactions, 128
tests for late-onset diseases, 124–125
tests for parent-of-origin and maternal effects, 127
tests for X-linked association, 126–127
tests using haplotypes, 126
rationale for studies, 119–121
transmission disequilibrium test (TDT), 122–125
sib TDT (S-TDT), 124–125
as test of association and linkage, 122–124
as test of linkage, 122
XRC-TDT, 127
XS-TDT, 127

Family-wise error rate (FWER), control of, 50–51
    Bonferroni method of correction, 50–51
    permutation test, 51
    problems with methods, 51
FAS-ESS, 178
FASTLINK package, 19
FatiGO, 198
FBAT (Family-Based Association Tests) program, 125
FDR. *See* False discovery rate (FDR)
Feedback of results, to study participants, 27
Fiers, Walter, 2
Fisher, Ronald, 2
Fluorescence in situ hybridization (FISH), 145, 147
Franklin, Rosalind, 2
FTO gene, 109, 111
Fusion proteins, from translocations, 146
FWER. *See* Family-wise error rate (FWER), control of

G

GABRB1 gene, 200–201
Gain-of-function mutations, 147
Galaxy, 198
γ-aminobutyric acid receptor (GABRA) signaling pathway, 200
GEEs (generalized estimating equations), 128
GEI (Genes, Environment, and Health Initiative), 114
Geiringer, Hilda, 89
Gene-based mutational burden analysis, 136
Gene–environment interaction, 28, 105–117
    candidate gene study, 115
    in diabetes, 107, 109, 110
    epidemiological definition, 106
    evidence for, in common disease, 106–107
    family-based association tests, 128
    genome-wide study, 115–116
    large-scale studies, 113, 114
    meta-analysis, 115
    in obesity, 106, 107
    overview, 105, 116–117
    perspectives on
        genetic, 107, 108
        public health, 108, 109
    power in studies, 113, 114
    reasons for study of, 109–111
    replication of study results, 115
    sample size in studies, 113–115
    study design, 111–113
        case-only study, 112
        cross-sectional case–control study, 112
        prospective cohort study, 112–113
    in variance component analysis framework, 45–46

Gene expression
    copy-number variation, impact of, 133
    gene set analysis (GSA), 197–198
Gene Expression Omnibus (GEO), 152, 154
Gene–gene interaction
    family-based association tests, 128
    in variance component analysis framework, 46
GENEHUNTER program, 22
Gene Ontology, 198
Generalized estimating equations (GEEs), 128
Genes, Environment, and Health Initiative (GEI), 114
Gene set analysis tools, 198
Gene set enrichment analysis (GSEA) algorithm, 198
Gene signature study, 151
Genetic association study. *See* Association study
Genetic code, 2
Genetic distance, 18
Genetic heterogeneity
    description, 24
    effect on mapping complex diseases, 24–25
Genetic imprinting, 127
Genetic map
    genetic heterogeneity in mapping complex diseases, 24–25
    linkage map creation, 21–22
    multipoint analysis, 22
Genetic power calculator, 58
Genetics
    central dogma of, 2
    history of modern, 1–3
    origin of term, 2
    why it matters, 1
GENEVA Gene Environment Association Studies, 114
GENEVAR, 198
GenGen Package, 198
Gene set analysis (GSA)
    application of, 199, 200–201
    challenges and perspectives, 202
    gene expression analysis, 197–198
    goal of, 196
    key steps, 197
    limitations of, 197
    mapping genetic associations to genes, 198–199
    tools for exploiting GWAS, 198
Genome Structural Variation Consortium, 132
Genome-wide association studies (GWAS)
    chips and platforms used for, 76–77
    common disease/common variant hypothesis, 73–74, 81, 85, 134–135
    copy-number variation (CNV), 85, 135–137
    data analysis methods, 81–82

disease associations uncovered by GWAS
  age-related macular degeneration, 84
  type 1 diabetes, 85
epistasis analysis, 200, 201
eQTL analysis, 198, 199, 202
follow-up studies, 188
gene–environment interaction, 115–116
gene set analysis (GSA), 196–202
haplotype analysis, 82–84
  E-M algorithm, 82–84
  haplotype blocks, 84
  identification of haplotypes, 82
high-throughput genotyping, 187–188
limitations of approach, 135
linkage disequilibrium, 74–76
meta-analysis, 95–101
negative study, 85
network analysis, 196–197, 199–202
overview, 83
pooling of samples, 187
power of, 14
preparation for analysis, 77–81
  data handling, 77, 78
  population stratification, 79–81
  quality handling, 77, 78
problems in analysis
  missing genotypes, 68–69, 95
  population stratification, 69
quality control, 189–190
rare variants, search for, 85–86
tag SNPs, 74, 134–135
tumor-associated SNPs, 149
Genome-wide mutational burden analysis, 136
Genomic control, 69, 80–81
Genomic disorders, 135
Genomic inflation factor λ, 97
Genomic rearrangements, finding, 146–147
Genomics, 2. *See also* Oncogenomics
Genotype frequency
  Hardy–Weinberg equilibrium, 6–7, 67–68
  sib transmission disequilibrium test (S-TDT), 124–125
Genotyping
  laboratory methods for high-throughput, 183–191
  quality, problems with, 67–69
Genotyping error, 79
GEO (Gene Expression Omnibus), 152, 154
Glioblastoma, 156
Goodness of fit, 9, 62, 67–68
GSA. *See* Gene set analysis (GSA)
GSEA (gene set enrichment analysis) algorithm, 198
GSEA-SNP, 198
GWAS. *See* Genome-wide association studies (GWAS)

H

Haldane, J.B.S., 2, 22
Haldane function, 22
Haplotype
  analysis of, 82–84
  ancestral mutation and, 75–76
  case–control studies, 28
  definition, 6
  E-M (expectation–maximization) algorithm, 82–84
  family-based association tests, 126
  identifying, 82
  linkage equilibrium, 6
  mutation and formation of, 75–76
  use as unit of analysis, 67
Haplotype association tests, 126
Haplotype-based HRR test (HHRR), 121–122
Haplotype blocks, 84
Haplotype relative risk (HRR), 121–122
Haploview software, 76, 77, 126, 185
HapMap
  as design study resource, 66
  International HapMap Project, 66, 73, 149
  mapping linkage disequilibrium using, 90–91
  number of SNPs, 74, 184
  populations sampled by, 91, 96
  use in tagging SNPs, 185
Hardy–Weinberg equilibrium
  deviation from as measure of genotyping error, 79
  probability theory and, 6–7
  testing for departure, 67–68
HBAT program, 126
Hepatocellular carcinoma, 156
Heredity, particle theory of, 2
Heritability
  choosing data transformations that maximize, 44
  definition, 38
  estimating, 38–39
Heterogeneous nuclear ribonucleoprotein (hnRNP) proteins, 168, 172–174, 176, 177
HGMD (Human Genome Mutation Database), 170
HHRR (haplotype-based HRR test), 121–122
Hidden Markov Models (HMM), for imputation of untyped loci, 92
High-throughput genotyping, 183–191
  next-generation sequencing, 190–191
  overview, 183–184
  parallel methods, 183, 187
  pooling of samples, 187
  quality control of SNP data, 188–190
  reasons for SNP use, 184–185
  serial methods, 183, 185, 186

High-throughput genotyping (*continued*)
  types of studies
    candidate genes, 185–186
    genome-wide association studies (GWAS), 187–188
    linkage, 186–187
Histone modifications, 149–151
HMGCR gene, 177
HMM (Hidden Markov Models), for imputation of untyped loci, 92
hnRNP. *See* Heterogeneous nuclear ribonucleoprotein (hnRNP) proteins
Hospital biobanking, 34
Hotelling's T² test, 82
Household matrix, 40
HRR (haplotype relative risk), 121–122
Human Epigenome Project, 150
Human genome, linkage map of, 21–22
Human Genome Mapping Project, 73
Human Genome Mutation Database (HGMD), 170
Human Genome Project, 2
Human leukocyte antigen (HLA) genes, 201
Hypothesis
  alternative, 11, 12, 14, 49
  null, 11–14, 49–50
  research, 11
Hypothesis testing, 11–14
  common disease/common variant hypothesis, 73–74, 81, 85
  description, 49–50
  examples, 12–13
    ABO blood type and ulcer susceptibility, 12–13
    female births, 12, 13
  genome-wide association studies, 73–74, 81, 85
  likelihood ratio (LR), 13
  likelihood ratio test (LRT), 13
  logarithm of the odds (LOD), 13
  multiple testing, 49–55
  power, 14
  p values, 11–12

I

IARC TP53 Mutation Database, 152
Identity by descent (IBD), 23, 42, 46, 122
Identity-by-state (IBS) measures, 79
Identity matrix, 39
Illumina chips, 76, 186, 187
Imputation
  accuracy and quality, 96–97, 98, 100
  chi-square test correction, 97–99
  description, 91–92
    incorporating uncertainty into meta-analysis, 99–101
    meta-analysis of genome-wide association studies, 95–101
    software tools, 96
IMPUTE program, 96
Independence
  definition, 6
  linkage equilibrium as example of, 6
Independent assortment, Mendel's second law of, 17
Inflation factor, 80
Inflection points, 8
Informed consent, 27
In situ hybridization techniques, 147
InterAct, 114
Interaction, analysis of, 67
International Cancer Genome Consortium, 153, 157
International HapMap Project, 66, 73, 149. *See also* HapMap
IntOGen (Integrative OncoGenomics), 153, 157–158
Intronic splicing enhancers (ISEs), 168, 173, 177
Intronic splicing silencers (ISSs), 168
Introns. *See also* Splicing
  splicing-regulatory elements (SREs) in, 165
  U2 type, 166
Invader assay, 186
IRGM gene, 135
Iterative gene counting algorithm, 7

K

Kadoorie Study of Chronic Disease in China (KSCDC), 34
KEGG, 198
Kinship coefficient, 39
Kruglyak, Leonid, 90
Kurtosis, 43

L

Lab functional neurogenomics, 198
Large-scale data resources (LSDRs)
  biorepositories (biobanks), 33–34
  description, 33
  integrated fungible studies, 35
  remote study, 34–35
Late-onset diseases, tests for, 124–125
LCRs (low-copy repeats), 133, 135
LD. *See* Linkage disequilibrium (LD)
LDL-C (low-density lipoprotein cholesterol), 177
LDLr gene, 177

Leptokurtosis, 43
Liability threshold model, 41
Lifestyle, effect of. *See* Gene–environment
        interaction
LightTyper, 185, 186
Likelihood
    definition, 10
    maximum estimation, 10–11
Likelihood ratio (LR), 13
Likelihood ratio test (LRT), 13, 44
Linkage
    description, 17
    fine-mapping by high-throughput SNP genotyp-
        ing, 186–187
    phase, 18
    transmission disequilibrium test (TDT)
        association in presence of linkage, 122–124
        linkage in presence of association, 122
Linkage analysis
    accommodating variations, 21
    description, 17–18
    of discrete traits, 17–25
    of a disease phenotype, 19–21
    genetic heterogeneity in mapping complex dis-
        eases, 24–25
    goal of studies, 18
    identity by descent (IBD), use of, 23
    linkage map creation, 21–22
    LOD score interpretation, 23–24
    misspecification of genetic model, effects of,
        22–23
    nonparametric, 23
    parametric, 20, 23
    penetrance, 20–21, 22
    utility in an era of genome-wide association stud-
        ies, 25
    in variance component method, 42
Linkage disequilibrium (LD)
    as basis of genome-wide association studies,
        74–76
    between copy-number variants (CNVs) and SNP
        haplotypes, 134
    correlation between SNP alleles measured
        by, 65
    description, 6, 74
    generation of, 75–76
    in genetic association testing, 65–66
    history of concept, 89
    identification of potentially functional quantita-
        tive trait locus (QTL) variants, 46–47
    imputation, 91–92, 95–96
    mapping using International HapMap Project,
        90–91
    patterns in human genome, 90–91
    statistic D', 89–90
    statistic r^2, 89–90

use in transmission disequilibrium test, 122
    visualizing using Haploview software, 76
Linkage equilibrium, as example of indepen-
        dence, 6
LINKAGE package, 19
Liver carcinoma, 156
Locus-specific recurrence risk ratio, 23
LODLINK package, 19
Logarithm of the odds (LOD) score
    correction constant for nonnormality, derivation
        of, 43
    description, 13, 18–19
    how to interpret, 23–24
    human linkage map creation, 21–22
    linkage analysis
        disease phenotype, 19–20
        sensitivity to model misspecification,
            22–23
    as measure of effect size in case–control
        study, 49
Logistic regression, 92, 100–101
    conditional (CLR), 128
    multiple, 66
Log-likelihood, 10–11
LOKI program, 22
Low-copy repeats (LCRs), 133, 135
Low-density lipoprotein cholesterol (LDL-C), 177
LSDRs. *See* Large-scale data resources (LSDRs)
Lymphoma, 146, 156

M

Machine-learning methods, 128
MACH program, 97
MAPH (multiplex amplifiable probe hybridization),
        copy-number variant (CNV) study
        using, 137
MAPT gene, 171–172
Markers, genetic
    analysis of multiple, 66
    for genetic map creation, 21–22
    selection of, 28–29
    sliding window of, 22
Maternal effects, tests of, 127
Maternal–fetal incompatibility test, 127
Mating, assortative, 68
Maximum Entropy, 178
Maximum likelihood estimation (MLE)
    description, 10–11
    female birth estimation (example), 11
    of recombination fraction, 17, 18, 20
MCAD (medium-chain acyl-CoA dehydrogenase)
        gene, 173, 174, 178
MDR (multifactor dimensionality reduction),
        67, 128

Measured genotype test, 44–45
Measurement precision, in gene–environment
    interaction studies, 113, 114
Medium-chain acyl-CoA dehydrogenase *(MCAD)*
    gene, 173, 174, 178
MeInfoText, 151
Melanoma, 156
Melting curve, 185, 186
Mendel, Gregor, 1–2
MENDEL package, 19
MERLIN program, 22, 23
Meta-analysis
    comparative genomic hybridization (CGH), 146
    description, 33
    fixed-effects, 101
    gene–environment interaction, 115
    of genome-wide association studies, 95–101
    imputation-based, 95–101
    noncertainty parameter computation for a given
        disease model, 101–102
    random-effects, 101
MethDB (DNA Methylation Database), 151
MethyCancer, 151, 152
mfold, 178
Microarray. *See also specific chips/platforms*
    array (aCGH), 131–132, 137, 145, 156
    assessing DNA methylation, 150–151
    cancer research, use in, 144–145
    ChIP-on-chip, 150, 151
    copy-number variation study, 131–132, 137–138,
        184, 187
    custom-designed genotyping, 187
    genome-wide association studies, use for, 76–77
    genome-wide chips/arrays, 187–188
    genomic regions of interest, use to capture, 191
    SNP genotyping, 185
Microhomology-mediated break-induced replication
    model (MMBIR), 133
Micro RNA (miRNA), 151, 154
Microsatellites, linkage studies using, 21
Migration studies, 107
Million Women Study, 34
Missing genotypes, 79
Mitelman Database of Chromosome Aberrations in
    Cancer, 146, 152
MLPA (multiplex ligation-dependent probe
        amplification), copy-number variant
        (CNV) study using, 137
MMBIR (microhomology-mediated break-induced
        replication model), 133
MOD (LOD score maximized), 23
Module maps, 154
Molecular concept maps, 154
Morgan, Thomas Hunt, 2
MQLS (more powerful quasi-likelihood score test),
    125

MRV/CD (multiple rare variant model of
        common disease) hypothesis,
        135, 136
MSigDB, 198
Mullis, Kary, 2
Multifactor dimensional reduction (MDR),
        67, 128
Multinomial distribution, 10
Multiple rare variant model of common disease
        (MRV/CD) hypothesis, 135, 136
Multiple testing, 49–55
    correction methods
        Bayesian analysis, 51–52
        control of family-wise error rate (FWER),
            50–51
            Bonferroni method of correction,
                50–51
            permutation test, 51
            problems with methods, 51
        false discovery rate (FDR), 52–55
            description, 52–53
            example, 54
            modification for dependent tests, 54
            parametric methods, 55
            π_0 estimation, 53–54
            problems with methods, 55
            q value, 53
            weighted, 54
    false positives, 50
Multiplex amplifiable probe hybridization (MAPH),
        copy-number variant (CNV) study
        using, 137
Multiplex ligation-dependent probe amplification
        (MLPA), copy-number variant
        (CNV) study using, 137
Multipoint analysis, 22
Multivariate analysis, 44
Mutation
    in cancer, 147–149
    copy-number variation, 133, 134, 137
    driver *versus* passenger, 148
    gain-of-function, 147
    mechanisms of structural, 133
    SNP formation and, 75–76
    splicing disregulation, 170–175
Mutational burden analysis, 136
MYC gene, 146

N

NAHR (nonallelic homologous rearrangement),
        133, 135
National Human Genome Research Institute
        (NHGRI) structural variation
        initiative, 132

NCBI/NCI's Cancer Chromosomes Database, 146, 152
NCI-CGAP, 153
NCP. *See* Noncentrality parameter (NCP)
Network analysis
    application of, 200, 201
    challenges and perspectives, 202
    combining GWAS and eQTL analysis with, 202
    goals of, 196, 197
    key steps, 197
    limitations of, 197
Network Edge Orienting (NEO) software, 202
Neural networks, for splice-site strength | prediction, 178
Neurofibromin 1 *(NF1)* gene, 172
NF1 gene, 172
NHEJ (nonhomologous end joining), 133
NHGRI (National Human Genome Research Institute) structural variation initiative, 132
NimbleGen arrays, 76, 191
Nonallelic homologous rearrangement (NAHR), 133, 135
Noncentrality parameter (NCP)
    computing for a given disease model, 100, 101–102
    description, 90
    relationship to power, 58, 90
Nonhomologous end joining (NHEJ), 133
Nonnormality, dealing with, 43–44
Normal distribution, 8–9
Null hypothesis
    description, 11
    effect size, 49
    examples, 12–13
    power and, 14
    p values and, 11–12, 49–50

O

OAS1 gene, 176
Obesity
    altered splicing as additive risk for, 177
    gene–environment interaction in, 106
Observed distribution, 7, 8
Odds ratio (OR)
    calculating confidence intervals on, 64
    for common variants, 86
    for rare variants, 86
    in rheumatoid arthritis gene association study, 64–65
Oligoadenylate synthetase enzyme, 176
Oncogenomics, 143–158
    copy number changes, 145–146
    epigenomic alterations in cancer, 149–151

        databases, 151
        methods for detecting, 150–151
        types of changes, 150
    genetic basis of cancer, 143–145
    genomic rearrangements, finding, 146–147
    integration of oncogenomic data types, 155–158
        Cancer Genome Atlas, 157
        Cancer Genome Project, 157
        International Cancer Genome Consortium, 157
        IntOGen, 157–158
        study approaches, 156–157
    prioritization of candidate cancer genes, 154–155
        cancer gene census, 154–155
        computational, 155
    resources and databases, 152–153
    somatic mutations
        databases, 148
        difficulties in predicting candidate cancer genes, 148–149
        mutational patterns, 147
        sequencing and mutational screens, 148
    transcriptomic changes in tumors, 151–154
        databases, 154
        expression analysis, 151, 154
        methods for detecting, 151, 154
        module maps, 154
        molecular concept maps, 154
    variants in cancer, common, 149
Oncomie, 153, 154
1000 Genomes Project, 101, 132
"Out of Africa" hypothesis for human evolution, 91

P

Paired-end mapping, 137, 147
Pancreatic cancer, 156
Parameter, definition of, 7
Parametric methods, for false discovery rate (FDR), 55
Parent-of-origin, tests of, 127
Parent-of-origin effects (POEs), 127
Parkinson's disease, 201
Pathway definition resources, 198
PBAT, 77
p53 Database, 148
PDT (pedigree disequilibrium test)
    multifactor dimensionality reduction (MDR), 128
    Tsp test, 124
    XPDT test, 127
Pearson, Karl, 2

Pearson's chi-square test, 62–63, 67–68
Penetrance
    confounding in linkage analysis, 20–21, 22
    rare variants and, 86
Perlegen, 76
Permutation testing
    for multiple-testing correction, 51
    use in genome-wide association study
        (GWAS), 82
PESX, 178
Pfam protein domains, 155
Pharmacogenetics, gene set analysis and, 200
Phase of linkage, 18
Phenotypic correlation between two traits, 44
Philadelphia chromosome, 146
Phosphorylation/dephosphorylation cycles,
        splicing and, 177
Pima Native Americans, diabetes mellitus in,
        107, 120
Plate effect, 190
Pleiotropy, 44
PLINK program, 77, 79, 81, 82, 85, 96, 198
POEs (parent-of-origin effects), 127
Point estimate, 30
Point mutations, rate of, 134
Polypyrimidine tract (PPT) sequence, 166, 173
Pontocerebellar hypoplasia, 172
Population admixture, 120
Population correlation coefficient, 8
Population covariance, 7
Population mean, 7
Population standard deviation, 7
Population stratification, 32–33, 45
    in case–control study, 69
    correction for, 80–81
        genomic control, 80–81
        principal component analysis, 81
    departure from Hardy–Weinberg equilibrium
        and, 68
    description, 79–80
    detection and correction, 69
    family-based association studies, 119–120
    hidden, 119–120
    Simpson's paradox, 79
Population variance, 7
Positional cloning, 144
Postgenomics era, 2–3
Power, statistical
    in association studies, 30
        improving using multiple markers, 66
        in quantitative trait study, 65
        replication study lacking power, 70
    calculation
        example, 56–57
        in identification of novel risk variants, 59
        in replication of a previous finding, 58

        sample size relationship, 57–58
        steps in, 55–56
        using genetic power calculator, 58
    description, 14, 49, 55
    in gene–environment interaction study, 113, 114
    increase with sample size, 57–58
    of liability threshold approach, 41
    linkage, 42
    noncentrality parameter (NCP), 58, 90
    quantitative trait association study, 65
    sample size, effect of, 14, 30
    in variance component analysis, 41, 42, 43
PPT (polypyrimidine tract) sequence, 166, 173
Pre-mRNA splicing. *See* Splicing
Principal components analysis
    detection and correction of population stratifica-
        tion, 69
    use in correction for population stratification,
        81
Probability density, 7, 10
Probability distribution, 7
Probability theory, 5–7
    conditional probability, 6
    definitions, 5–6
    frequency of ABO blood type by affection
        status (example), 7
    Hardy–Weinberg equilibrium (example), 6–7
    independence, 6
    linkage equilibrium (example), 6
    rules of probability, 6
Progenetix Database, 146, 152
Prospective cohort study, 112–113
Prostate cancer, 149
Proto-oncogenes, 144, 155
Proxy-haplotype test, 82
Pseudoexons, 168, 172
Pseudo-splice sites, 167, 172
Publication bias, 30, 117, 185
Public health perspective, on gene–environment
        interaction, 109
Public Population Project in Genomics (P3G), 34
PubMed, 151
p value
    Bonferroni method of correction, 50–51, 70
    conversion to *Z* scores, 99–100
    correction by permutation testing, 51
    definition, 49–50
    meta-analysis *Z* scores converted back to, 100
Pyrosequencing, 185, 186

Q

QTDT program, 126
Quality control
    for association study, 31

genome-wide association study (GWAS), 77–79
high-throughput SNP genotyping, 188–190
Quantile–quantile (Q-Q) plot, 80
Quantitative polymerase chain reaction (PCR),
 copy-number variant (CNV) study
 using, 137
Quantitative trait locus (QTL)
 detection by linkage analysis, 42
 identifying potentially functional variants, 46–47
Quantitative trait nucleotide (QTN), 46–47
Quantitative traits
 association analysis for, 44–45, 65
 family-based tests, 126
 liability threshold model, 41
 transmission disequilibrium tests, 45
 variance component analysis and, 37–48
q value, 53

R

Random error, 30
Random variable, definition of, 7
Rare variants, search for, 85–86
RARS2 gene, 172
Recombination
 hidden, 90
 hot spots, 90
 linkage and, 17–19
 punctate in sperm, 91
Recombination fraction θ
 confounding with penetrance, 20–21, 22
 description, 17–18
Regression
 conditional logistic regression (CLR), 128
 with interaction terms, 67
 linear, 100–101
 logistic, 66, 81, 92, 100–101
 weighted, 67
Relatedness, statistical measure of, 79
Relative risk, haplotype (HRR), 121
Remote study, 34–35
Replication of association studies
 gene–environment interaction, 115
 pitfalls and problems of, 69–70
Representational oligonucleotide microarray analysis
 (ROMA), 137
RESCUE-ESE, 178
Research hypothesis, 11
Retinoblastoma, 156
Retrieval, 31
Rheumatoid arthritis (RA), gene association study
 example, 64–65
RNA polymerase II, transcription elongation rate
 and, 170
RNA-recognition motifs (RRMs), 168

ROMA (representational oligonucleotide microarray analysis), 137

S

SABP (schizoaffective bipolar) disorder, 201
SAGE (serial analysis of gene expression), 156
Sample correlation coefficient, 8
Sample covariance, 8
Sample selection, for association study, 29–30
Sample size
 effective, 100
 in gene–environment interaction studies,
  113–115
 relationship of power, 57–58
Sample standard deviation, 8
Sample variance, 7
Sampling mean, 9
San Antonio Family Heart Study (SAFHS), 37–38
Sanger, Fred, 2
SBCAD gene, 170–171, 178
Schizoaffective bipolar (SABP) disorder, 201
Schizophrenia, role of copy-number variants
 in, 136, 137
Scn8a gene, 176
Scnm1 gene, 176
Segmental duplications (SDs), 132, 133
Selection, and departure from Hardy–Weinberg
 equilibrium, 68
SELEX (Systematic Evolution of Ligands by
 Exponential Enrichment), 179
Sequencing
 chromosomal aberrations, detection of, 147
 copy-number variant identification, 137
 exon resequencing of human tumors, 145
 focused on signaling pathway, 148
 massively parallel, 190
 next-generation high-throughput, 190–191
 "$1000 genome" target, 185, 190
 pyrosequencing, 185, 186
 SNP detection by, 184
Sequenom MassARRAY, 185, 186–187, 188
Serial analysis of gene expression (SAGE), 156
Serine/arginine-rich (SR) proteins, 168, 177, 179
Set of tools for gene ontology, 198
Set theory, basics of, 5
SF1 (splicing factor 1), 166
Shared environmental effects, 39–40
Shared environmental variance, 37
Short/branched-chain acyl-CoA dehydrogenase
 deficiency, 170, 178
Sib TDT (S-TDT), 124–125
Sidak correction, 50–51
Simpson's paradox, 79
SimWalk program, 22

Single-base extension, 185, 186
Single major locus (SML) model, 19–23
Single Nucleotide Polymorphism Database (dbSNP),
    148, 176
Single-nucleotide polymorphisms (SNPs)
    conditional splicing defects, 165, 176
    copy-number variation (CNV) compared, 134
    density of, 21
    detection by direct sequencing, 184
    DNA microarray chips, 76–77
    formation of SNPs, 75–76
    genetic association studies and, 61–70
    genome-wide association studies and, 73–79,
        81–82, 84
    HapMap project and, 90–91
    identification of potentially functional QTL
        variants, 46–47
    imputation of untyped loci, 91–92, 95–101
    laboratory methods for high-throughput genotyp-
        ing, 183–191
    linkage disequilibrium and, 65–66, 74–75,
        91–92
    linkage map construction using, 21
    meta-analysis of genome-wide association stud-
        ies, 95–101
    number of, 90, 176, 184
    quality control of data, 188–190
    real *versus* sequencing artifacts, 184
    tag, 74, 134–135
Skewness, 43
SKY (spectral karyotyping), 147
Small nuclear ribonucleoprotein (snRNP), 166,
    169–170, 176
Small nuclear RNAs (snRNAs), 166, 170–172,
    177–178
SML (single major locus) model, 19–23
SNAP, 198
SNaPshot Multiplex System, 186
SNPlex Genotyping System, 186
SNPs. *See* Single-nucleotide polymorphisms (SNPs)
SNPTEST, 77
Sodium channel modifier 1 *(Scnm1)* gene, 176
SOLAR program, 125, 126
Somatic mutations
    databases, 148
    difficulties in predicting candidate cancer genes,
        148–149
    mutational patterns, 147
    sequencing and mutational screens, 148
Spectral karyotyping (SKY), 147
Spliceosome
    common genetic variation in components
        of, 176
    components of, 166
    splicing mechanisms, 166–167
SpliceRack, 178

Splicing, 165–179
    alternative, 165, 168, 173
    conditional splicing alleles, 165, 176
    degeneracy of splice-site sequences, 167
    environmental influence, 176–178
    missplicing as disease mechanism, 165–166,
        170–175
    mutations
        splice-site, 170–172
        in splicing-regulatory elements (SREs),
            173–175
    process of, 166–170
        enhancers and silencers, actions of, 167–170
        steps in, 166–167
        transesterification reactions, 166
    pseudo-splice sites, 167
    regulation by splicing-regulatory elements
        (SREs), 167–170
    in silico tools for evaluating splicing effects,
        178–179
Splicing factor 1 (SF1), 166
Splicing-regulatory elements (SREs)
    functions of, 167–170
    in introns, 165
    mutations, 173–175
    in silico tools for analysis of, 178–179
    types of, 168
SR (serine/arginine-rich) proteins, 168, 177, 179
Stability genes, 144
Stanford Microarray Database, 154
Statistics, 5–14
    distributions
        binomial, 10
        chi-square, 9
        discrete uniform, 8
        multinomial, 10
        normal, 8–9
        observed, 7, 8
        theoretical, 7, 8
    hypothesis testing, 11–14
        examples, 12–13
        likelihood ratio (LR), 13
        likelihood ratio test (LRT), 13
        logarithm of the odds (LOD), 13
        power, 14
        p values, 11–12
    maximum likelihood estimation, 10–11
    parameters, 7
    probability theory, 5–7
        conditional probability, 6
        definitions, 5–6
        frequency of ABO blood type by affection
            status (example), 7
        Hardy–Weinberg equilibrium (example),
            6–7
        independence, 6

linkage equilibrium (example), 6
rules of probability, 6
set theory, basics of, 5
variables, 7
Supernormal controls, 62
Sutton, Walter, 2
Systematic error, 30, 77
Systematic Evolution of Ligands by Exponential
Enrichment (SELEX), 179

T

Tagger program, 185
Tag SNPs (single-nucleotide polymorphisms), 74,
134–135
Tamal, 198
TaqMan SNP Genotyping Assays, 185, 186
Tardive dyskinesia (TD), 200
t-distribution, 43
TDT. *See* Transmission disequilibrium test (TDT)
Temozolomide, 156
Temperature, effect on splicing, 177–178
Theoretical distribution, 7, 8
Thyrotropin β-subunit *(TSHB)* gene, 172
Tissue banks, 34
Transcription factor 7-like 2 gene *(TCF7L2),* 109,
110
Transcriptomic changes in tumors, 151–154
databases, 154
expression analysis, 151, 154
methods for detecting, 151, 154
module maps, 154
molecular concept maps, 154
Transesterification reaction, in splicing mechanism,
166
Transformation of cells, 143–144
Transformations, use to normalize data, 43–44
Transient neonatal diabetes mellitus (TNDM),
127
Translocation, chromosomal, 146
Transmission disequilibrium test (TDT), 69, 122–
125
illustration, 121
sib TDT (S-TDT), 124–125
as test of association and linkage, 122–124
as test of linkage, 122
variance component network, 45
XRC-TDT, 127
XS-TDT, 127
TRANSMIT program, 125, 126
TSHB (thyrotropin β-subunit) gene, 172
Tsp test, 124
Tumor suppressor genes, 144, 147, 155
Twin studies
gene–environment interaction and, 107

shared environmental effects, 39
Type 1 diabetes, gene associations with, 85

U

UCSC Cancer Genomics Browser, 153
UK Biobank, 33–34
Ulcers, ABO blood type by affection status, 7
Ulcers, blood type association with, 7, 12–13
Uniform distribution, 8
Unique environmental variance, 37
Unordered genotype, 7
UNPHASED test, 125, 126
Unshared environmental variance, 37, 38

V

Variance
additive, 37, 38–39
defined, 7
dominance, 37
environmental, 37, 38, 39–40, 45–46
epistatic, 37, 46
genetic, 37, 38
locus-specific QTL, 42
variance component methods for analysis of
complex phenotypes, 37–48
Variance component methods, 37–48
ascertainment scheme, effect of, 42–43
association analysis for quantitative traits, 44–45
covariate selection effect on, 40–41
description, 37
gene–environment interactions, 45–46
gene–gene interactions, 46
heritability estimation, 37–38
identification of potentially functional
variants, 46–47
liability threshold model, 41
linkage analysis, 42
measured environmental factors as covariates, use
of, 40
multivariate analysis, 44
nonnormal distributions, dealing with, 43–44
pleiotropy, test for, 44
shared environmental effects, dealing with, 39–40
Venter, Craig, 190
VITESSE package, 19

W

Watson, James D., 2, 190
Wellcome Trust Case Control Consortium
(WTCCC), 33, 84, 101, 200

Whole-genome shotgun (WGS) sequence data, mining for copy-number variants, 132
Wilkins, Maurice, 2
Woolf method, of confidence interval calculations, 64

X

XAPL test, 127
X-linked association, tests for, 126–127
XLRT test, 127
XPDT test, 127

XQTL test, 127
XRC-TDT test, 127
XS-TDT test, 127

Z

Z score
comparison between imputed genotypes and experimentally observed genotypes, 98–99
converting meta-analysis back to *p* values, 100
p values converted to, 99–100